Janvier 200.

Forensic Speaker Identification

Taylor & Francis Forensic Science Series
Edited by James Robertson
Forensic Sciences Division, Australian Federal Police

Firearms, the Law and Forensic Ballistics
T A Warlow
ISBN 0 7484 0432 5
1996

Scientific Examination of Documents: methods and techniques, 2nd edition
D Ellen
ISBN 0 7484 0580 1
1997

Forensic Investigation of Explosions
A Beveridge
ISBN 0 7484 0565 8
1998

Forensic Examination of Human Hair
J Robertson
ISBN 0 7484 0567 4
1999

Forensic Examination of Fibres, 2nd edition
J Robertson and M Grieve
ISBN 0 7484 0816 9
1999

Forensic Examination of Glass and Paint: analysis and interpretation
B Caddy
ISBN 0 7484 0579 9
2001

Forensic Speaker Identification

Philip Rose

London and New York

First published 2002 by Taylor & Francis
11 New Fetter Lane, London EC4P 4EE

Simultaneously published in the USA and Canada
by Taylor & Francis Inc,
29 West 35th Street, New York, NY 10001

Taylor & Francis is an imprint of the Taylor & Francis Group

© 2002 Taylor & Francis

Typeset in Times by Graphicraft Ltd, Hong Kong
Printed and bound in Great Britain by TJ International Ltd, Padstow, Cornwall

All rights reserved. No part of this book may be reprinted or reproduced or utilised in any form or by any electronic, mechanical, or other means, now known or hereafter invented, including photocopying and recording, or in any information storage or retrieval system, without permission in writing from the publishers.

Every effort has been made to ensure that the advice and information in this book is true and accurate at the time of going to press. However, neither the publisher nor the author can accept any legal responsibility or liability for any errors or omissions that may be made. In the case of drug administration, any medical procedure or the use of technical equipment mentioned within this book, you are strongly advised to consult the manufacturer's guidelines.

British Library Cataloguing in Publication Data
A catalogue record for this book is available from the British Library

Library of Congress Cataloging in Publication Data
A catalogue record has been requested

ISBN 0-415-27182-7

To my mum, who died on the day before she could see this book finished.
To my dad, who taught me how to observe.

Contents

	Acknowledgements	xiii
1	**Introduction**	1
	Forensic speaker identification	2
	Forensic phonetics	2
	Readership	3
	The take-home messages	3
	Argument and structure of the book	4
2	**Why voices are difficult to discriminate forensically**	9
	Between-speaker and within-speaker variation	10
	Probabilities of evidence	13
	Distribution in speaker space	14
	Multidimensionality	14
	Discrimination in forensic speaker identification	17
	Dimensional resolving power	18
	Ideal vs. realistic conditions	18
	Lack of control over variation	19
	Reduction in dimensionality	21
	Representativeness of forensic data	22
	Legitimate pooling of unknown samples	30
	Chapter summary	31
3	**Forensic-phonetic parameters**	33
	Types of parameters	34
	Acoustic vs. auditory parameters	34
	Traditional vs. automatic acoustic parameters	41
	Linguistic vs. non-linguistic parameters	43
	Linguistic sources of individual variation	45
	Forensic significance: linguistic analysis	48
	Quantitative and qualitative parameters	50
	Discrete and continuous parameters	50
	Requirements on forensic-phonetic parameters	51
	Chapter summary	53

Contents

4	**Expressing the outcome**	55
	The likelihood ratio	57
	Combination of likelihood ratios	60
	Prior odds	63
	Alternative hypothesis	64
	Bayesian inference	66
	Pros and cons	69
	Pro-Bayesian arguments	69
	Anti-Bayesian arguments	73
	A reality check	76
	Chapter summary	78
5	**Characterising forensic speaker identification**	81
	Speaker recognition	81
	Speaker identification and verification	82
	Relationship between forensic speaker identification and speaker identification/verification	87
	Summary: Verification and identification	91
	Naive and technical speaker recognition	92
	Technical speaker recognition	92
	Conditions on forensic-phonetic speaker identification experiments	96
	Naive speaker recognition	97
	Naive speaker recognition: Discussion	105
	Familiarisation in auditory forensic analysis	106
	Aural-spectrographic (voiceprint) identification	107
	Linguistic controversy	111
	Legal controversy	120
	Summary: Aural-spectrographic method	122
	Chapter summary	122
6	**The human vocal tract and the production and description of speech sounds**	125
	The vocal tract	126
	The basic dichotomy	126
	Vocal cords	127
	Forensic significance: Vocal cord activity	131
	Summary: Vocal cord activity	131
	The supralaryngeal vocal tract	131
	Forensic significance: Nasals and nasalisation	135
	Squeezing the supralaryngeal vocal tract tube: Vowels and consonants	135
	Place and manner description of consonants	137
	Place of articulation	140
	English consonants	142
	Vowels	144
	Primary parameters of vowel description	144
	Secondary parameters of vowel description	148
	English vowels	150
	The componentiality of speech sounds	152

Summary: Consonants and vowels	153
Suprasegmentals: Stress, intonation, tone and pitch accent	153
Stress	154
Forensic significance: Stress	156
Intonation	156
Tone	158
Pitch accent	160
Typology of suprasegmentals	161
Forensic significance: Suprasegmentals	161
Timing of supralaryngeal and vocal cord activity	162
Non-linguistic temporal structure	166
Rate	166
Forensic significance: Rate of utterance	169
Continuity	170
Forensic significance: Continuity	172
Summary: Non-linguistic temporal structure	173
Chapter summary	173

7 Phonemics — 175

Speech sounds in individual languages	175
Phonemic contrast	176
Phonemes and allophones	176
Comparison with English: The phonemic insight	179
Types of distribution	180
The reality of phonemes	181
Phonotactics	182
English phonemes	183
Forensic significance: Phonemic structure	185
Establishing between-sample comparability	185
Between-speaker and within-speaker differences in phonemic structure	186
Morphophonemics	192
Between-speaker and within-speaker allomorphic differences	193
Accent imitation	194
Chapter summary	194

8 Speech acoustics — 195

Speech sounds	195
Speech waves	196
Spectral representation	199
The acoustic theory of speech production	207
Source	207
Filter	209
Formants	211
Interaction of source and filter	212
Forensic significance: Vocal tract length and formant frequencies	215
Summary: Source–filter theory	220
Spectrograms	220

Contents

Spectrograms of other vowels	225
Acoustic vowel plots	228
Between-speaker variation in vowel acoustics	230
Within-speaker variation in vowel acoustics	232
Forensic significance: Differential effects of phonological environment	235
Forensic significance: Within-speaker variation in vowel acoustics	237
Higher-frequency formants	237
Forensic significance: Higher-frequency formants	237
A whole-word example	238
Forensic significance: Vowel acoustics	243
Fundamental frequency	244
Forensic significance: Fundamental frequency	246
Long-term fundamental frequency distributions	248
Fundamental frequency distributions and probability	253
Modelling fundamental frequency distributions	257
Long-term spectrum	259
Forensic significance: Long-term spectrum	261
The cepstrum	262
Forensic significance: Cepstrum	265
Differential software performance	265
Chapter summary	267

9 Speech perception — 269

Speech-specific perception	270
Forensic significance: Speech-specific perception	273
Expectation effect	273
Forensic significance: Expectation effect	273
Perceptual integration	274
Forensic significance: Perceptual integration	275
Chapter summary	275

10 What is a voice? — 277

The meaning of 'voice'	277
Voice quality and phonetic quality	278
Forensic significance: Phonetic quality and voice quality	281
Tone of voice (1)	282
Forensic significance: Tone of voice	282
The need for a model	283
Voice as 'choice' and 'constraint'	284
Vocal mechanism	286
Linguistic mechanism	286
Tone of voice (2)	290
Communicative intent	291
Forensic significance: Affect	293
Forensic significance: Social intent	294
Forensic significance: Regulatory intent	295
Intrinsic indexical factors	296

	Forensic significance: Intrinsic indexical factors	300
	Chapter summary	302
11	**The likelihood ratio revisited: A demonstration of the method**	303
	Calculation of likelihood ratio with continuous data	306
	A likelihood ratio formula	310
	Applications	312
	The likelihood ratio as a discriminant distance	318
	Problems and limitations	319
	Chapter summary	325
12	**Summary and envoi**	327
	Requirements for successful forensic speaker identification	329
	In the future?	333

Glossary — 337
References — 343
Index — 353

Acknowledgements

In getting to the stage where I could even contemplate writing this book, and then actually writing it, I have benefited from the help of many people, in many different ways. Now I can thank them formally for their contributions, which is a great pleasure. When I think of this book I think of the following people.

Hugh Selby, Reader in Law at the Australian National University. A long time ago, in commissioning a chapter on forensic speaker identification for his legal reference series *Expert Evidence*, Hugh introduced me to the Bayesian evaluation of evidence, with the implicit suggestion that I apply it to forensic speaker identification. Things seemed then to fall into place. How much is a key conceptual framework worth? That much I have to thank Hugh. Hugh also acted as my legal guinea-pig and read several early drafts, and he was also my source of information on legal matters. I also thank **Mie Selby**, who was apparently never quite able to escape in time from having Hugh read some sections to her.

Dr Yuko Kinoshita, soon to be Lecturer in the School of Communication at the University of Canberra. Yuko gave me permission to use, in Chapter 11, a large portion of the results from her PhD research into forensic speaker identification.

Dr Ann Kumar, Reader in Asian History at the Australian National University. Ann has contributed in several significant ways. She let me have some of her long Australian vowels for acoustic analysis in Chapter 8, and also compiled the glossary. Apart from reading and correcting several drafts, Ann was always there to listen to me try to formulate and express ideas, and offered continual encouragement. She said she found it very stimulating, and I'm sure it satisfied her intellectual curiosity. But I think she really should have been spending the time studying the comparative manufacture of Javanese Krises and Japanese ceremonial swords, or the agreement in mitochondrial DNA between Japanese and Indonesians. Now she can.

Drs Frantz Clermont and **Michael 'Spike' Barlow** of the Department of Computer Science at the University College of New South Wales. Frantz and Spike have been this poor linguistic phonetician's window into the world of automatic speaker recognition. They also provided the superb if somewhat slightly extra-terrestrially disconcerting three-dimensional figure of the vocal tract tube in Chapter 6. Frantz has patiently tried to make the mathematics of the cepstrum understandable to me, without confusing me with what he says are its 'dirty bits'. He spent a lot of his time creating the two *Matlab*™ figures of the cepstrum in Chapter 8, and then getting them just right for me.

My wife **Renata**. This book shows how command of the background statistical information is essential for the proper evaluation of differences between forensic samples. As well as offering never-ending support and encouragement, Renata has been in total command of all the onerous background work and thus ensured that I have always been free to devote my time to writing. Unfortunately, unless I can do some really quick thinking, I now no longer have an excuse for not mowing the lawn.

Dr Francis Nolan, Reader in Phonetics at the University of Cambridge. As is obvious from the number of citations in the text, this book owes a tremendous intellectual debt to Francis' extensive work on the phonetic bases of speaker recognition and its application to forensic phonetics. Francis also read an earlier draft and made many invaluable suggestions for improvement.

Alison Bagnall, Director of Voicecraft International. I have Alison to thank for making available the beautiful xero-radiographs of my vocal tract, and video of my vocal cords, which appear in Chapter 6. We made them at the (now) Women's and Children's Hospital in Adelaide, as part of an investigation for their marvellous craniofacial unit into variation of vowel height and velopharyngeal port aperture.

My colleagues and students in the Linguistics Department at the Australian National University, and elsewhere. In particular, **Professor Anna Wierzbicka**, for her help with questions of semantic analysis; **Professor Andrew Butcher**, of the Department of Speech Pathology at Flinders University, South Australia, for encouragement and ideas and keeping me posted on his forensic-phonetic case-work in the context of many informative discussions on forensic speaker identification; **Belinda Collins** and **Johanna Rendle-Short** for their help with conversation analysis; **Jennifer Elliott** for keeping me up-to-date with between-speaker and within-speaker variation in *okay*.

John Maindonald, formerly of the Statistical Consulting Unit of the Australian National University. John patiently fielded all my questions on statistical problems and then answered them again, and often again, when I still wasn't quite sure.

James Robertson, head of forensic services for the Australian Federal Police. James was my series editor, and made many useful comments and corrections to drafts of several chapters of the book.

Len Cegielka, my eagle-eyed copy-editor. This book has greatly benefited from Len's experienced and professional critique. I'm sure he is well aware of the considerable extent of his contribution, and doesn't need me to tell him that he really has made a material improvement. So instead of that I will just express my thanks for a job very well done.

Professor Stephen Hyde in the Research School of Physical Sciences' Department of Applied Mathematics at the Australian National University. Stephen contributed the 3-D figure in Chapter 2.

The book is ultimately the fault of **Dr** (now Pastor) **Mark Durie**, who was responsible for my embarking on forensic speaker identification. There have been times when I'm really not sure whether he actually deserves thanks for this.

All have given their knowledge and expertise to help make the book better, but also the most valuable thing of all: their time. They are in the happy, and deserved, position of partaking of all of the credit but none of the criticism. That, of course, is down to me alone.

I want to close with a kind of anti-acknowledgement. It comes as no news in what now remains of Australian higher education that the past decade has witnessed a dramatic decline in – some would say jettisoning of – academic standards (Coady 2000: 11–24). This has been particularly the case since 1996, when the newly-elected government initiated a series of savage cuts by slashing 600 million dollars from the tertiary education budget. All the while incanting inane alliterations like 'clever country' and 'knowledge nation', we are teaching less and less to more and more. Sadly, I don't suppose this is much different from many other countries.

That modern master of the essay, Stephen Jay Gould recently wrote this about the current jeopardy of scholarly enquiry:

> I just feel that the world of commerce and the world of intellect, by their intrinsic natures, must pursue different values and priorities – while the commercial world looms so much larger than our domain that we can only be engulfed and destroyed if we make a devil's bargain of fusion for short-term gain.
>
> Gould (2000: 26)

Much the same sentiment is expressed in a recent P. D. James detective novel:

> People who, like us, live in a dying civilisation have three choices. We can attempt to avert the decline as a child builds a sand-castle on the edge of the advancing tide. We can ignore the death of beauty, of scholarship, of art, of intellectual integrity, finding solace in our own consolations. . . . Thirdly, we can join the barbarians and take our share of the spoils.
>
> James (2001: 380)

Unfortunately, the person who wrote these pessimistic words in the novel is in fact the brutal murderer! Well, I too have often felt like murdering in the past few years, as change after change, cut after cut, has made it increasingly difficult to teach, to supervise, to do case-work and meaningful research, and to write this book.

The point I wish to make, however, is this. This book is about an aspect of applied scholarly endeavour, forensic phonetics, that carries with it very serious social responsibilities. The book makes it clear that forensic speaker identification requires scholarly expertise, and in several disparate areas. Expertise, like forensically useful fundamental frequency, is a long-term thing. It requires an enormous amount of dedication on the part of students and teachers to get someone to the stage where they may responsibly undertake forensic case-work. This simply cannot be realised, in both senses of the term, under an ideology so clearly obsessed with the bottom line. Most people can learn to do a variety of things – I think the result is what is called multiskilling – but whether they will do them with sufficient skill and expertise is another matter. With the topic of this book that is precisely the question. It is about acquiring and exercising expertise. As Francis Nolan (1997: 767) has written: Justice will depend on it.

On a more optimistic note, then: I hope that this book will go a little way to helping all interested parties excel.

References

Coady, T. (2000) 'Universities and the ideals of inquiry', in Coady, T. (ed.) *Why Universities Matter*: 3–25, St. Leonards: Allen and Unwin.

Acknowledgements

Gould, S. J. (2000) 'The lying stones of Marrakech', in Gould, S. J. *The Lying Stones of Marrakech: Penultimate Reflections in Natural History*: 9–26, London: Jonathan Cape.

James, P. D. (2001) *Death in Holy Orders*, London: Faber and Faber.

Nolan, F. (1997) 'Speaker Recognition and Forensic Phonetics', in Hardcastle, W. J. and Laver, J. (eds) *The Handbook of Phonetic Sciences*: 744–67, Oxford: Blackwell.

1

Introduction

Paul Prinzivalli, an air freight cargo handler in Los Angeles, stood trial for telephoning bomb threats to his employer, Pan Am. He was suspected because he was known to be a disgruntled employee, and because some Pan Am executives thought that the offender's voice sounded like his. Defence was able to demonstrate with the help of forensic-phonetic analysis that the offender's voice samples contained features typical of a New England accent, whereas Prinzivalli's accent was unmistakably from New York. To untrained West Coast ears, the differences between New York and Boston accents are not very salient; to ears trained in linguistic and phonetic analysis, the recordings contained shibboleths galore. Reasonable doubt was established and Prinzivalli was acquitted (Labov and Harris 1994: 287–99).

In a case of the kidnapping and murder of an 11-year-old German girl (Künzel 1987: 5, 6), considerable agreement was found between the voice samples of a suspect and that of the kidnapper and, on the basis of this and other evidence, the suspect was arrested. Subsequently, a more intensive comparison between offender and suspect voice samples yielded yet more similarities. The man confessed during his trial.

In another American case involving a telephoned bomb threat (Hollien 1990: 51), the defendant had been identified by his voice. However, it was clear to forensic phoneticians even from an auditory comparison that the voices of the defendant and the offender were very different. For example, the offender's voice had features typical of someone who spoke English as a second language. The case was dismissed.

In the late 1990s in Australia, in a case concerning illegal drug trafficking (Q. vs. Duncan Lam 1999), police intercepted 15 incriminating telephone conversations containing 31 voice samples in Cantonese. Forensic-phonetic analysis was able to assign these 31 voice samples to three different speakers. Since the police, but not the analyst, knew the identity of some of the samples, not only could two of the speakers be identified, but the accuracy of the identification could also be checked.

Another recent Australian case involved the interception of telephone conversations between two brothers, one of whom was charged with drug-related matters. Defence claimed that their voices were so similar that the incriminating recordings could not be attributed to the suspect. A forensic-phonetic analysis was able to show that the brothers' voices were distinguishable. Although their voices were indeed acoustically very similar in many respects, they still differed in others, and in particular they both had different ways of saying their 'r' sound.

Forensic speaker identification

Expert opinion is increasingly being sought in the legal process as to whether two or more recordings of speech are from the same speaker. This is usually termed forensic speaker identification, or forensic speaker recognition. As the examples above show – and many more could be cited – forensic speaker identification can be very effective, contributing to both conviction and elimination of suspects. Equally importantly, the examples also demonstrate the necessity for expert evaluation of voice samples, since three of them show how the truth actually ran counter to the belief of naive listeners.

Forensic speaker identification of the type described in this book – that is, using a combination of auditory and acoustic methods – has been around for quite a long time. Germany's Bundeskriminalamt was one of the first institutions to implement it, in 1980 (Künzel 1995: 79), and the first conference on forensic applications of phonetics was held in the United Kingdom in 1989. Yet there is still a considerable lack of understanding on the part of law enforcement agencies, legal practitioners, and indeed phoneticians and linguists, as to what it involves, what constitutes appropriate methodology, what it can achieve, and what its limitations are. In a 1995 Australian Court of Appeal report, for example, (Hayne and Crockett 1995: 2, 3) some perceived 'weaknesses in the science' [of speaker identification] were explicitly listed, some of which were inaccurate (Rose 1996d). The aim of this book is to make explicit and explain in detail for the relevant professionals what is involved in forensic speaker identification (FSI), and to clarify the problems of inferring identity from speech under the very much less than ideal conditions typical in forensics.

Forensic phonetics

Forensic speaker identification is a part of forensic phonetics. Forensic phonetics is in turn an application of the subject of phonetics. Different experts have slightly differing opinions on the exact subject matter of phonetics, and even whether it constitutes a discipline (Kohler 2000; Laver 2000; Ohala 2000). However, the following characterisation will not be controversial. Phonetics is concerned primarily with speech: it studies especially how people speak, how the speech is transmitted acoustically, and how it is perceived.

As well as forensic speaker identification, forensic phonetics includes areas such as speaker profiling (in the absence of a suspect, saying something about the regional or socioeconomic accent of the offender's voice(s)); the construction of voice line-ups; content identification (determining what was said when recordings are of bad quality, or when the voice is pathological or has a foreign accent); and tape authentication (determining whether a tape has been tampered with) (French 1994: 170; 182–4; Nolan 1997: 746). This book will not be concerned with the latter three areas of voice line-ups, content identification and tape authentication, and only very indirectly with speaker profiling.

Readership

It is a good idea to be specific early on about the intended audience. I have had several types of readers in mind when writing this book. As just mentioned, I have written it to help members of the legal profession, the judiciary, and law enforcement agencies understand what forensic speaker identification is about. This will help them when requesting a forensic-phonetic investigation and help them understand and evaluate forensic-phonetic reports and evidence. I have also thought of linguistic phoneticians, budding and otherwise, who might be lured away from describing how *languages* differ phonetically to do valuable forensic-phonetic research into the complementary area of how their *speakers* differ. This book is also intended to be useful to those contemplating a professional career in forensic phonetics. I hope that the speech science community at large, including those working in the area of automatic speaker identification, will find much that is of use in the book. I hope, too, that the book will be accessible to both graduate and advanced undergraduate students in all related disciplines. Finally, I hope that the book will help stimulate the interest of statisticians who are thinking of researching, or supervising, forensic-statistical topics: there is especially much to be done in this area!

The take-home messages

The most important points that the book will make are these:

- The forensic comparison of voice samples is extremely complex.
- In the vast majority of cases the proper way to evaluate forensic speech samples, and thus to evaluate the weight of the forensic-phonetic evidence, is by estimating the probability of observing the differences between them assuming that the same speaker is involved; and the probability of observing the differences between them assuming that different speakers are involved. This method is thus inherently probabilistic, and as such will not yield an absolute identification or exclusion of the suspect.
- The two main problems in evaluating differences between samples are (1) differential variation in voices, both from the same speaker and from different speakers, and (2) the variable and generally poor degree of control over forensic speech samples.
- Speech samples need to be compared both acoustically and auditorily, and they also need to be compared from the point of view of their linguistic and non-linguistic features.
- Forensic speaker identification requires expert knowledge of not just one, but several different specialist areas related to speech-science. These include sub-areas of linguistics, acoustics and statistics. This is in addition to the knowledge of how to interpret the results forensically.

Argument and structure of the book

Speaker identification in the forensic context is usually about comparing voices. Probably the most common task involves the comparison of one or more samples of an offender's voice with one or more samples of a suspect's voice. Voices are important things for humans. They are the medium through which we do a lot of communicating with the outside world: our ideas, of course, but also our emotions and our personality:

> The voice is the very emblem of the speaker, indelibly woven into the fabric of speech. In this sense, each of our utterances of spoken language carries not only its own message, but through accent, tone of voice and habitual voice quality it is at the same time an audible declaration of our membership of particular social regional groups, of our individual physical and psychological identity, and of our momentary mood.
>
> Laver (1994: 2)

Voices are also one of the media through which we (successfully, most of the time) recognise other humans who are important to us – members of our family, media personalities, our friends and enemies. Although evidence from DNA analysis is potentially vastly more eloquent in its power than evidence from voices, DNA can't talk. It can't be recorded planning, carrying out or confessing to a crime. It can't be so apparently directly incriminating. Perhaps it is these features that contribute to the interest and importance of FSI.

As will quickly become evident, voices are extremely complex things, and some of the inherent limitations of the forensic-phonetic method are in part a consequence of the interaction between their complexity and the real world in which they are used. It is one of the aims of this book to explain how this comes about.

Because of the complexity of the subject matter, there is no straightforward way to present the information that is necessary to understand how voices can be related, or not, to their owners. I have chosen to organise this book into four main parts:

- The first part, in Chapters 2–5, introduces the basic ideas in FSI: Why voices are difficult to discriminate forensically, Forensic-phonetic parameters, Expressing the outcome, Characterising forensic-phonetic speaker identification.
- The second part, Chapters 6–9, describes what speech sounds are like: The human vocal tract and the production and perception of speech sounds, Phonemics, Speech acoustics, Speech perception.
- The third part, in Chapter 10, describes what a voice is.
- The fourth part, in Chapter 11, demonstrates the method using forensically realistic speech.

A more detailed breakdown of the contents of these four main parts – Basic ideas, What speech sounds are like, What is a voice? Forensic speaker identification demonstrated – is given below. But before that it is important to flag an additional feature of the overall organisation, namely that it reflects the strongly cumulative nature of the topic. The concepts in some early chapters constitute the building blocks for those to come. Thus, for example, it is difficult to understand how speakers differ in phonemic structure without understanding phonemics in Chapter 7, and very difficult to understand phonemics without a prior understanding of speech sounds from Chapter 6. The demonstration of the method in Chapter 11 will be much easier to understand with knowledge of speech acoustics in Chapter 8.

Because this book is intended to be of use to so many disparate groups, I have adopted an inclusive approach and erred on the side of detail. This means that some readers may find more detail in some chapters than they need. I can imagine that members of the legal profession will not be able to get quite as excited as I do about formant frequencies, for example, and readers with some background knowledge of articulatory phonetics do not need to learn about vocal tract structure. Feel free to skip accordingly, but please be aware of the cumulative structure of the book's argument.

A brief characterisation of the contents now follows.

Basic ideas

In Chapters 2 to 5 are introduced the ideas that are central to the problem of FSI. There are four themes, one to each chapter. The first theme, in Chapter 2, describes what it is about voices that makes FSI difficult. This includes the existence of within-speaker as well as between-speaker variation, and its consequences for discriminating between speakers; what conditions variation, and our lack of control over it in the real world. The second theme, in Chapter 3, is forensic-phonetic parameters. Here are discussed the different types of parameters used to compare speech samples forensically. The third theme describes the proper conceptual framework and way of expressing the outcome of a forensic-phonetic identification. This is in Chapter 4. The fourth theme, in Chapter 5, shows how FSI relates to other types of speaker recognition.

What speech sounds are like

FSI is performed on recordings of human vocalisations – sounds made exclusively by a human vocal tract. Although other vocalisations, for example laughs or screams, may from time to time be forensically important (Hirson 1995), most of the vocalisations used in FSI are examples of speech. Speech is the primary medium of that supremely human symbolic communication system called Language. One of the functions of a voice – perhaps the main one – is to realise Language, by conveying some of the speaker's thoughts in linguistic form. Speech is Language made audible.

Moreover, when forensic phoneticians compare and describe voices, they usually do so with respect to linguistic units, especially speech sounds, like vowels or consonants. They might observe, for example, that the *ee* vowels in two samples are different; or that the *th* sound is idiosyncratically produced in both. It is therefore necessary to understand something of the structure of speech sounds, and how they are described. A large part of this book, Chapters 6 to 9, is accordingly given over to a description of the nature of speech sounds.

How speech sounds are produced – articulatory phonetics – is covered in Chapter 6. Since speech sounds are traditionally described in articulatory terms, their description is also covered in this chapter.

Chapter 7 (Phonemics) is concerned with how speech sounds are functionally organised in language. Phonemics is a conceptual framework that the forensic description of speech sounds usually assumes, and within which it is conveniently, and indeed indispensably, presented.

Phonemics regards actual speech sounds (called *phones*) as realisations of abstract sounds, called *phonemes*, whose function is to distinguish words. Thus the vowel in the word *bead*, and the vowel in the word *bid* are said to realise different phonemes in English because the phonetic difference between them – one vowel is, among other things, shorter than the other – signals a difference between words. The consonant sounds at the beginning of the words *red* and *led* realise different phonemes for the same reason. Phonemics is indispensable, because it supplies the basis for comparison of speech sounds in the first place. It allows us to say, for example, that two speech sounds are potentially comparable across samples because they are realisations of the same phoneme. This would enable us to legitimately compare, acoustically and phonetically, the vowel in the word *car* in one sample, say, with the vowel in the word *far* in another, because they are realisations of the same phoneme.

It is generally assumed that similarities and differences between forensic speech samples should be quantified acoustically, and acoustic comparison is an indispensable part of forensic-phonetic investigation. It is therefore necessary to describe speech acoustics, especially those that are assumed to be the ones in which speaker identity is optimally encoded. Speech acoustics are described in Chapter 8.

Several aspects of speech perception – how humans decode the speech acoustics to hear speech – are relevant for forensic phonetics. In particular, one argument for the necessity of analysing forensic speech samples both auditorily and acoustically is based on speech perception. It is also important to know about the expectation effect (you hear what you expect to hear), another aspect of speech perception. Both are discussed in Chapter 9: *Speech perception*.

The contents of Chapters 6 to 9, then, describe some of the basic knowledge that informs forensic-phonetic work. For example, it would be typical for a forensic-phonetic expert to listen to two speech samples, decide what is comparable in terms of occurrences of the same phoneme, describe and transcribe the phonemes' realisations, and then quantify the differences acoustically.

Each of these chapters on its own constitutes a vast area of scholarship. Scholarship, moreover, that is informed by several disciplines. Thus the multidisciplinary basis of phonetics comprises at least linguistics, anatomy, physiology, acoustics and statistics. Although I cannot hope to cover the contents of each chapter in a depth appropriate to the complexity of subject matter, I have tried to avoid the Charybdis of cursory treatment: to present *here* the figure of a spectrogram as an example of speech acoustics; *there* a word transcribed phonetically as an example of articulatory phonetics. Instead I have tried to include enough speech acoustics to help the reader appreciate how one speaker can differ acoustically from another, at least in some respects, and to understand what they are actually looking at in a spectrogram. I have also included enough traditional phonetics and phonology to help them understand at least some of what might otherwise be arcane parts of a forensic-phonetic report. As a result of all this, the reader will find that they are able to follow important arguments demonstrated with the appropriate acoustics and linguistic analysis.

Introduction

These aims have their downside. It would be silly to pretend that all of the specialist topics I have tried to explain in some detail are easy. They are quite definitely not, and the reader must be prepared to find at least some of the chapters difficult. Like many difficult passages, however, they become much more understandable on repeated readings.

What is a voice?

A voice is more than just a string of sounds. Voices are inherently complex. They signal a great deal of information in addition to the intended linguistic message: the speaker's sex, for example, or their emotional state or state of health. Some of this information is clearly of potential forensic importance. However, the different types of information conveyed by a voice are not signalled in separate channels, but are convolved together with the linguistic message. Knowledge of how this occurs is necessary to interpret the ubiquitous variation in speech, and to assess the comparability of speech samples.

Familiar things like voices we tend to take for granted. In this case familiarity breeds false understanding. We assume they are simple, and that we know about them. This is especially typical for language, and phenomena that are intimately connected with language (Lyons 1981: 38). Language is absolutely fascinating, and nearly everyone has an unreflected opinion on aspects of their own language. Although not all languages have a separate word for voice, all languages have words for describing the way speakers sound when they talk: in English, for example, they can be harsh, kind, sexy, masculine, gruff, melodious, sibilant, booming, staccato, etc. (Laver 1991f).

A voice is, however, an extremely complex object, and a large part of this complexity lies in its relationship with its owner. When comparing voices, it is imperative that one knows about what one is comparing. Consequently, it is very important to have a model for the information content in a voice, and how these different components interrelate and interact. A model for the voice is presented in Chapter 10: *What is a voice?*

The likelihood ratio revisited: a demonstration of the method

A book like this could obviously not be taken seriously if it could not demonstrate that the method worked – that forensically realistic speech samples from the same speaker can be distinguished from speech samples from different speakers. In Chapter 11, then, will be found such a demonstration, together with an explanation of the basic mechanics of the statistical approach used (called the likelihood ratio), and its shortcomings.

Chapter 12 contains a summary of the book, and addresses questions of what, in terms of data, method, and qualifications of practitioner, constitutes the requirements for a successful forensic speaker identification. It also asks what developments can be expected in the future.

I have throughout written from the perspective of the expert in an adversarial system, and referred, for example, to strength of evidence supporting the defence or prosecution. The application of the ideas remains the same in an inquisitorial context.

I hope this book will inform; will stimulate interest; and will suggest research agendas in a field where much fascinating and valuable work remains to be done.

2

Why voices are difficult to discriminate forensically

As flagged in the introduction, the aim of this and the following three chapters is to introduce in as non-technical way as possible ideas and distinctions that are central to the understanding of forensic speaker identification. In this chapter I describe what it is about voices that makes identifying them or telling them apart under realistic forensic conditions problematic.

Probably the most common task in forensic speaker identification involves the comparison of one or more samples of an unknown voice (sometimes called the questioned sample(s)) with one or more samples of a known voice. Often the unknown voice is that of the individual alleged to have committed an offence (hereafter called the offender) and the known voice belongs to the suspect. Both prosecution and defence are then concerned with being able to say whether the two samples have come from the same person, and thus being able either to identify the suspect as the offender or to eliminate them from suspicion. Sometimes it is important to be able to attach a voice to an individual, or not, irrespective of questions of guilt.

In order to tell whether the same voice is present in two or more speech samples, it must be possible to tell the difference between, or *discriminate* between, voices. Put more accurately, it must be possible to discriminate between samples from the voice of the same speaker and samples from the voices of different speakers (Broeders 1995: 158). So identification in this sense is the secondary result of a process of discrimination. The suspect may be identified as the offender to the extent that the evidence supports the hypothesis that questioned and suspect samples are from the same voice. If not, no identification results. In this regard, therefore, the *identification* in forensic speaker identification is somewhat imprecise. However, since the term identification is commonly found (*recognition* is another common term), I have chosen to use it in this book. It can also be noted here that the word discrimination is usually used in a sense that differs somewhat from that just described. This point will be taken up later.

One of the important messages in this book will be that it is possible to discriminate between voices extremely well under ideal circumstances, but there are good reasons to assume that this ability diminishes considerably under the very much less than ideal conditions normally present in forensic case work. One of the aims of this chapter is therefore to clarify the reasons for this – why discrimination becomes more problematic in the real world of forensic phonetics.

It stands to reason that if entities are to be discriminated by one of their attributes, those entities must differ in that attribute. If everyone had the same voice, voices would be no use for discrimination. Thus, if individuals are to be discriminated by their voices, individuals must differ in their voices. They do, but the voice of the same speaker can also show considerable variation. This variation must now be discussed.

Between-speaker and within-speaker variation

Under ideal conditions, speakers can be identified reasonably easily by their voices. We know this from the excellent performance of automated speaker recognition systems (to be discussed below). And we probably suspected it already because we have all had the experience of hearing identity: we humans recognise familiar voices, fairly successfully, all the time. This probably entitles us to assume that different speakers of the same language do indeed have different voices, although as already foreshadowed the extent to which this can be assumed forensically is a separate issue. We thus have to deal with variation between speakers, usually known as *between-speaker* (or *inter-speaker*) *variation*.

Although it is a general assumption that different speakers have different voices, it is crucial to understand that the voice of the same speaker will always vary as well. It is a phonetic truism that no-one ever says the same thing in exactly the same way. (For a demonstration of some of the acoustic differences that occur between repeats of the same word, see Rose 1996a). It is therefore necessary to come to terms with variation within a speaker. This is usually termed *within-speaker* or *intra-speaker* variation. (The latter term, with *intra-*, sounds confusingly similar to the *inter-* of inter-speaker variation and, as I also always have trouble in remembering which is which, is accordingly avoided in this book.)

Probably the most important consequence of this variation is that there will always be differences between speech samples, even if they come from the same speaker. These differences, which will usually be audible, and always measurable and quantifiable, have to be evaluated correctly if forensic speaker identification is to work. To put it another way, forensic speaker identification involves being able to tell whether the inevitable differences between samples are more likely to be within-speaker differences or between-speaker differences.

Given the existence of within-speaker variation, it is obvious that the other logical requirement on the feasibility of forensic speaker identification (or indeed any identification system) is that variation between speakers be bigger than the variation within a speaker. Again, this appears to be the case: providing that one looks at the right things, there is greater variation between the voices of different speakers than within the voice of the same speaker. It is also intuitively obvious that the greater the ratio of between-speaker to within-speaker variation, the easier the identification.

To illustrate this, Figure 2.1 shows three speakers differing in one particular aspect of their voice. It does not matter what the aspect is, as long as it is quantifiable – one possibility that is easy to envisage might be the speaker's 'average pitch'. Average

Why voices are difficult to discriminate forensically

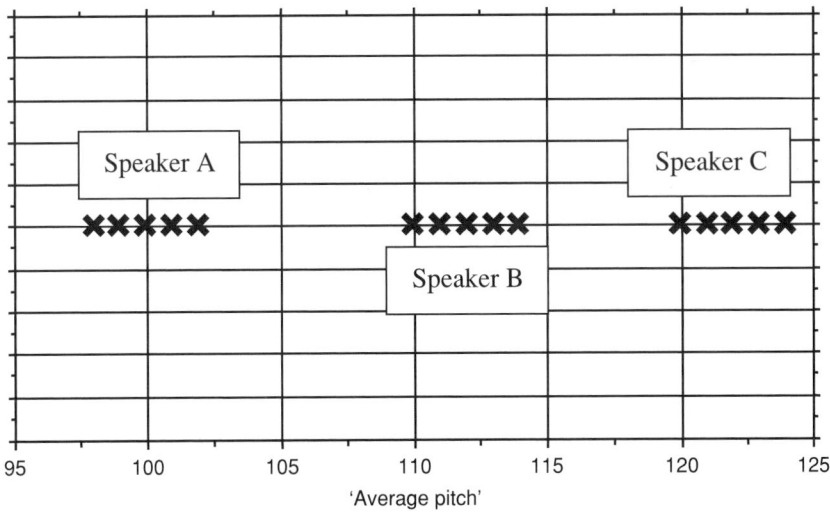

Figure 2.1 Three speakers in one dimension

pitch would then be termed a dimension. (Other commonly found synonyms are parameter, or feature.) In the figure, '✕' marks a particular value within this dimension (or parameter, or feature). For the moment, let us assume that these values were measured from the speakers on one particular occasion: perhaps measurements were taken once at the beginning and end of their speech, once in the middle, and twice somewhere in-between, thus making five measurements per speaker. (Measurements are often referred to as observations, and the terms are used interchangeably in this book.)

Figure 2.1 shows how speakers vary along the single 'average pitch' dimension – speaker B's lowest value is 110 'average pitch' units, for example. It can be seen firstly that there is between-speaker variation: each speaker has different values for the dimension. Within-speaker variation is also evident: each speaker is characterised by a set of different values for the dimension. It can also be seen that variation within a speaker for the 'average pitch' dimension is small compared to the variation between speakers. For example, the largest difference between two observations for the same speaker is smaller than the smallest difference between any two speakers.

It can be appreciated that, given this kind of distribution, if we take any two values for this dimension, it would be easy to correctly discriminate them – to say whether the two values had come from the same speaker or not. If we took values of 100 and 104, for example, we could say that they came from the same speaker; if we took 100 and 110, we could say they came from different speakers. The reader might like to formulate the discriminant function for these data: the condition under which any two observations could be said to come from different speakers.[1]

---

1 A magnitude of difference between any two observations greater than 4 average pitch units would indicate that they came from different speakers.

Forensic Speaker Identification

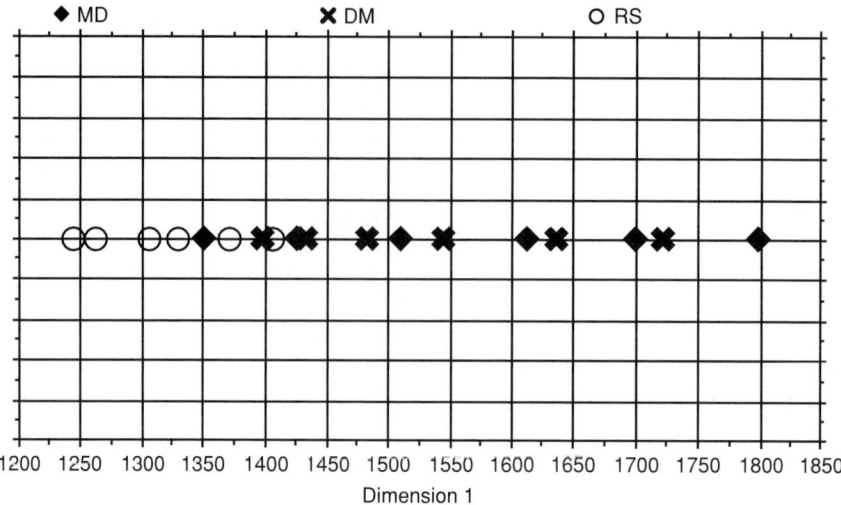

Figure 2.2 Three speakers with overlapping values in one dimension

In reality, however, speakers overlap in their distributions for a dimension. Sometimes they overlap a little, sometimes a lot, and sometimes completely. This situation is illustrated in Figure 2.2, which, in contrast to the made-up data in Figure 2.1, now shows real data from three speakers: MD, DM, RS. The dimension involved is an easily measured acoustic value in the second syllable of the word *hello* (the data are from Rose 1999a). Six *hello*s from each of the three speakers were measured, and the values plotted in the figure (their unit is hertz, abbreviated Hz). It can be seen that the speaker with the lowest observation was RS, whose lowest value was about 1240 Hz. The highest value, of about 1800 Hz, belongs to speaker MD.

The most important feature of Figure 2.2 is that the speakers can be seen to overlap. Speakers MD and DM overlap a lot: DM's range of values is from 1399 Hz to 1723 Hz, and is covered completely by MD, whose range is from 1351 Hz to 1799 Hz. Speaker RS, whose range is from 1244 Hz to 1406 Hz, encroaches a little on the lower ends of the other two speakers' ranges.

The consequences of this overlapping are drastic. Now it is no longer possible to give a value for magnitude of difference between samples that will *always* correctly discriminate between the same and different speakers. This is because the smallest difference between two different speakers, e.g. the difference of 7 Hz between DM's lowest value of 1399 Hz and RS's highest value of 1406 Hz, is smaller than the smallest within-speaker difference. Therefore, although we could say that a difference between any two observations that is greater than 448 Hz always indicates different speakers (because 448 Hz is the biggest within-speaker difference between the lowest and highest observations in the data) we cannot stipulate a threshold, values smaller than which will always demonstrate the presence of the same speaker.

Even though it is now not possible, because of the overlap, to correctly discriminate all between-speaker and within-speaker pairs in the figure, it is nevertheless possible to estimate a threshold value such that a certain percentage of the pairs are correctly discriminated. For example, if we chose a threshold value of 150 Hz, such that differences between paired observations greater than 150 Hz were counted as coming from different speakers, and differences smaller than 150 Hz counted as the same speaker, we could actually correctly discriminate about 60% of all 153 pairs of between-speaker and within-speaker comparisons. (With six tokens each from three speakers, there are 45 possible within-speaker comparisons; and 108 possible between-speaker comparisons.) For example, the difference of 35 Hz between speaker RS's highest value of 1406 Hz and his next highest value of 1371 Hz is less than 150 Hz, and is therefore correctly discriminated as a within-speaker difference. Likewise the difference of 155 Hz between RS's lowest value of 1244 Hz and DM's lowest value of 1399 would be correctly discriminated (just!) as a between-speaker difference. However, the difference between RS's highest value of 1406 and his lowest value of 1244 Hz is greater than 150 Hz and would be incorrectly classified as coming from two different speakers. This would count as a false negative: deciding that two observations come from different speakers when in fact they do not.

More significantly, the difference between RS's highest value of 1406 and DM's lowest value of 1399 Hz will be incorrectly discriminated as coming from the same speaker. This would count as a judicially fatal false positive: DM, assuming he is the accused, is now convicted of the crime committed by RS.

If we choose this threshold (of 150 Hz), then we get about the same percentage of between-speaker comparisons incorrect as within-speaker comparisons. This is called an equal error rate (EER); since we get about 60% correct discriminations, the EER in this case will be about 40%.

The threshold can, of course, be set to deliver other error rates. If we wished to ensure that a smaller percentage of false positives ensue (samples from two different speakers identified as coming from the same speaker), then we set the threshold lower. This would, of course, have the effect of increasing the instances where the real offender is incorrectly excluded.

Where does the figure of 150 Hz for the equal error rate threshold come from? It can be estimated from a knowledge of the statistical properties of the distribution of within-speaker and between-speaker differences, or it can be found by trial and error by choosing an initial threshold value, noting how well it performs, and adjusting it up or down until the desired resolution, in this case the equal error rate, is achieved.

Probabilities of evidence

The typical fact that only a certain percentage of cases can be correctly discriminated – for these data 60% – is another consequence of the overlap, and introduces the necessity for probability statements. For example, we would expect that a decision based on the application of the '150 Hz threshold' test to any pair of observations from the above data would be correct 60% of the time and incorrect 40% of the time.

You are obviously not going to stake your life on the truth of such a call, but this way of looking at things will nevertheless turn out to be of considerable importance. This is because it allows us to compare the probability of observing the magnitude of difference between two observations assuming that they come from the same speaker (60%), and assuming they come from different speakers (40%). Note that it is the *probabilities of the evidence* as provided by the test that are being given here and not the *probability of the hypothesis* that the same speaker is involved. This thus gives us a measure of the strength of the test. In other words, whatever our estimate was of the odds in favour of common origin of the two samples before the test, if the two observations are separated by less than 150 Hz, we are now 60/40 = one and a half times more sure that they do have a common origin. And if they are separated by more than 150 Hz, we are one and a half times more sure that they do not. Not a lot, to be sure, but still something.

Note that this also introduces the notion of probability in the sense of an expectation, or as a measure of how much we believe in the truth of a claim (here, that the two observations come from the same speaker). The importance of assessing the probabilities of the evidence and not the hypothesis is a central notion and will need to be addressed in detail in Chapter 4.

Distribution in speaker space

The three speakers' monodimensional data in Figure 2.2 also illustrate two more important things about voices. It is possible to appreciate that RS differs from MD and DM much more than the latter two differ. This is illustrative of the fact that *voices are not all distributed equally in speaker space: some are closer together than others*. This could also possibly be guessed from the fact that some voices *sound* more similar than others. Not so easy to guess is the fact that *speakers can differ in the amount by which they vary for a dimension*. This is also illustrated in Figure 2.2, where it can be seen that RS's individual observations distribute more compactly than DM's and MD's. The fact that speakers vary in their variability is an additional complication in discrimination.

The unequal distribution of voices is one reason why our illustrative discrimination, with its 'average equal error rate', is unrealistic. It will not be the case that all pairs of speakers will be discriminable to the same extent, because a small amount of really good discrimination between speakers who are well separated in speaker space, for example between RS and DM, or between RS and MD, can cancel out a large amount of poor discrimination, for example between MD and DM. This means that an average equal error rate can be misleading. It is because of this that sophisticated statistical techniques, for example linear discriminant analysis, are required to give a proper estimate of the inherent discriminating power of dimensions.

Multidimensionality

It now needs to be explained that it is unrealistic to envisage a comparison on a single dimension only, like our notional 'average pitch', or the acoustic dimension in *hello*. This leads into the next main message: that *voices are multidimensional objects*. A speaker's voice is potentially characterisable in an exceedingly large number of different dimensions. This is illustrated below.

Why voices are difficult to discriminate forensically

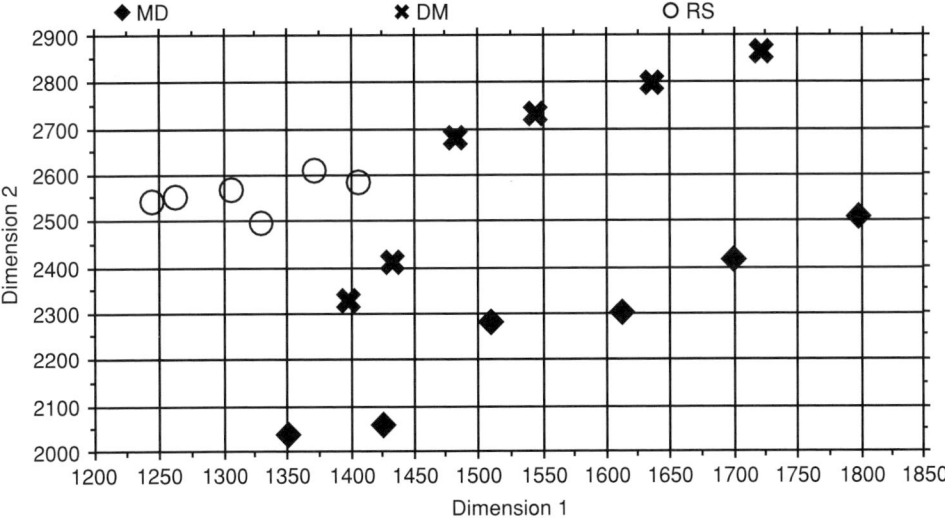

Figure 2.3 Three speakers in two dimensions

Two-dimensional comparison

Figure 2.3 shows the same three speakers as Figure 2.2, this time positioned with respect to two dimensions. The original dimension from Figure 2.2 is retained on the horizontal axis, and another acoustical dimension, this time having to do with the 'l' sound in the word *hello*, and also quantified in Hz, is shown vertically.

Now it can be seen that, although DM and MD largely overlap along the first dimension, they overlap much less on the second: DM's lowest value is 2328 Hz, MD's highest value is 2509, and two of DM's values lie within MD's range. RS's values, which have a narrower distribution in the second dimension than the other two, are overlapped completely by DM and only minimally by MD.

As a result of quantifying speakers with respect to these two dimensions, the speakers can be seen to occupy more different positions in two-dimensional than one-dimensional speaker space.

It should be clear that, if it is possible to take into account differences in two dimensions, and the dimensions separate out speakers a little, there is the possibility of achieving a better discrimination than with a single dimension. This can be demonstrated with the data in Figure 2.3.

As with Figures 2.1 and 2.2 we are still concerned with distances between individual observations. There are different ways of calculating a distance between two observations in two dimensions. The most straightforward way is by using the theorem of Pythagoras: calculating the square-root of the sum of the squared magnitudes of the differences between the observations in both dimensions (this is called a Euclidean distance). So, in Figure 2.3, the distance between speaker RS's leftmost observation, which is at the coordinates 1244 Hz and 2543 Hz, and his next nearest observation (1262 Hz, 2556 Hz) is

Forensic Speaker Identification

$$\sqrt{(1244 - 1262)^2 + (2543 - 2556)^2}$$

$$= \sqrt{(-18)^2 + (-13)^2}$$

$$= \sqrt{(324) + (169)}$$

$$= \sqrt{493}$$

$$= 22.2$$

It can be seen from Figure 2.3 that there are still many different-speaker pairs that are separated by about the same amount as same-speaker pairs. Nevertheless, the overall greater separation results in a slight improvement in discrimination power (Aitken 1995: 80–2). Now about 70% of all 153 pairs can be discriminated, compared to 60% and 62% with the first and second single dimensions respectively. An improvement in discrimination performance will only happen if the dimensions are not strongly correlated, otherwise the same discrimination performance occurs with two well-correlated dimensions as with one. For the data in Figure 2.3, it can be seen that there is in fact a certain degree of correlation, since the second dimension increases as the first increases. The question of correlation between dimensions is an important one, to be examined in greater detail later.

Figure 2.3 shows two configurations typical for voice comparison: that one pair of speakers can be well separated in dimension 1 but not in dimension 2, whereas another pair might be well separated in dimension 2 but not in dimension 1. The other two logical possibilities are also found: that a pair of speakers can be well separated in both dimensions, and not separated in either dimension. Note that the second dimension in Figure 2.3, just like the first, will not correctly separate out all three speakers: there is no single dimension that will correctly discriminate all speakers.

Multidimensional comparison

It is of course possible, theoretically, to carry the comparison to as many dimensions as desired. (The Euclidean distance between two observations in three dimensions now becomes the square-root of the sum of the squares of the differences between the observations in all three dimensions.) Figure 2.4 gives an idea of how three speakers might position in three-dimensional space (different dimensions and values are used from the previous examples).

The idea is, therefore, that as dimensions are added – as speakers are characterised in more and more dimensions – the possibility increases that a given speaker's range of variation will occupy a different position from all other speakers of the language in multidimensional speaker space. As a consequence, the speaker will eventually no longer overlap with other speakers, and can be absolutely discriminated from them as in the original one-dimensional example in Figure 2.1. Recall that, due to the existence of within-speaker variation for a dimension, speakers will not occupy a point in this multidimensional space, but a volume (in three-dimensional space) or hypervolume (in more than three dimensions).

Why voices are difficult to discriminate forensically

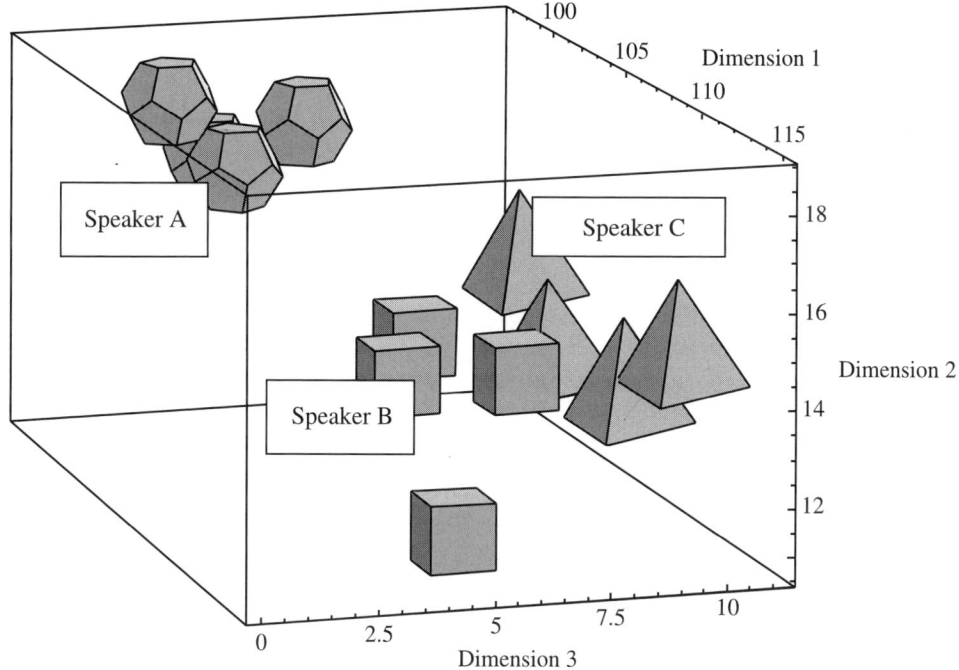

Figure 2.4 Three speakers in three dimensions

Discrimination in forensic speaker identification

In forensic speaker identification, then, the task is to discriminate between same-speaker voice samples and different-speaker voice samples. It was observed earlier in this chapter that this is using the word *discrimination* in a somewhat different sense from that encountered generally in statistics. The difference between the two senses is important and is explained here.

The way discrimination is usually construed in statistics – let us call this classical discriminant analysis – can be illustrated once again using data from the three speakers in Figure 2.3. A typical classical discriminant analysis on these three speakers would involve deriving a discriminant function from the existing data, *such that any new tokens from any of the three speakers could be correctly attributed to the correct speaker*. For example, we might take another of DM's *hello*s, find that it measured 1575 Hz on dimension one, and 2650 Hz on dimension two. The discriminant function that had been established with the old data would then hopefully identify this new token as having come from DM. It can be seen that an identification is involved here: the new token is identified by classical discriminant analysis as belonging to DM.

In this way, the individual-identifying power of different sets of dimensions can be tested. Powerful dimensions will allow a higher proportion of new tokens to be assigned to the correct speaker; weak ones will have a high proportion of incorrect decisions. If many new tokens from DM were identified as belonging to MD and vice versa, we would say that DM could not be well discriminated from MD with these dimensions.

Classical discrimination is clearly useful for finding out whether a particular dimension might be of use in forensic speaker identification, and there are many experiments in the literature where discriminant analysis of this type is used to demonstrate that a certain set of dimensions can correctly discriminate a predetermined set of speakers. However, it is of limited applicability in circumstances typical of forensic speaker identification. This is because in forensic speaker identification one does not normally have a predetermined set of known individuals to which the questioned sample is to be attributed on the basis of a set of dimensions. All one has is two (sets of) samples: questioned and suspect, and one does not know (of course!) whether the suspect's voice is the same as that of the offender.

If one wants to look at the problem from the point of view of discriminant analysis, only two categories are involved in forensic speaker identification: same-speaker pairs and different-speaker pairs. The task is then to examine the pair of samples from offender and suspect and determine to what extent the evidence supports whether the pair is to be attributed to the category of same-speaker pair or different-speaker pair. It is part of the aim of this book to show that this can be done, and how it is done.

Dimensional resolving power

Although a speaker's voice can be characterised in many different dimensions, not all dimensions are equally powerful. They vary according to their discriminatory power. For example, the discriminating power of the first dimension in Figure 2.3 on its own was 60%, whereas that of the second was 62%. Obviously, it is of no use to compare samples along dimensions that have little or no resolving power, and it is part of general forensic-phonetic knowledge to know which are the more powerful dimensions. This is a very important point, since time is usually limited and cannot be wasted on comparing samples with respect to weak dimensions.

The most powerful dimensions are naturally those that show a small amount of variation within a speaker and a large amount of variation between speakers, so a common way of selecting potentially useful forensic speaker identification dimensions has been to inspect the ratio of between-speaker to within-speaker variation (Pruzansky and Mathews 1964; Wolf 1972; Nolan 1983: 101; Kinoshita 2001). This ratio is called the F-ratio, and is usually a by-product of a statistical procedure called Analysis of Variance or ANOVA ('ann-over') (the proper statistical term for variation in this sense is *variance*). The discriminant power of the dimensions is then determined in discriminant analyses, as mentioned above.

The reader should now be able to understand that it is always important to establish that any dimension that is being used to compare speakers forensically is known to show at least greater between-speaker variation than within-speaker variation.

Ideal vs. realistic conditions

It has been explained so far that speaker identification works by (1) using sets of dimensions that (2) show relatively high ratios of between-speaker to within-speaker variance.

So far, all of this applies to any speaker recognition system. However, it is important to understand that there are very big differences between speaker recognition under ideal conditions and forensic speaker identification, where the real-world circumstances make conditions very much less than ideal. This is of course the individuality postulate in another guise, whereby no two objects (here: speakers' voices) are identical, although they may be indistinguishable (Robertson and Vignaux 1995: 4). These less than ideal conditions make forensic speaker identification difficult for the following two reasons: lack of control over variation, and reduction in dimensions for comparison. These will now be discussed.

Lack of control over variation

It was pointed out above that variation is typical of speech, and that there is variation both between and within speakers. Some of the variation in speech is best understood as random, but a lot of it can be assumed to be systematic, occurring as a function of a multitude of factors inherent in the speech situation. For example, between-speaker variation in acoustic output occurs trivially because different speakers, for example men and women, have vocal tracts of different dimensions (why this is so is explained in Chapter 6). Within-speaker variation might occur as a function of the speaker's differing emotional or physical state.

Since we usually do not have total control over situations in which forensic samples are taken, this is one major problem in evaluating the inevitably occurring differences between samples. This point is so important that it warrants an example. A concrete, but simplified, example with our 'average pitch' dimension will make this clear.

Let us assume that some speakers differ in the pitch of their voice. The pitch difference, which can be heard easily between men's and women's voices, also exists between speakers of the same sex, and is due to anatomical reasons connected with the size and shape of the vocal cords, which are responsible for pitch production.

However, the pitch of a speaker also varies as a function of other aspects of the situation. These will include, but not exhaustively: the speaker's mood, the linguistic message, and their interlocutor. The pitch can change as a function of the speaker's mood: it might be higher, with a wider range, if the speaker is excited; lower, with a narrow range if they are fed-up. The speaker's pitch will also change as a function of the linguistic message. In English it will normally rise, and therefore be higher, on questions expecting the answer yes or no. (Try saying 'Are you going?' in a normal tone of voice, and note the pitch rise on *going*.)

Not quite so well known is the fact that the speaker's voice might also change as a function of their interlocutor. Human beings typically 'move towards' or distance themselves from each other in many ways as a part of their social interaction – so-called postural echo is one form of convergence, for example (Morris 1978: 83–6). This also occurs in speech (Nolan 1983: 67, 134), where there is the opportunity for speakers to converge on, or diverge from, their interlocutor in many different ways. Pitch is one way in which they might move towards them. For example, Jones (1994: 353) notes that a famous sporting identity was demonstrated to significantly covary their pitch with that of 16 different interlocutors, and Collins (1998) also demonstrates pitch convergence. Speakers might even change basic aspects of their sound patterns. One subject of a recent forensic-phonetic investigation has been observed using an *r* sound after their vowels in words like *car* and *father* when speaking with inmates of an American prison, but leaving the postvocalic *r* sounds out when speaking with Australians (J. Elliott, personal communication). This behaviour suggests that he was accommodating to either real or perceived features in his interlocutors' voices. This is because one of the ways in which most (but not all) varieties of American English differ from Australian English is in the presence vs. absence of postvocalic *r*.

Imagine a situation where an offender is recorded on the video surveillance system during an armed hold-up. The robber shouts threats, is deliberately intimidating: their voice pitch is high, with a narrow range. A suspect is interviewed by the police in the early hours of the morning. He almost certainly feels tired, and perhaps also alienated, intimidated, cowed and reticent. As a result, if he says anything useful at all, his pitch is low, with a small range. Although the pitch difference between the questioned and suspect samples will almost certainly be bigger than that obtaining between some different speakers, it cannot be taken to exclude the suspect, because of the different circumstances involved.

Now imagine a situation where an offender with normally high pitch is perpetrating a telephone fraud. The suspect, who has a normally low-pitched voice, is also intercepted over the telephone talking to their daughter: they are animated, congenial, happy: their pitch is high, and very similar to that of the offender. Although the observed pitch difference between the samples is less than that which is sometimes found within the same speakers, it cannot be taken as automatically incriminatory because of the different circumstances involved.

Individuals' voices are not invariant but vary in response to the different situations involved. Depending on the circumstances, this can lead to samples from two different speakers having similar values for certain dimensions; or samples from the same speaker having different values for certain dimensions. It is also to be expected that samples from the same speaker will be similar in certain dimensions, and samples from different speakers will differ in certain dimensions.

Differences between voice samples, then, correlate with very many things. The differences may be due to the fact that two different speakers are involved; or they may reflect a single speaker speaking under different circumstances – with severe laryngitis, perhaps, on one occasion and in good health on another. One common circumstance that is associated with within-speaker variation is simply the fact that a long time separates one speech sample from another: there is usually greater within-speaker variation between samples separated by a longer period of time than between speech samples separated by a short period (Rose 1999b: 2, 25–6).

The effect of all this will be to increase the variation, primarily within speakers, and thus reduce the ratio of between-speaker to within-speaker variance of dimensions. This will result in a decrease in our ability to discriminate voices.

In order to evaluate the inevitable differences between speech samples that arise from ubiquitous variation, therefore, it is necessary to know the nature of the situational ('real-world') effects on the voice dimension under question. For example, assume that we measure the pitch from two voice samples: one a telephone intercept of the offender's voice, the other from a police interview of the suspect. If it is the case that speakers have a higher pitch when talking over the telephone, and there are a few data to suggest that this is so (Hirson *et al.* 1995), a higher pitch for the phone voice will be expected and not be exclusionary. However, a lower pitch for the telephone voice, or perhaps even the same pitch, will suggest an exclusion of the suspect.

In order to be able to evaluate all this, we need to know about how voices differ in response to different circumstances. Situations also need to be comparable with respect to their effect on speech. There may be situations where it is not possible to compare forensic samples because the effect of the situations in which they were obtained is considered to be too different. This basic comparability is one of the first things that a forensic phonetician has to decide.

Reduction in dimensionality

The second effect of the real-world context on forensic speaker comparison is a substantial reduction in dimensionality of the voices to be compared. This commonly occurs in telephone speech, for example, where dimensions that may be of use – that is dimensions with relatively high resolution power – are simply not available because they have been filtered out. It can also occur as the result of many other practical factors, like noisy background, someone else speaking at the same time, and so on.

A not insignificant factor in this respect is time (Braun and Künzel 1998: 14). The identification, extraction, and statistical comparison of speech dimensions take a lot of time, especially for investigators who are not full-time forensic phoneticians. Whereas the enlightened governments of some countries, e.g. Holland, Germany, Sweden, Austria, Spain and Switzerland, employ full-time forensic phoneticians, in others, e.g. Australia and the United Kingdom, a lot of this work is done by academic phoneticians who already have heavy research, teaching, supervisory and administrative loads. Time might therefore simply not be available to compare samples with respect to all important dimensions. Since our ability to discriminate voices is clearly a function of the available dimensions, any reduction in the number of available dimensions constitutes a limitation.

In addition to reduction, distortion of dimensions is a further problem associated with the real world. This occurs most commonly in telephone transmission (Rose and Simmons 1996; Künzel 2001). For example, the higher values of the second dimension that I adduced in Figure 2.3 to improve discrimination would in all probability be compromised by a bad telephone line. Distortion can also result from inadequate recording conditions, a low-quality tape recorder and/or microphone, echoic rooms, etc. Again, it is necessary to know what the effects are on different dimensions, and which dimensions are more resistant to distortion.

An interesting point arises from this practically imposed reduction in dimensionality. As a result, dimensions in which samples from different speakers differed a lot could be excluded. Then samples from two voices that in reality differed a lot would appear not to differ so much.

It was pointed out above that there will always be differences between forensic speech samples. These differences can be small, however, either because pre-existing differences have been excluded or distorted or because the voices from which they were taken were indeed similar in the first place. Two samples, therefore, whether they come from different speakers or the same speaker, cannot get any closer than minimally different, whereas there is no upper limit on the amount by which two samples can differ. There thus exists an asymmetry in the distribution of differences between samples, with a wall of minimal differences. (For a useful discussion on the effect of the wall phenomenon see Gould (1996, esp. ch.13)).

At first blush, this sounds like good news for defence. It looks as though it could be argued that, although two samples do not differ very much, they may still be from different speakers, since it is possible that the dimensions strongly differentiating the speakers were not available for comparison. Moreover, a *mutatis mutandis* argument – although there are large differences between the voice samples, they still come from the same speaker – is clearly not available to the prosecution.

Such a defence argument is, in fact, logically possible if discussion is confined to *differences* between samples. As will be demonstrated in detail below, however, discussion cannot be thus restricted, since the magnitude of *differences* between samples constitutes only half the information necessary to evaluate the evidence: it will be shown below that the question of how *typical* the differences are is equally important for its evaluation.

Representativeness of forensic data

A final way in which real-world conditions make it difficult to forensically discriminate speakers has to do with the representativeness of the data that are used to compare voices forensically. By representativeness is meant how representative the actual observations are of the voice they came from. The more representative the data are of their voices, the stronger will be the estimate of the strength of the evidence, either for or against common origin. Below we examine the main factors influencing the representativeness of forensic speech data. First, however, it is necessary to outline a model showing how typical forensic speaker identification data are structured. This model is shown in Figure 2.5.

Typically, speech samples in forensic speaker identification come from two sources. One source is the voice of the offender. The other is the voice of the suspect. These two sources are marked in Figure 2.5 as 'questioned voice' and 'suspect voice'.

It is worthwhile trying to conceptualise what these two sources must represent. One possible thought is that they must involve the totality of the speech of the offender, and the totality of the suspect's speech. But how could such a concept be delimited? Would it include their speech as children, or the speech that they are yet to utter? A more sensible interpretation would be to assume some kind of temporal delimitation, namely the totality of the suspect/questioned speech at the time relevant to the committing of the crime and its investigation.

In statistics, the idea of a totality is termed a population. Thus, for example, we could talk of the population of a speaker's *ah* vowels. In the forensic case, this would be *all* the speaker's *ah* vowels spoken at the time relevant to the committing of the crime and its investigation. We cannot know this totality, of course, and we cannot therefore determine, say, what the average value for an acoustic feature from *the totality* of the offender's *ah* vowels is. (The population value is often called the 'true' value.) We do, however, have access to the totality by taking *samples* from it. Moreover, since it is one of the basics of statistical inference that large samples tend to reflect

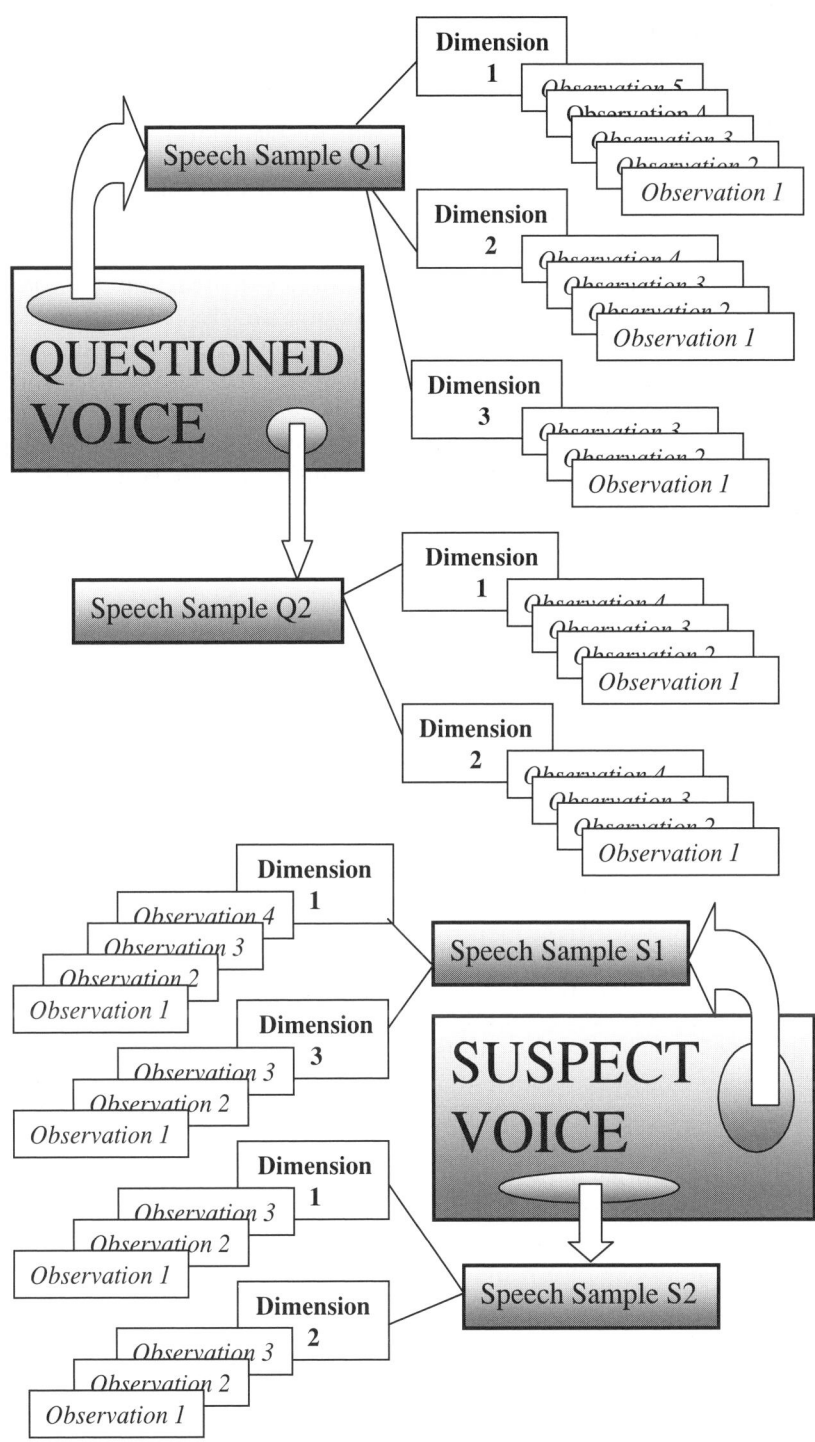

Figure 2.5 Architecture of observation data in forensic speech samples

Forensic Speaker Identification

the values of the population they are taken from (Elzey 1987: 42), we can in fact estimate the true population values from large samples taken from it.

It may or may not be the case that the questioned source and the suspect source are the same. If the former, the suspect is the offender; if the latter, the suspect is not the offender. In reality, either the questioned source and the suspect source are the same, or they are not. However we cannot know this, as we only have access to the samples from the sources, and, as demonstrated above, the outcome of the investigation is thus probabilistic, not categorical.

The point to be made from all this is a simple one. The strength of the voice evidence depends on how well the questioned and suspect samples reflect their respective sources: how well the questioned samples reflect the offender's voice and how well the suspect samples reflect the suspect voice. How well they do this depends on the amount of data available. Therefore the more data available, the more representative the samples are of their sources, and the stronger will be the estimate of the strength of the evidence, either for or against common origin.

What does 'more data' mean? Three factors influence the representativeness of forensic speech samples:

- the number of the questioned and suspect samples available (called sample number);
- the number of dimensions involved (dimension number, also called dimensionality); and
- the size of the dimensions (dimension size).

Sample number, dimension number, and dimension size are explained below.

Sample number

Let us assume that four speech samples are available: two of the offender, call these the questioned samples, and two of the suspect, the suspect samples. Then the questioned samples reflect a part of the totality of the questioned source, and the suspect samples reflect part of the totality of the suspect source.[2] Figure 2.5 symbolises these speech samples, marked 'Speech Sample Q(uestioned)' 1 and 2, and 'Speech Sample S(uspect)' 1 and 2, as having come from the questioned and suspect sources respectively. The number of speech samples for each source is called the *sample number*. So in this example, the sample number for the questioned voice is two, and the sample number for the suspect's voice is also two. It might be the case that there are 20 questioned samples (from 20 intercepted phone-calls perhaps) and 3 suspect samples (2 from phone-calls and 1 from a record of police interview). Then the questioned sample number would be 20 and the suspect sample number would be 3.

---

2 I am aware that I am not using the term *sample* in the proper statistical sense here. Strictly speaking, what I have denoted as the two questioned samples are in fact *two observations from one sample* from the population of the questioned voice. Quite often the reference of terms in applied areas of statistics diverges from that in pure statistics, and this is an example. Talking about speech samples in the way I have done is the accepted usage in forensic phonetics and speech science in general, and it is sensible to adopt the usage in this book.

Dimension number

As explained above, a voice is characterisable in very many dimensions. For example, the three voices in Figure 2.3 were quantified with respect to two acoustic dimensions in the word *hello*. The number of dimensions per sample is called the sample's dimension number, or dimensionality.

Voice samples are also characterisable in very many dimensions, and a few of these dimensions are represented, numbered, in Figure 2.5. Thus it can be seen that one of the speech samples from the offender (Q1) has been quantified for three different dimensions numbered 1, 2 and 3. As a concrete example, we could imagine that dimension 1 was an acoustic value in the offender's *ah* vowel in the word *fucken'*; dimension 2 might have something to do with the offender's voice pitch; and dimension 3 could be a value in the offender's *ee* vowel. The other speech sample from the offender (Q2) has been shown quantified for dimensions 1 and 2. Dimension 3 is not shown in the second speech sample: it might have been that there were no usable examples of *ee* vowels in Q2.

As far as the suspect's speech samples are concerned, the first (S1) obviously had some *ah* vowels in *fucken'* (dimension 1), and some *ee* vowels (dimension 3), but no usable pitch observations. The second suspect sample S2 lacks the *ee* vowel in dimension 3.

The fact that not all samples in Figure 2.5 contain the same dimensions is typical. It can be seen, however, that questioned and suspect samples can still be compared with respect to all three dimensions.

Dimension size

Each dimension in Figure 2.5 can be seen to consist of a set of measurements or observations. Thus the questioned sample Q1 had five measurable examples of *ah* vowels from *fucken'* (perhaps he said *fucken'* eight times, five tokens of which had comparable and measurable *ah* vowels). The five observations for the *fucken'* vowel dimension in Q1 might be the values 931, 781, 750, 636 and 557.

By *dimension size* is meant the number of observations contributing to each dimension. So in the above example the dimension size for dimension 1 in the questioned speech sample is five, and the dimension size for dimension 2 in the suspect's speech sample is three.

The different numbers of observations per dimension, of dimensions per sample, and of samples per source, will affect the degree of representativeness of the data. Below we examine how this occurs, starting with dimension size. Before this, however, a short digression is necessary to introduce the concept of mean values.

Mean values

Probably the most important thing needing to be quantified in comparing forensic speech samples is how similar or different their component dimensions are. This is because it is this difference that constitutes the evidence, the probability of which it is required to determine.

It was shown above with Figure 2.2, in discussing discriminability between speech samples, that if two sets of measurements from different speakers are compared, the

Forensic Speaker Identification

difference between the individual observations will vary. Sometimes the difference between the observations will be big and sometimes small. Thus in Figure 2.2, RS's highest value of 1406 and MD's lowest value of 1351 differ by only 55, but RS's lowest value of 1244 differs from MD's highest value of 1799 by 555. How different *is* RS's speech sample from MD's in this dimension then? If differences between individual observations vary like this, it is clearly not possible to answer this question by comparing individual observations, and dimensions must be compared with respect to values that are more representative of their sets of measurements as a whole (Moroney 1951: 34).

One such commonly used representative value is the arithmetical mean of the set of measurements. The arithmetical mean, symbolised $\bar{x}$ (pronounced x-bar), is commonly known as the average, and usually called just the mean, or sample mean. One of the values that are used to represent and compare the dimensions of forensic samples is therefore the dimension mean. In Figure 2.5, then, the means of the two sets of observations for dimension 1 in questioned samples Q1 and Q2 would be compared with the means from the two dimension 1 sets in suspect samples S1 and S2.

Effect of dimension size

It can be appreciated that, when comparing the dimensions of speech samples with respect to their means, if the dimensions only consist of a single, or few, observations, this will not necessarily provide an accurate estimate of the difference that would be obtained if the dimensions consisted of many more observations.

To illustrate this, Table 2.1 shows a sample of 17 observations of an acoustic feature from *hello*, taken from one speaker on one occasion. This acoustic feature is therefore the dimension. The mean of all the observations, indicated $\bar{x}_{1-17}$, is 1605. (The standard deviation of 65, after the overall mean, is a measure of the spread of the individual measurements.) The means of several sub-groups of observations are also given, as well as the magnitude of the difference of these sub-group means from the mean of all the observations (these differences are indicated by Δ). Thus the mean of the first three observations, indicated $\bar{x}_{1-3}$, is 1591, and the magnitude of the difference between the mean of this sub-group and the overall mean is (1605 − 1591 =) 14. It can be seen that the single observation 7 in Table 2.1 (1468, $\Delta = 137$), or single observation 17 (1714, $\Delta = 109$), both give a very poor estimate of the overall mean of 1605, and the sub-groups with fewer observations do not tend to give such good estimates as those with more. (Note that one sub-group with three observations (1–3) does perform better than one sub-group with nine, however.)

What this means if we are comparing speech samples is shown graphically in Figure 2.6. Figure 2.6 replots, as unfilled symbols, the values of the *individual* tokens in the two-dimensional comparison of RS, DM and MD of Figure 2.2, but also plots each speaker's *mean* position in two dimensions (shown by solid symbols). It can be seen that the speakers' mean values sit in the centre of gravity of their individual observations.

It is easy to see in Figure 2.6 that the mean positions as derived from all six observations of the three speakers are well separated. Imagine now that DM was the suspect and RS was the offender, and that the only tokens in the questioned sample were RS's rightmost two tokens in Figure 2.6, and the only two tokens in the suspect

Why voices are difficult to discriminate forensically

Table 2.1 A set of measurements of an acoustic value from 17 *hello*s produced by one speaker on one occasion

1. 1645	4. 1562	7. 1468	10. 1649	13. 1564	16. 1671
2. 1566	5. 1585	8. 1608	11. 1723	14. 1563	17. 1714
3. 1562	6. 1581	9. 1637	12. 1639	15. 1545	

Overall mean $\bar{x}_{1-17}$ = **1605** (standard deviation = 65)

$\bar{x}_{7-8}$ = 1538	$\bar{x}_{16-17}$ = 1693	$\bar{x}_{1-3}$ = 1591	$\bar{x}_{13-15}$ = 1557	$\bar{x}_{1-9}$ = 1579	$\bar{x}_{1-12}$ = 1602
Δ = 67	Δ = 88	Δ = 14	Δ = 48	Δ = 26	Δ = 3

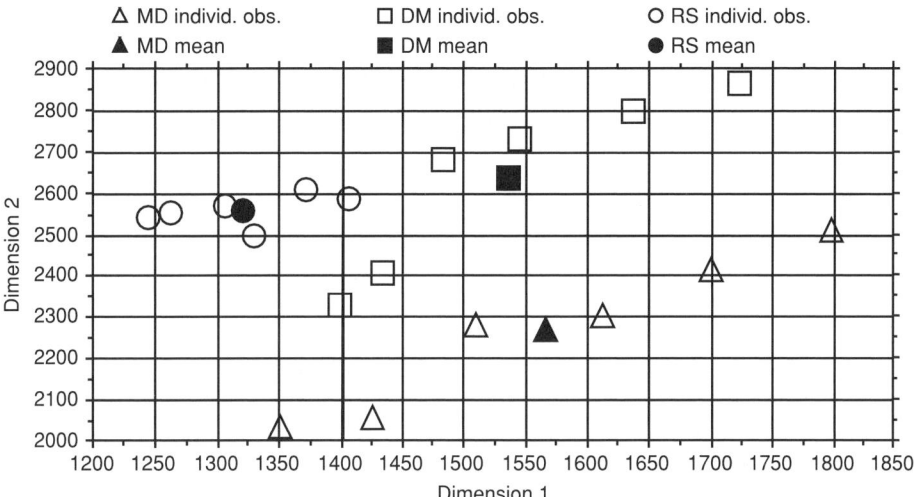

Figure 2.6 Three speakers' mean and individual values in two dimensions

sample were DM's leftmost two. Now the questioned and suspect samples would appear much more similar than is indicated by their mean values as based on six observations. This is because, as can be seen from the data in Figure 2.6, these individual tokens are not very representative of the respective speakers' means. Thus with fewer observations per dimension it might be more difficult to estimate the true difference between samples.

The effect of dimension size on the degree of representativeness of the mean is to alter the width of the interval wherein the true mean can be expected to be found at a given confidence level. This interval is called the confidence interval. Although confidence intervals are not strictly speaking part of the statistical approach espoused in this book, they are of use in giving an idea of how the number of observations affects the representativeness of the data. To illustrate with the data in Table 2.1, it can be calculated that the 95% confidence interval for the mean is ±33. This means that we can be 95% sure that the true mean lies in the interval of 66 between 1638 (i.e. the overall mean of 1605 + 33) and 1572 (i.e. 1605 − 33).

The confidence interval is a function, among other things, of the number of observations in the dimension: the greater the number, the smaller the confidence interval, and vice versa. By taking, say, four times as many observations one could approximately double the reliability of the average as an estimate of the true value (Gigerenzer *et al.* 1989: 82). What we are more interested in, however, is the way our degree of confidence changes with the number of observations. Using data similar to those in Table 2.1, a confidence interval of 35, for example, would correspond to a confidence level of 98% if the mean and standard deviation are derived from 20 observations, but ca. 68% if based on about five observations – a drop in confidence of about 30%. You could be 98% sure that the true mean lies within ±35 of the sample mean with 20 observations, but only ca. 70% sure if there were only five.

The message from this is that to get an accurate idea of the dimensional differences between speech samples, the number of observations in the dimension – the dimension size – needs to be large. With few observations it is less easy to be confident that the true dimension mean is being reflected, and that the real difference between the dimensions of the questioned and speech samples is being estimated.

It is usually assumed that 30 or more observations are required to give a good estimate of the population dimension (e.g. Elzey 1987: 66–7). Quite often, however, the dimension size in forensic speech samples is very restricted: there are unlikely to be 30 or more tokens of *ee* vowels in a short incriminating telephone call, for example, to compare with 30 or more *ee* vowels in a suspect sample. It will be explained in Chapter 4 how the effect of smaller dimension size is to lessen the strength of the evidence either in favour of, or against, a common origin.

Effect of dimension number

It was suggested above that speech samples from different voices will separate out and be discriminable in proportion to the number of dimensions that are used to compare them. In the forensic comparison of speech samples, the question of dimension number also arises, because it contributes to the accuracy of the evaluation of the true difference between speech samples, and hence the strength of the evidence. The way this happens will be explained in Chapter 4. Thus the more dimensions used in comparing forensic speech samples the better.

Effect of sample number

Small dimension size is not the only statistical restriction resulting from the typical lack of data in forensic comparison. It was pointed out above that a major source of within-speaker variation is the fact that a speaker speaks differently on different occasions. This means that in order to estimate a speaker's true mean value, and to get an accurate picture of how different samples relate, we need to take into account their mean values not just from one occasion, but many. This is illustrated in Figure 2.7.

Figure 2.7 shows some values from our three speakers, using the same two dimensions as in Figure 2.3. These data are from a study into long-term and short-term variation in the acoustics of similar-sounding speakers (Rose 1999b). Two things need to be noted in this figure. Firstly, and most importantly, the values represent data elicited on three separate occasions. Two constitute short-term variation: the

Why voices are difficult to discriminate forensically

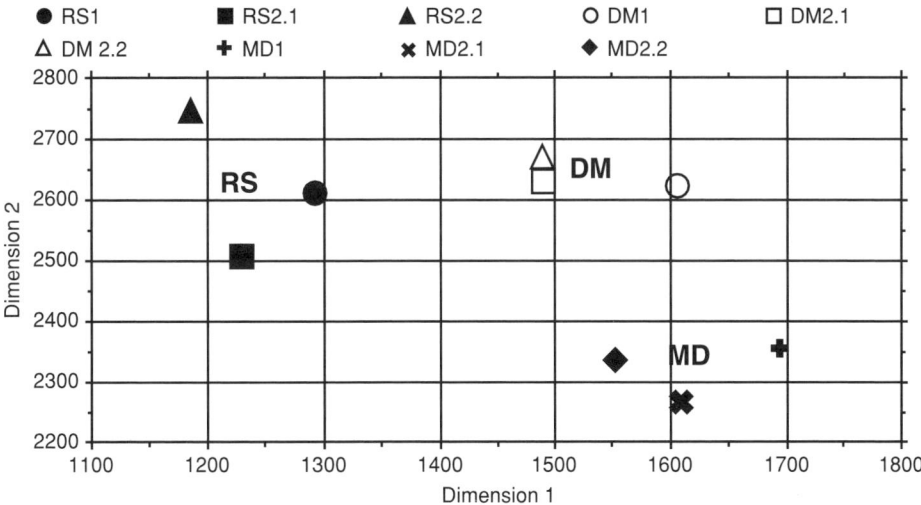

Figure 2.7 Three speakers' two-dimensional mean values on three different occasions

occasions were separated by only a few minutes. This might correspond to a situation where two telephone calls separated by a couple of minutes were intercepted. Two of the occasions represent long-term variation, being separated by at least a year. Thus the solid square and the solid triangle in the left of the figure (notated RS 2.1 and RS 2.2 in the legend) represent values for speaker RS taken from two occasions separated by a few minutes, and the solid circle (RS 1) represents a recording made one year earlier than RS 2.1 and RS 2.2.

Secondly, in keeping with the desideratum of the preceding section, each symbol represents the mean of a set of values. (There are only about ten observations per occasion, which is still well short of the value of 30 for a large sample, but fairly typical for forensic samples.) Thus, for example, DM's 2.1 sample contained values measured from 9 different *hello*s, and MD's sample 1 contained values from 12 different *hello*s. Figure 2.7 thus gives us a more realistic picture of how the three speakers actually differ, at least for these two dimensions on these three occasions.

The within-speaker variation in figure 2.7 is clear: the mean values for each speaker's three different occasions differ. Variation as a function of different occasions was mentioned above as an important element of within-speaker differences, and it can be seen that for two speakers (DM and MD) the short-term variation – the distance between the samples notated 2.1 and 2.2 in the legend – is smaller than the long-term variation – the distance between the samples marked 1 and the other two samples marked 2.1 and 2.2.

Thus just in the same way that a single dimension mean is not necessarily well reflected by one, or even a small number, of its constituent observations, it is likely that each speaker's true mean will not necessarily be well estimated from one mean from one occasion, or even three as shown in Figure 2.7. Although the speaker's true mean *might* lie somewhere near the centre of gravity of the means for the three occasions in Figure 2.7, samples from many more separate occasions would be required to be confident of having something like the true relationship between them.

Once again, the fewer the samples the less is the strength of evidence for or against common origin.

The data in Figure 2.7 must also be used to make a more positive methodological point. It can be seen that the three speakers are nicely spaced out, such that the between-speaker variation exceeds the within-speaker. There would obviously be much less difficulty discriminating these three speakers on the basis of their mean data. In fact, since the largest within-speaker distance is smaller than the smallest between-speaker distance, all pairs of tokens could be correctly discriminated as either same-speaker or different-speaker pairs. Remember, however, that there are considerably more than three speakers in the world, among whom there are bound to be many who are not discriminable in these two dimensions from one or more of these three speakers.

This section has shown that if one is able to calculate means of questioned and suspect samples that contain many dimensions, themselves each containing large numbers of observations, *and* over several different occasions, as in Figure 2.7, a better estimate can be obtained of where the samples' true means lie. Consequently, a better picture is obtained of the extent to which the samples actually differ, and the strength of the evidence for or against common origin of the samples will increase accordingly.

Legitimate pooling of unknown samples

There is one further important consideration, which arises from the desirability of combining means from different occasions. It has just been demonstrated that, in order to have a good idea of the true relationship between forensic speech samples, it is necessary to have means from samples from different occasions. Then, if we could combine the means from a speaker's separate occasions into one mean figure for a given speaker the resulting super-mean would give an even better estimate of that speaker's true mean, and might lead to an even clearer separation from the other speakers. Having a lot of samples from different occasions might also offset to a certain extent the restrictions from small dimension size: it is probably better to have a lot of small-sized samples from different occasions than one large-sized sample from one occasion.

However, caution is required here. Imagine that each of the mean values for a given speaker in Figure 2.7 represents a separate sample (as indeed they do). Imagine further that the RS samples were from the suspect, and that the DM samples were tendered by the police as samples of the offender's voice. The RS/suspect individual sample means can be combined to get one super-mean, because it is known that they do in fact all come from the same speaker. Before the means of the individual questioned samples can be combined, however, we need to know that they can be assumed to have come from the same speaker.

There might well be such evidence, independent of the phonetic comparison, which makes it clear that the same speaker is involved in each of the questioned samples. (Note that if it is known that the same speaker is involved, this does not necessarily imply that the speaker's identity is known.) However, if there is no independent evidence that they do (the police's assertion that they do does not of course count), then each individual questioned/DM sample would have to be separately compared with the combined RS/suspect sample. Such separate comparisons would usually not evince the same strength of evidence as a single comparison between the combined RS/suspect sample and a combined questioned/DM sample, because they would involve smaller dimension size (on the part of DM). This is therefore a further restriction often imposed from the real-world situation, and the question of to what extent questioned/unknown forensic speech samples can legitimately be pooled is an important one that must always be addressed.

Chapter summary

Forensic speaker identification is about attempting to discriminate between same-speaker and different-speaker voice samples. Voices are characterisable in an exceedingly large number of dimensions, and, because it is likely that they occupy separate regions in multidimensional speaker space, it is also likely that speakers of a given language all have different voices. However, the reduction in dimension number, and increase in within-speaker variation that is imposed by the realities of forensic-phonetic investigation mean that discrimination is not absolute, but must be probabilistic. The probabilities refer to the evidence. That is, one attempts to say how much more probable the observed difference between questioned and suspect samples is, assuming that they have come from the same speaker, and assuming they have come from different speakers. Further limitations obtain as a function of the amount of questioned and suspect data available for comparison, and whether separate unknown samples can be pooled. These limitations all affect the strength of the evidence.

3

Forensic-phonetic parameters

In Chapter 2 it was shown how voice samples are quantified and compared forensically along different dimensions, or parameters. Thus the voice samples of speakers in Figure 2.1 were compared with respect to values along a notional 'average pitch' dimension; and the voice samples of speakers in Figures 2.2 and 2.3 were compared using actual acoustic dimensions. This chapter is about aspects of the different dimensions with respect to which voices are compared. Since the term 'parameter' appears to be preferred to 'dimension' in this context, it will be used in this chapter.

Although there are some preferred parameters – acoustic values associated with vocal cord vibration and vowels are two that will be compared if possible – there is no predetermined set of parameters to use, as choice is so often dictated by the reality of the circumstances. Sometimes the existence of a prior auditory difference, for example a perceived difference in overall pitch range or a perceived similarity in overall vowel quality, will determine the choice, as the investigator seeks to quantify and check the acoustic bases of their observations.

The choice of parameters will also depend in part on the language of the samples, since variation between speakers and within speakers is found in different parameters in different languages. If comparison were between samples of some varieties of English, for example, then one of the things one would pay attention to would be how the *r* sounds are produced; whereas if the samples were in Cantonese this would be difficult, since Cantonese does not have *r* sounds. Instead one would be paying attention to, among other things, aspects of the *ch*- and *ng*-like sounds at the beginning of words, the behaviour of which is different from English. It can be appreciated from this that, because of the cross-linguistic differences in the way speaker identity is signalled in speech sounds, the forensic phonetician must have specialist knowledge of the speech sounds in the language(s) under investigation (French 1994: 174). Thus, in general, it constitutes part of forensic-phonetic expertise to know in which parameters speaker identity is best to be sought given the real-world conditions of the investigation.

It is important to know two general things about forensic-phonetic parameters: their different types, and their ideal properties. These two aspects are addressed below.

	LINGUISTIC	NON-LINGUISTIC
AUDITORY	**AUDITORY–LINGUISTIC**	**AUDITORY–NON-LINGUISTIC**
ACOUSTIC	**ACOUSTIC–LINGUISTIC**	**ACOUSTIC–NON-LINGUISTIC**

Figure 3.1 Primary classification of forensic-phonetic parameters

Types of parameters

The phonetic parameters that are used to compare speech samples forensically can be categorised with respect to two main distinctions: whether they are *acoustic* or *auditory*, or whether they are *linguistic* or *non-linguistic*. This gives rise to four main types of parameter: linguistic auditory-phonetic, non-linguistic auditory-phonetic, linguistic acoustic-phonetic and non-linguistic acoustic-phonetic (Figure 3.1; the *phonetic* will be left off below). Some further sub-classification is possible. Parameters can be *quantitative* or *qualitative*, and acoustic parameters can be either traditional or automatic. These distinctions will now be explained, and some important points connected with the distinctions will be discussed. I first discuss the fairly straightforward difference between acoustic and auditory parameters, then the distinction between linguistic and non-linguistic parameters, where examples of the four main types are given.

Acoustic vs. auditory parameters

Acoustic parameters are self-explanatory, and comparing voice samples with respect to their acoustic properties extracted by computer is perhaps what first comes to mind when one thinks of forensic speaker identification. However, it should not be forgotten that it is also possible to describe and compare voices with respect to their auditory features: how the voices and speech sounds *sound* to an observer trained in recognising and transcribing auditory features.

The earlier history of forensic-phonetic activity tended to be divided into acoustic and auditory factions, the controversial activity known as voiceprinting, despite its being a visual process, being especially associated with the former (Nolan 1997: 759–65; French 1994: 170). Three radically different positions can be identified with respect to the difference between auditory and acoustic parameters, and all of them will probably be encountered in the presentation of forensic speaker identification evidence:

- that auditory analysis is sufficient on its own (Baldwin 1979; Baldwin and French 1990: 9);

- that auditory analysis is not necessary at all: it can all be done with acoustics; and
- that auditory analysis must be combined with other, i.e. acoustic, methods (Künzel 1987, 1995: 76–81; French 1994: 173–4).

The third, hybrid, approach, sometimes called the phonetic–acoustic approach, is now the accepted position:

> Given the complementary strengths of the two approaches, it would be hard to argue coherently against using both in any task of [forensic] speaker identification, assuming the quality of the samples available permit it.
>
> Nolan (1997: 765)

> ... nowadays the vast body of professional opinion internationally recommends a joint-auditory-acoustic phonetic approach [to forensic speaker identification].
>
> French (1994: 173)

Today it is thus generally recognised that both approaches are indispensable: the auditory analysis of a forensic sample is of equal importance to its acoustic analysis, which the auditory analysis must logically precede. It is important to know the arguments that support this position, so that forensic-phonetic reports that are exclusively acoustic or auditory can be legitimately assessed and their inherent limitations recognised.

Even at the simplest level, it is necessary to first listen to forensic samples to decide whether their quality warrants proceeding further with analysis. A sensible heuristic is that if it is not possible to understand what has been said, forensic comparison is not possible. The point was mentioned above – and will be taken up in Chapter 6 – that we tend to hear what we expect to hear. Therefore a judgement on the understandability of a sample should also be made without recourse to any transcript or indication of what other interested parties think has been said.

Also, of course, one has to listen to samples in order to identify the variety of language(s) spoken and select the appropriate parameters that can be used to compare them in the further auditory and subsequent acoustic analysis. One might note, for example, that both samples contained many words with the same vowel in similar phonological environments, the acoustic features of which could be compared.

As mentioned above, it is generally assumed that an approach that combines acoustic and auditory analysis is required. This is because both approaches on their own can exhibit significant shortcomings. An auditory approach on its own is inadequate because, owing to one of the ways the perceptual mechanism works, *it is possible for two voices to sound similar even though there are significant differences in the acoustics*. This effect is explained in detail in Nolan (1990). This paper is very important because it describes a case in which two phoneticians had claimed identity between the questioned and suspect samples on the basis of auditory analysis, but it could be shown that the samples still differed consistently and sufficiently in their acoustics to warrant reasonable doubt that they came from the same speaker. We return to this case in Chapter 9 on *Speech perception*.

An acoustic analysis on its own is inadequate because, as already explained, without some kind of control from an auditory analysis indicating what is comparable,

Forensic Speaker Identification

samples can differ acoustically to any extent. Furthermore studies have shown that it is possible for two speakers to be effectively indistinguishable acoustically because of near identity in vocal tract dimensions, and yet still show linguistic differences that would be obvious to the trained ear. Nolan and Oh's 1996 paper 'Identical twins, different voices' reports, for example, how identical twins who could be assumed to have the same vocal tract dimensions and who differed minimally in general acoustic output, nevertheless differed consistently in the type of *r* sound they used.

Having explained why a forensic speaker identification report must contain both acoustic and auditory descriptions, some important aspects of both these types of analysis can be described.

Auditory analysis

The auditory analysis will be predominantly concerned with comparing samples linguistically, especially with respect to aspects of the sound system that is assumed to underlie the speech. An auditory-phonetic analysis should provide a summary of the similarities and differences between the samples of the sound system used and the way it is realised. Thus typical auditory-phonetic statements might be that the *t* sound in both samples is realised as a glottal stop at the end of certain words; or that the *ah* sound is further back in one sample than in another; or that, in a tone language, the tone in a particular word in one sample regularly has a mid level rather than the expected rising pitch.

It is important to describe what auditory-phonetic analysis implies; this is not generally well known, even if you have seen *Pygmalion / My Fair Lady*. Auditory analysis is part of the traditional training of phoneticians, those who study how speech is produced, transmitted acoustically, and perceived. It involves learning how to listen to speech in such a way as to permit its phonetic transcription. Phonetic transcription uses a conventional set of symbols that can of course be interpreted by anyone else who shares the same convention. The best-known and most pervasive convention is that of the *International Phonetic Association*, or *IPA*. A comprehensive and up-to-date guide to its use, with exemplification in several languages, is the *Handbook of the International Phonetic Association* (IPA 1999). IPA conventions also exist for the description of disordered speech.

Here is the sentence *This is an example of phonetic transcription* transcribed phonetically with IPA symbols:

[ˠˈðɪsəz nəɡˈzaːmpləy̰ ˈfn̥ɛdɪk˺ t̪ɹ̪æn˺ˈskɹɪpʃn̩]

As can be seen, it consists of a string of symbols, some of which (e.g. s, f, a) will be familiar to the reader from the English alphabet, and some of which (e.g. ə, ð, ɹ, æ) might not. There are also some diacritics (₀ , ˺ , ̪ , ˈ). The meaning of many of these symbols will be explained in Chapter 6, but for the moment it is enough to point out that some indicate consonants and vowels, and some indicate other features of the sentence such as its rhythm (stress patterns) and intonation (pitch).

One important feature is that the whole thing is enclosed in square brackets, which means that the symbols have a fixed phonetic interpretation. This interpretation is usually understood to refer to the way the sounds are produced. For example the [ð] symbol, representing the first sound in *this*, is interpreted as a sound made with the vocal cords vibrating, and with a constriction formed between the front of the tongue and the upper teeth narrow enough to make the air flowing through it turbulent. The sound also has a name, which is useful, given the mouthful of the preceding definition: it is a voiced dental fricative. What all this means is that a phonetician will be able to reproduce from the symbols the way the sentence was originally said in some detail: not only the vowels and consonants of its constituent words, but also its intended rhythm and intonation.

Importantly for forensic phonetics, as has already been demonstrated with [ð], phonetic transcription incorporates and implies *description*. A phonetician can objectively describe the speech sounds they hear – *any* speech sounds they hear. (And, equally important forensically, they can also say to what extent they *cannot* hear aspects of speech sounds.) This means that the description of the speech sounds in forensic samples can be checked and verified by other phoneticians, an indispensable feature for forensic evidence (cf. Robertson and Vignaux 1995: 6).

Finally, and most importantly, a phonetic transcription implies *theory* and *analysis*:

> ... the descriptive labels and transcriptional symbols are only the visible part of a large section of phonetic theory about the substance of speech. Each label, with its associated symbol, is a reflection of an underlying network of theoretical concepts and assumptions. A complex analytical framework underlies the description of speech sounds.
>
> Laver (1994: 4)

It is important to draw attention to the fact that there is a complex analytical framework underlying the description of speech, because non-linguists often do not understand that speech is the realisation of structure (Labov and Harris 1994: 291), and very complicated structure at that. This structure, and the analytic framework that reflects it, informs all forensic-phonetic work, and some of it is described in Chapters 6 and 7.[1]

The technique of phonetic transcription finds its primary application in descriptive linguistics, where the sounds of a previously undescribed language, and the way they combine, have to be worked out before the rest of its grammar – the structure of words, how the words combine into sentences and what meanings are expressed – can be described. It is therefore possible in principle to describe the differences and similarities in the speech sounds of different samples of speech using auditory techniques. To the extent that speakers differ in the auditory quality and arrangement of their speech sounds (and they do) this expertise can be used forensically.

Descriptive and classificatory schemes are also available for the non-linguistic aspects of a voice: those aspects that do not have to do directly with individual speech

---

1 Readers who already have some familiarity with linguistic analysis will note that the IPA transcription of the sentence actually contains symbols that imply two different levels of representation. Thus the vowel and consonant symbols represent a less abstract level nearer to the actual sounds, whereas the symbols representing the stress are of a higher order of linguistic abstraction. Since stress is realised mainly by pitch, its less abstract representation should be in terms of pitch.

sounds. Thus the auditory description will normally include a characterisation of the (non-linguistic) voice quality using some kind of accepted framework. One very thorough system is Laver's (1980) book *The Phonetic Description of Voice Quality*. A typical statement using this framework might be that 'phonation type is predominantly whispery, pitch range wide, and velopharyngeal setting nasal'. (*Whispery phonation type* means that it sounds as though the speaker's vocal cords typically vibrate in a particular, whispery, way when they speak. A *nasal velopharyngeal setting* means that it sounds as if the speaker has their soft palate down, thus sounding nasal. These terms will be explained in greater detail in Chapter 6.) If symbols representing these settings were incorporated within the square brackets of the example sentence above, thus:

[{ v̥ ṽ ↘ 'ðɪsəz nəg'zaːmpləy̥ 'fn̥nɛdɪk˺ tɹ̥ɹæn˺'skɹɪpʃn̥ }]

it would indicate that the whole sentence was said with whispery voice (v̥) and a nasal setting (ṽ). The curly brackets indicate the extent of the settings (Laver 1994: 420–1).

Acoustic analysis

As just mentioned, the acoustic analysis will usually be done after comparable items have been selected from the auditory analysis. It is in the acoustic analysis, perhaps, where the border between linguistic and individual information is most clearly crossed. This is because speech sounds have to be produced with an individual's vocal tract, and *the acoustic output of a vocal tract is uniquely determined by its shape and size*. Thus individuals with differently shaped vocal tracts will output different acoustics for the same linguistic sound. This is one obvious source of between-speaker variation mentioned above. The converse of this must not be forgotten either: individuals who do *not* differ very much in vocal tract dimensions will also tend to differ *less* in overall acoustics.

All this can now easily be illustrated with some acoustic data from Cantonese. Cantonese is a tone language. That means it uses pitch to signal word identity just as much as consonants and vowels, and each word has a fixed pitch shape in the same way as it has a fixed shape in terms of consonants and vowels. For example the words for *elder sister*, *cause*, and *ancient* are all said with a consonant that sounds similar to an English *g* followed by a *oo* vowel (this is transcribed phonetically as [ku]). Each of the three words sounds different, however, because they are said on different pitches. *Elder sister* is said with a high level pitch thus: [ku 55]; *cause* is said on a mid level pitch: [ku 33]; and *ancient* on a pitch which starts low and then rises to high in the speaker's pitch range: [ku 24]. As the reader will probably have worked out, pitch can be transcribed with integers, using a scale from 5 [= highest pitch] to 1 [= lowest pitch]. The three different pitches in the examples above ([55], [33], [24]) represent three of the six different Cantonese tones.

Speakers produce different pitches by controlling the rate of vibration of their vocal cords, and these different vibration rates are signalled acoustically by an easily measurable acoustic parameter called fundamental frequency, often referred to as F0

Forensic-phonetic parameters

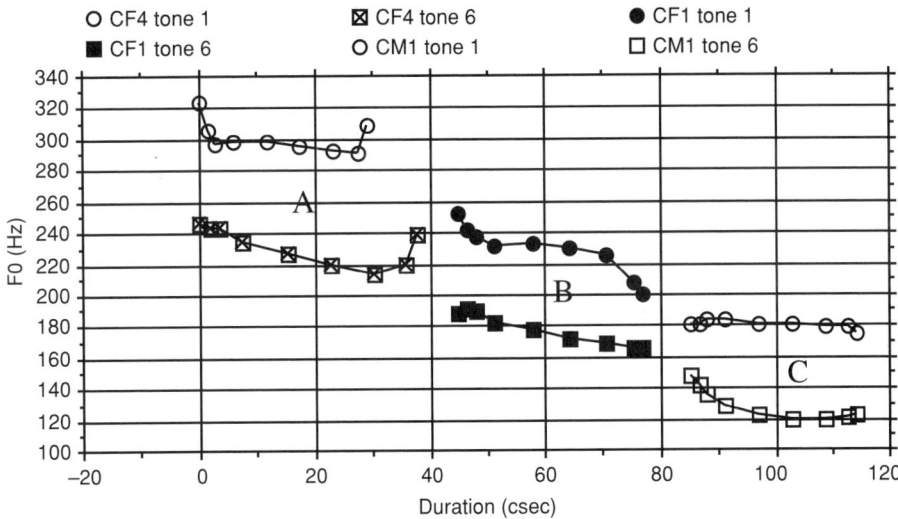

Figure 3.2 Three Cantonese speakers' high and low level tones

(eff-oh)[2]. However, the rate of vocal cord vibration depends, among other things, on the size of the speaker's cords. Big cords vibrate at lower frequencies; small cords at higher. Since females have on average smaller cords than males, females will have higher rates of vocal cord vibration and correspondingly higher F0 values than males for the same linguistic sound (in this case the same Cantonese tone). Individual differences can be found within the same sex too. This is illustrated in Figure 3.2.

Figure 3.2 shows the fundamental frequency for two Cantonese tones – the high level [55] tone and the low level [22] tone – as spoken by three speakers: two females (A and B) and a male (C). The data are from Rose (2000). The speakers' values have been plotted alongside each other for ease of comparison. The fundamental frequency scale is shown vertically; duration (in hundredths of a second, or centiseconds, csec) is shown horizontally. It can be seen that each speaker has two F0 curves, the top one is for the high level tone, the bottom for the low level.

Figure 3.2 shows that the acoustic values for the same tones across speakers are different. In particular, the between-speaker differences are big enough to swamp the linguistic differences. Thus the female speaker A's low tone is very close in frequency to the female speaker B's high tone (both lie between ca. 240 and 220 Hz), and the female speakers B's low tone is very close in frequency to the male speaker C's high tone (at ca. 180 Hz).

This Cantonese example demonstrates several points. Apart from showing how speakers can differ in their acoustics for the same linguistic sounds (the two Cantonese tones), this example also illustrates the importance for comparison of a prior

2 In this book, 'F0' will be used both to refer to the 'name' of the fundamental frequency as a concept or entity and to represent a particular value of the variable of fundamental frequency (e.g. F0 = 100 Hz). Strictly speaking, the latter mathematical use should be notated with italic type thus: $F0$, or F_0. This lack of a notational distinction will also apply to other parameters.

Forensic Speaker Identification

linguistic analysis. If we were not able to identify the tones, and compared the acoustics of speaker A's low tone with speaker B's high, for example, we might not conclude that they were very different, when in fact they are.

Consideration of the data in Figure 3.2 will help understand how much of the knowledge of between-speaker acoustic differences has come primarily from work on *speech*, as opposed to *speaker*, recognition: from the need to extract acoustic correlates of speech sounds from raw speech acoustics. In real speech, as Figure 3.2 shows, the differences between speech sounds are swamped by the 'noise' from the differences between speakers. From Figure 3.2 it is obvious that in order for a Cantonese listener to recognise a tone as high level or low level from its F0 (and thus help recognise the word), their brain must somehow factor out the speaker-dependent attributes of the sound. Likewise, for a machine to recognise a *speech* unit, it must also be able to remove the *individual* content.

There are many ways of doing this. One of the simplest is to use what is called a *z*-score normalisation, where a speaker's average value and range of values are calculated, then the values for their tones are normalised by subtracting them from the speaker's average value and dividing by their range (Rose 1987, 1991). For example, speaker CF4's average F0 value was 266 Hz and her F0 range was 37 Hz. The first F0 value for speaker CF4's low tone was 246 Hz (see Figure 3.2), so this value is normalised to $(246 - 266)/37 = -0.54$.

The result of such a normalising transform is shown in Figure 3.3, where the three speakers' normalised tonal values have now been superimposed. (The reader can check that the first normalised value of CF4's low tone (−0.54) lies indeed just below the −0.5 value.) In Figure 3.3 it is clear that very much of the between-speaker differences in the tonal acoustics has been removed, leaving three very similar normalised F0 curves for each tone. It would now be possible to recognise a tone as being either high or low from its normalised curve: any tone with a normalised value greater than 0 for any

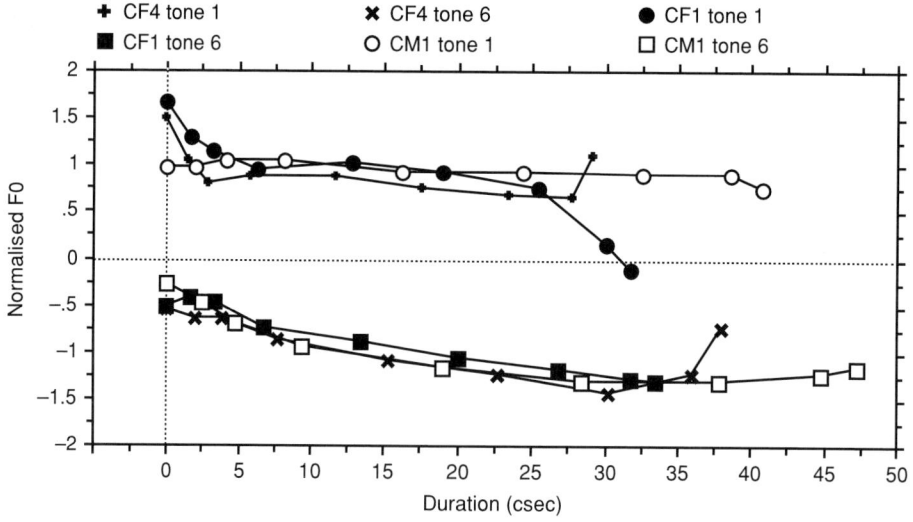

Figure 3.3 Three Cantonese speakers' normalised tones

except its final point is *high*; any with normalised value less than 0 is *low*. Note that not all the original between-speaker differences have been eliminated. The curves still differ in duration, for example, and in the details of their onset and offset perturbations. (The latter reflect the different ways speakers have of letting their vocal cords stop vibrating. But as always the magnitude of within-speaker to between-speaker variance for this feature has to be known before it can be considered for use in FSI.)

The residue of this normalisation process is the speaker-dependent features: in this case the features of average F0 and F0 range that differentiated the speakers before normalisation. F0 average and F0 range emerge therefore, although they are noise to be excluded in *speech* recognition, as potential acoustic parameters for *speaker* discrimination (cf. van der Giet 1987: 121).

It is parameters like these that were used to illustrate discrimination in the earlier sections, and that can legitimately be considered as features encoding the individual, as opposed to the linguistic, content. Of course, in order to contribute to successful speech recognition, these parameters must also have small within-speaker variation. That is, the same speaker must differ from occasion to occasion in their average F0 and F0 range less than different speakers do.

Traditional vs. automatic acoustic parameters

It is important to be aware of a distinction between two different types of acoustic parameters: traditional and automatic.

Traditional acoustic parameters

The acoustic cues that relate to differences between language sounds – either within a language or between languages – can be called traditional acoustic parameters (another term is *natural* parameters; Braun and Künzel 1998: 15).

When, in the late 1940s, it first became practically possible to study the acoustic patterns of speech, interest naturally focused on what the acoustic correlates of speech sounds were (e.g. Jakobson *et al.* 1952). What in the acoustics signalled the difference between an *r* sound and an *l* sound, for example; what signalled the difference between the *a* vowel as in *had* and the *ea* vowel as in *head*; between *inSULT* (verb) and *INsult* (noun); what signalled the differences in pitch that correlate with statements and questions?

Since then, an enormous amount of research has been concentrated on these questions and as a result our knowledge in this area is considerable. We now have a very good idea indeed of the range of sounds in the world's languages (Ladefoged and Maddieson 1996: 1–2), and of the acoustic features that help cue the linguistic-phonetic differences between them. A highly recommended text for the layman is now available, together with a CD, which explains and exemplifies much of this: Ladefoged's (2001) *Vowels and Consonants – An Introduction to the Sounds of Languages*.

We also know how the acoustic cues are affected by being produced by vocal tracts of different shapes and sizes (Fant 1960; Stevens 2000). Finally, we know what factors mediate the perception of the cues. To use the example of Cantonese tones just discussed, we know that pitch height is a linguistically relevant feature because, in a tone language for example, it signals the difference between different tones, which help distinguish different words. We know further that the main acoustic correlate of pitch is fundamental frequency. We also know that, due to the different sizes of speakers' vocal cords, the same tone can be encoded in different F0 values for different speakers. Thus we know that the acoustic correlate of tonal pitch (as opposed to pitch in general) is normalised F0. We also know that other factors mediate the perception of normalised F0 as tonal pitch, like the quality and temporal alignment of vowels or consonants said in the same syllable as the tone (Rose 1989).

Automatic acoustic parameters

A different development in the approach to speech acoustics occurred with the advent of computer technology. This approach was motivated by non-linguistic aims: how well can computers be programmed to *automatically* recognise speakers and speech? It was able to implement and develop powerful digital signal-processing pattern recognition algorithms, backed up by intensive probabilistic statistical processing, to accomplish with high degrees of success the tasks of automatic speaker and speech recognition.

Humans and computers recognise speech patterns in different ways (Ladefoged 2001: 78–95), and the computerised approaches, which for obvious reasons can be called automatic, make use of acoustic parameters some of which differ considerably from the traditional acoustic parameters that humans use to recognise speech sounds. One of the reasons why automatic ones are used is that it often proves difficult to automatically identify and extract the traditional ones in normal running speech. Some of the automatic parameters are of considerable mathematical complexity and beauty. One of the most important of them – the cepstral coefficients and their cepstrum – is exemplified in Chapter 8.

Thus there are two different types of acoustic parameters: traditional and automatic. Two main differences between the two types are important for forensic phonetics. One is that, because they are more robust under controlled conditions than the traditional parameters, automatic parameters are more powerful. Experiments comparing the performance of automatic and traditional parameters in speaker recognition, for example, generally show that the automatic parameters outperform the traditional ones by about 10% (Hansen *et al.* 2001; Rose and Clermont 2001). It is as yet unknown, however, how far their robustness extends. Broeders (2001: 8), for example mentions their 'extreme sensitivity' to transmission channel effects (for example from different handsets and telephone lines), and to environmental factors, such as extraneous noise, and reminds us that such effects are typical in forensic samples. For a useful summary of these various environmental and channel distortion effects, see Furui (2000: 20).

The other difference is that the automatic parameters do not relate in any straightforward way to the linguistic auditory or articulatory properties of speech sounds (Braun and Künzel 1998: 15). Thus whereas linguistic pitch can be clearly related to the traditional acoustic parameter F0, as demonstrated with Cantonese above, the fourth cepstral coefficient, for example, is a mathematical abstraction of a high order that is difficult to conceptualise at all, let alone relate to something phonetic: 'Ein solcher "abstrakter" Merkmalvektor entzieht sich jedoch jeder Anschaulichkeit' (van der Giet 1987: 125). Some tenuous relationships do exist (for example, Clermont and

Itahashi (1999) have shown that certain aspects of vowel quality relate to the second and third cepstral coefficients) but these relationships are not strong enough to be of any forensic use.

Neither do the automatic parameters refer directly to what is known about the way speakers actually differ in their speech. One of the leading figures in automatic speaker and speech recognition has said:

> Recent advances in speaker recognition are mainly due to improvements in techniques for making speaker-sensitive feature measures and models, and they have not necessarily come about as an outgrowth of new or better understanding of speaker characteristics or how to extract them from the speech signal.
>
> Furui (1994: 8)

It is worth noting, however, that Furui adds: 'It can be expected that better understanding of speaker characteristics in the speech signal can be applied to provide more effective speaker recognition systems.'

The tenuous relationship between automatic parameters and phonetic features has certain important consequences for forensic phonetics. Suppose, firstly, that one hears an important difference, or similarity, between two speech samples. Perhaps they differ in the quality of a vowel in the same word, or they show very similar, but unusual, pitch shapes for a tone. The acoustic quantification of the difference or similarity is then relatively easy with traditional acoustic parameters, because it is known in which traditional acoustic dimensions these auditory differences are likely to reside. This is not the case with some automatic parameters.

Moreover, thanks to the received Acoustic Theory of Speech Production, the behaviour of some traditional acoustic parameters is articulatorily interpretable. This means that we can infer certain aspects of the production of sounds from the acoustics, which is of obvious use when one is ultimately trying to make inferences from acoustic patterns to the vocal tract(s) that produced them (Rose 1999b: 7). Again, this is not the case to any useful extent with some automatic parameters.

Finally, it has also been suggested that, because they are difficult to conceptualise, there would be problems in explaining automatic parameters to the judiciary and juries.

The above must not be taken to imply that automatic parameters cannot, or must not, be used forensically: '. . . the factors limiting the applicability of automatic procedures do not in fact always constitute absolute impediments' (Broeders 1995: 157). Given their power, and the fact that they circumvent some of the problems associated with traditional parameters, it would be foolish not to try to make use of them when possible, and not to continue to research the extent of their forensic applicability (Rose and Clermont 2001). Certainly the ideal, if for time and financial constraints currently unrealistic, situation would be for forensic phoneticians to be able to apply both traditional and automatic methods wherever possible.

Linguistic vs. non-linguistic parameters

The second main distinction within forensic speaker identification parameters concerns whether they are *linguistic* or *non-linguistic*.

Linguistic parameters

As yet, inadequate information has been given to indicate what is meant by a linguistic (phonetic) parameter. A linguistic parameter can be thought of as any sound feature that has the potential to signal a contrast, either in the structure of a given language, or across languages or dialects. This could be for example a contrast between words in a language: the difference in sound between the *a* and *ea* vowels in the words *had* and *head* is a linguistic parameter; as is the difference in sound between the way a language signals a question and a statement. A linguistic parameter could also be whether two words like *lore* and *law* were homophonous.

This characterisation includes an enormous number of possibilities. For example, it might be observed that in one sample the word *car* was pronounced with a final *r* sound, whereas in another sample it was not. This is a linguistic difference firstly because *r* is one of the sounds English uses to distinguish words (cf. *road* vs. *load*), and secondly because words are also distinguished by the absence or presence of speech sounds (cf. *heart* vs. *art*). Thus the analyst will describe and characterise in their report both structural and realisational features of the sound systems in the samples that may have diagnostic value (these two aspects of a language's sound structure, especially the technical sense of *realisation*, are described in detail in Chapters 6 and 7).

A well-known case in which linguistic features played a major role was that of Prinzivalli, mentioned in the introduction (Labov and Harris 1994: 287–99). In this case, defence was able to demonstrate that the questioned samples contained linguistic features typical of a New England accent, whereas the suspect clearly had a New York accent. For example, in the word *that*, Eastern New England usually has a particular vowel quality that never occurs in New York (*ibid*.: 291–2). Another linguistic feature identifying the samples as coming from different dialects was whether they manifested the so-called COT–CAUGHT merger (*ibid*.: 292–7). In a large area of the United States, including Eastern New England, words like *cot* and *caught* are pronounced with the same vowel (so *cot* and *caught* are homophonous). But not in New York.

Non-linguistic parameters

A typical non-linguistic observation might be that voices in both samples sound to have lower than average pitch. This is a non-linguistic observation because, although it might signal the speaker's emotional state, a speaker's overall level of pitch is not used to distinguish linguistic structure: to signal the difference between words, for example.

Linguistic and non-linguistic observations can be made both in auditory and in acoustic terms. The examples given in the two preceding paragraphs would constitute auditory observations. An example of an acoustic–linguistic comparison would be a comparison of the acoustics of the same linguistic sound in different samples. This occurred in the Prinzivalli case when the acoustics of diagnostic vowels in the bomb threats were compared with the same vowels in Prinzivalli's voice (Labov and Harris 1994: 294–7). An example of a non-linguistic acoustic analysis might be when, in the case just mentioned with the two voice samples with lower than normal overall pitch, the acoustical correlate of overall pitch (called mean fundamental frequency) was quantified and compared. Like this example, most non-linguistic acoustic parameters fall under the heading of long-term parameters, which are now briefly discussed.

Long-term acoustic parameters

There is one class of acoustical quantification that involves finding statistical values (e.g. averages and ranges) for a particular parameter based on measurements made over relatively long stretches of speech. These are called long-term measurements, and the parameters from which they are made are called long-term parameters.

The idea behind long-term parameters is precisely to factor-out the contribution of individual sounds to a particular acoustic parameter so as to yield an overall value for a speaker that is independent of the contribution of individual sounds to the parameter. For example, the F0 of the voice changes in speech almost continuously from moment to moment as a function of many factors. If F0 is sampled every hundredth of a second over a minute of speech, however, an 'average F0' can be derived that is not a function of the local, short-term factors, but which is more likely to reflect individual characteristics like vocal cord size (Rose 1991). Long-term parameters do not appear to refer to prior linguistic analysis, since one simply measures the whole of the speech sample for the usually quasi-continuous variable.

Long-term parameters are important because they are considered among the more powerful indicators of individual voice quality (Hollien 1990: 239–40). Automatic long-term parameters are used with great success in automatic speech and speaker recognition. They are exemplified in Chapter 8. However, in forensics, some kind of prior estimate of comparability is still necessary, since long-term measures, just like any other measures, are never totally inert to real-world factors. For example long-term F0 can vary as a function of a speaker's state of health (e.g. if they have laryngitis), of their emotional state (whether they are shouting a stream of abuse at a football referee), or of their interlocutor.

Finally, it is important to point out that the distinction between linguistic and non-linguistic parameters is language-specific. In other words, a parameter may be non-linguistic in some languages and linguistic in others. A good example is the mode of vibration of a speaker's vocal cords, called phonation type. In English, this signals non-linguistic information like emotional state, or personal idiosyncrasy. In (Northern) Vietnamese, it serves, like consonants and vowels, to signal the difference between words and is therefore linguistic.

Linguistic sources of individual variation

It has just been pointed out that both linguistic and non-linguistic parameters are of use in comparing voice samples. It is important to emphasise this because it is tempting to see the non-linguistic parameters as those carrying the individual-identifying information. This is an important point and merits some discussion.

When we hear someone saying something in our own language, our naive response is that we can normally tell at least two types of thing. We hear of course what is being said: for example, that the speaker said 'Put it in the back', and not 'Put it in the bag'. This can be called the linguistic message. But we can invariably also say something about who is conveying the message: their sex almost certainly, possibly also their emotional state: 'That man sounds angry'. Sometimes we recognise the voice but not its owner: 'I've heard that voice before; don't know who it is though'. And sometimes we even identify the voice's owner: 'That's Renata's voice'. Thus a very simple model for the information content in a voice might be that it contains two types: *linguistic* and *individual* (a much more detailed model will be presented in Chapter 10). We can do this whether we see the speaker or not, and so these two types of information must be contained in the speech acoustics.

Although we do it with ease, to hear the linguistic and individual content separately is no mean cognitive feat since, as will be shown, the two types of information are not signalled in separate parts of the acoustics but are convoled and have to be disentangled by the perceptual mechanism. This fact has important consequences for speaker identification, since we need to understand how this happens in order to be able to interpret the variation inherent in speech.

It is important to realise, however, that although speakers' voices are probably most often thought of as differing in their individual content, voices can also differ in the structure encoding the linguistic message: '. . . a significant part of between-speaker variability exhibits regularities which are linguistic, rather than solely acoustic, in nature' (Barry *et al*. 1989: 355).

Regional accents

The most obvious example of this would be a situation where different regional accents of a language are involved. One voice for example might differ from another in pronouncing an *r* sound after the vowel in words like *car*, *hear*, *word*, and *fourth*. (The owner of the first voice might come from England's West Country, Scotland or Northern Ireland; such an accent with the post-vocalic *r* is termed 'rhotic'.) Failing truly bidialectal speakers, such differences, if they are consistent, can constitute the strongest evidence that different speakers are involved, and there are cases where a suspect has been exonerated on the basis of a careful dialect-geographical comparison of his speech with that of the offender (the above-mentioned Prinzivalli case is an example).

Of course, it is possible that two samples that differ strongly in geographical accent will not be confused in the first place. However, accentual differences can be subtle enough to elude the non-expert. In the Prinzivalli case, for example, the differences between a New York and a New England accent, although clear to an expert, were not salient to the Los Angeles police on the other side of the continent who were responsible for the arrest. Moreover, there are areas where there is very little regional accentual variation, as in Australia. Furthermore, non-accentual linguistic differences between speakers can often be a lot more subtle and complicated. For example, as mentioned above, even identical twins can differ in the small details of the implementation of their linguistic system, consistently using different types of *r* sounds. Or a speaker might have a slightly different way of pronouncing a *th* before an *r* in words like *throw* and *three*, as did my former flat-mate.

Social accents

Another example of the degree of subtlety involved in linguistic differences comes from sociolinguistic variation. Accents can differ socially as well as regionally, and one of the most important insights from modern sociolinguistics is that an individual does not have a single invariant linguistic system, but uses several variants in different

Table 3.1 Percentage of *r*-full forms in contexts of differing linguistic formality for six New York speakers

Speaker	Casual speech	Careful speech	Reading passage	Reading word list	Reading minimal pairs
BN	0	0	13	33	33
JP	0	3	23	53	50
SK	0	6	8	38	100
DH	0	31	44	69	100
AG	12	15	46	100	100
ML	32	47	39	56	100

sociologically defined circumstances (within-speaker variation again!). Thus our upwardly mobile West Country rhotic speaker might not pronounce her *r*s so frequently under different social circumstances, for example when she wants to make a favourable impression (since post-vocalic *r* in British English sounds rustic). Or an Australian speaker who says an *f* sound at the end of a word like *both* or *mouth* when talking informally with his mates will probably substitute the *th* sound on those very infrequent occasions when he actually gets to meet his bank manager.

One of the earliest and best-known demonstrations of this behaviour was carried out in New York's Lower East Side. The New York accent is traditionally classified as non-rhotic, or lacking *r* sounds after vowels in words like *beer, fire, flower, Saturday, November*. However, presumably under the influence of General American, which is fully rhotic, *r* sounds have made a comeback; thus the accent is described as having 'variable non-rhoticity' (Wells 1982: 503, 505–6). The extent to which speakers pronounced post-vocalic *r* was examined as a function of five different contextual styles by Labov (1972: 99–109). The contextual styles differed in the degree to which they focused the speakers' attention on their linguistic performance, and were: casual speech, careful speech, reading a passage, reading word lists, and reading pairs of similar words differing orthographically in the presence of *r*, like *guard* and *god, dark* and *dock*. Table 3.1 summarises the percentage of occurrences of *r* in the different styles for several New Yorkers.

It can be seen from Table 3.1 that the percentage of words with *r* generally increases with the linguistic formality of the situation. In casual speech, BN, JP and SK are non-rhotic, but they start to pronounce *r* sounds in more formal contexts. The *r* sound is said to be a sociolinguistic variable, and symbolised (r). There are many other factors besides linguistic formality that evince different speech behaviour. For example, the different image one presents in various social interactions has also been shown to correlate with different vowel qualitites in Philadelphia (Labov 1986: 409–11).

Within-speaker variation in pronunciation as a function of contextual factors like these has been demonstrated for many different languages, and must be assumed to be fairly ubiquitous. Of course, the details are different for different languages. For example, in the Chinese dialect of Pudong, spoken near Shanghai, speakers shift towards the local Shanghai standard in more formal styles, like reading, by changing word-initial *h* to *f* in words like hong *bee* and hoq *happiness* (Xu 1989: 92). Both

American and British English have *r* as a sociolinguistic variable, but in American English the percentage of *r* forms increases with formality, whereas for British English social accents it decreases (Trudgill 1978: 21).

An additional complication to this, and one that is forensically important because it involves between-speaker variation, is that speakers from different socioeconomic backgrounds display different incidences of this behaviour. So a working-class New Yorker will differ from a middle class speaker in retaining a lower incidence of *r*s in the same sociologically defined circumstances. There is evidence of this in Table 3.1, where BN is a truck driver with minimal education, and ML is a well-educated lawyer.

Speakers thus do not have an invariant accent. The consequences for forensic-phonetic comparison are obvious. When comparing samples with respect to linguistic features it is necessary to take into account any factors that may have sociolinguistic relevance. Imagine two forensic-phonetic samples of New York dialect that differed in their *r* incidence: one sample has 0% and one 32%. They would have to be evaluated very differently depending on comparability with respect to sociolinguistic factors. If they were comparable then that would suggest different speakers (cf. BN's and ML's casual speech). However, if the sample with 0% was from casual, and the sample with 32% from careful speech, the evidence would not be exclusionary (cf. BN's casual and reading word-list figures). Likewise, *mutatis mutandis*, two samples differing minimally in *r* incidence, for example 32% and 33%, are not necessarily a match under non-comparable sociolinguistic conditions (cf. BN's reading word-list and ML's casual speech figures).

Forensic significance: linguistic analysis

The foregoing examples have to do with the possible between-speaker and within-speaker differences in the realisation of speech sounds. In some, like the New York *r*, the realisation is conditioned primarily by non-linguistic context, in another by sound environment (my flatmate's *r* affects the realisation of his *th*), and the identical twin example represents (probably) free choice within the linguistic system. Apart from illustrating the important point that speakers can differ linguistically in exceedingly complex ways, these examples also show that their proper forensic evaluation must be informed by knowledge from the discipline of linguistics, that is, knowledge of the nature of language and the way it works. That is why above all forensic phoneticians need to be linguists, and it is important to devote part of this book to an explanation of the conceptual frameworks within which such things like speech sounds, their realisations and variation are understood and described in linguistics.

The areas of linguistics that deal with these questions are phonetics, phonology, sociolinguistics and dialectology. As already mentioned, phonetics is a subject that studies how humans speak; how the speech is transmitted acoustically; and how it is perceived. Phonology looks at how the speech sounds are organised. There are many different phonological approaches. The most important one for forensics is called phonemics, which looks at what sounds signal the difference between words in a language. Both sociolinguistics and dialectology look at variation within a language, the former as a function of sociological variables like income and education; and the latter as a function of geographical location.

One message in this section, therefore, is that it is important for forensic-phonetic analysis to incorporate a comparison based on linguistic features. It is worth rehearsing the three reasons for this. The first, and most important, reason has already been mentioned: that speakers can differ linguistically. The second reason is that it is the linguistic structure that specifies what is comparable in the first place. Common sense suggests that it is necessary to compare samples with respect to the same sounds. It is effectively of no value comparing the acoustics of an *ee* vowel in one speech sample with the acoustics of an *ah* vowel in the other; one has to compare *ee* vowels in both samples. Although the question of what is the same sound is by no means a trivial one (why this is so is explained in Chapter 7), we recognise what is an *ee* vowel by linguistic analysis.

The third reason for the primacy of a linguistic approach is a contingent, rather than principled one. It is clearly the case that the descriptive and analytic tools of phonetics are intended to make sense of the linguistic rather than the individual content in a voice. This is so because the primary application of phonetics has been linguistic: studying speech with reference to language. Also, one suspects, it is because the separation of speech sound from who is saying it is the way that humans function. Thus phoneticians are able to focus on and describe the speech sound independently of the voice it is being realised in, and the productional, acoustic and perceptual correlates of speech sounds are now generally very well-known indeed. We know, for example, what is common to the production of the speech sounds *p* and *a* in the English word *pat* when it is said normally; and we know how they differ from the *k* and *ih* sound in *kit*. These are linguistic data.

In contrast, the description of non-linguistic content, at least auditorily, is not so well advanced. To continue the previous example, we know less about how to describe the individual aspects of my *p* and *a*, if there are any, that make it sound different from yours and everybody else's. To be sure, descriptive frameworks and transcriptional conventions for the auditory description of voices, as opposed to speech sounds, exist. Laver's *The Phonetic Description of Voice Quality* has already been mentioned. However, such works do not constitute a necessary part of traditional phonetic training.

It can be noted in passing that this lack of focus on a phonetic model of between- (and within-) speaker differences has been one source of the above-mentioned controversy as to the applicability of *auditory*-phonetic analysis to forensic speaker identification. So much so that at the 1980 meeting of the British Association of Academic Phoneticians the motion was passed (30 to 12 with 8 abstentions) that 'phoneticians should not consider themselves expert in speaker identification until they have demonstrated themselves to be so' (Nolan 1983: 17).

Although auditory-phonetic description may have limitations when it comes to describing how one speaker's voice sounds different from another's, this is clearly now not the case with modern acoustic-phonetic description. An acoustic phonetician would indeed delight in quantifying how similar or different your *a* was from theirs. Being able to do this presupposes a knowledge of the dimensions in which acoustics differ, which itself presupposes a fully developed theory of the acoustics of speech production. (Aspects of this theory will be presented in Chapter 8.) The upshot of this is that it is possible to characterise voices acoustically extremely well – certainly well enough for them to be recognised automatically under ideal conditions.

Quantitative and qualitative parameters

A final distinction to be drawn is between quantitative and qualitative parameters (Aitken 1995: 14, 15). Quantitative parameters, as their name suggests, are parameters in terms of which samples can be given a numerical value. The dimensions used in Figures 2.1 to 2.4, and 2.6 and 2.7, are all quantitative since samples can be described and compared with numbers. In Figure 3.2, again, the fundamental frequency dimension is quantitative since all speakers' samples are specified in terms of F0 values. The refractive indices of glass fragments constitute another quantitative parameter in forensic science. Quantitative parameters can be further classified as either continuous or discrete, and this distinction is taken up below, after a brief discussion of qualitative parameters.

Qualitative parameters are those in terms of which samples can be described in terms of a quality, and which do not allow samples to be ordered numerically. Thus the apparent dialect or language of a speech sample, or the apparent sex of its speaker, or whether the first syllable of *OK* is left out, or the particular way a vowel sounds, are all qualitative parameters: they cannot be assigned numerical values, and simply allow qualities to be predicated of samples. The blood group of a blood stain found at the scene of a crime would be another example of a qualitative parameter.

Qualitative parameters as such are likely to be of relevance in the initial comparison of speech samples. Thus it may be noted that both samples are dissimilar in that one sounds as if it is an example of dialect *x*, whereas the other sounds like dialect *y*. Under these circumstances, the police might decide not to continue with the investigation. Or the samples may be initially similar in that they both sound to be of a male speaking Cantonese with a Mandarin accent and a speech defect, in which case the comparison is likely to proceed.

However, to contribute to the proper evaluation of the forensic-phonetic evidence, qualitative observations will ultimately need to be converted into quantitative data. This will usually be done by counting occurrences, and will be addressed in the section below on discrete and continuous data.

Discrete and continuous parameters

It was mentioned above that quantitative parameters can be either discrete or continuous. This distinction must now be clarified. Continuous parameters are those in which samples can be quantified to any degree of precision (although often within a particular interval). Thus, for dimension 1 in Figure 2.3, it is possible for a speaker to have any value (in the interval between about 1200 Hz and 1850 Hz). For example, a speaker may be measured to have a value of 1401 Hz in a particular token and 1402 Hz in another; another speaker may be measured to have 1206.2 Hz in one token and 1206.3 Hz in another. (In principle the value can be quantified to any degree of precision – 1402.1 ... 1402.13 ... 1402.134, etc. – but practical limitations and other considerations usually result in some kind of rounding-off.)

As might be appreciated, such continuous parameters are primarily involved where acoustic measurements (both linguistic and non-linguistic) are concerned. However, not all parameters are continuous, and some, most commonly qualitative linguistic parameters, can take only a fixed number of values – often only two.

A well-known English example is the first vowel in the word *economics*, which can either be like the vowel in *deck* or like that in *creek*. In this case, we would be dealing with a discrete linguistic parameter which could take just two values: *eh* (as in *deck*) and *ee* (as in *creek*). (It is a linguistic parameter because the difference between these two vowels is important for signalling the difference between words: cf. *head* vs. *heed*.) The pronunciation in Australian English of the *t* sound as a *d* in words like *city* or *phonetics* is a further example of a binary discrete linguistic parameter. Another example of a parameter with discrete values is the first vowel in the word *hello* in English. The reader might like to check whether they can say *hello* in their variety of English with its first vowel sounding like the vowel in *hut*; like the vowel in *head*; or like the vowel in *hat*. If that is the case, then this parameter has three discrete values.

Voices can of course be characterised and compared with respect to the value selected for such discrete parameters: whether for example *city* or *ciddy* is said. However, as with the continuous acoustic parameters, both within-speaker and between-speaker variation exists. This means that characterisation and comparison is typically more complicated, and will involve comparison between different *incidences* of the parameter. Thus in one sample there may be 20 words in which the *t* could have been said as *d*, but only two words where it in fact was; and in another sample there may be 30 *t* words, all except five said with a *d*. In this case, the difference between samples in incidence of (2/20 =) 10% and (25/30 =) 83% needs to be evaluated. The question: is the 73% difference in incidence more likely to be observed if the samples were from the same speaker, or from different speakers?

An additional complication in the evaluation of discrete auditory linguistic parameters is that they often involve so-called sociolinguistic variation, where values for the parameter are selected in certain social situations. For example, a speaker may be more likely to say *city* in a formal situation, but *ciddy* in less formal circumstances. Comparability therefore needs to be assured before they can be compared. This is not a trivial task, since the formality of a situation is quite often judged precisely by the values of the sociolinguistic variable selected, and hence there is a danger of an argument becoming circular.

Discrete parameters are not all auditory: it is possible to characterise voices according to the absence or presence of an acoustic feature, for example an extra resonance from a particular manner of vocal cord vibration. Needless to say, such features also show within-speaker as well as between-speaker variation.

As the preceding section has shown, speech samples can be compared with respect to many different types of parameter. Irrespective of the type of the parameter, however, the ultimate question remains the same. Given these speech samples, what is the probability of observing this difference for this parameter assuming the samples come from the same speaker, and what is the probability of observing the difference assuming different speakers?

Requirements on forensic-phonetic parameters

The following desiderata, from Nolan (1983: 11), have often been cited for an ideal acoustic parameter in forensic speaker identification, but are applicable to any parameter. According to Nolan, the ideal parameter should:

- show high between-speaker variability and low within-speaker variability (i.e. a high *F*-ratio: the necessity for this criterion has already been discussed);
- be resistant to attempted disguise or mimicry;
- have a high frequency of occurrence in relevant materials;
- be robust in transmission; and
- be relatively easy to extract and measure.

There is no single parameter that satisfies all five criteria.

There is one other desideratum for parameters as a group that should arise naturally when voices are being considered but that surprisingly does not often rate a mention. Suppose I were to cite as evidence that two speech samples came from the same speaker (1) that both samples had an Australian accent and (2) that they both had vowels with a particular quality characteristic of Australian. If I were to claim that this evidence is strong since it is derived from not just one but two parameters, it would legitimately be objected that, since the agreement in vowel quality can be predicted from the accent, or vice versa, the two parameters are not independent. The evidence is therefore not as strong as claimed, since there is actually only one item, not two. Whence the additional desideratum:

- each parameter should be maximally independent of other parameters.

The notion of dependence has been explicated by saying that two pieces of evidence are independent if the truth or falsity of one does not affect our assessment of the probability of the other. A phonetic example of this might be if one of the items of evidence was that both questioned and suspect voices had an idiosyncratic way of pronouncing a certain speech sound (as for example Michael Palin/Pontius Pilate's *r* in the Monty Python film *The Life of Brian*), and the other way was that both samples also differed by a certain amount in a certain acoustic feature of another speech sound, their *ee* vowel, perhaps. If, as is likely, the probability of observing the amount by which the two samples differ in the acoustic feature of the *ee* vowel sound is not predictable from a presence or absence of the *r* idiosyncrasy, these two items are independent.

The evaluation of dependence between forensic-phonetic parameters is very complicated, and, since it has not yet been adequately addressed in the literature, cannot be given a detailed treatment here. However, an idea of the complexity can be given by continuing with the apparently anodyne preceding example. The independence between funny *r* and *ee* vowel acoustics would certainly not obtain between the funny *r* and the first *ee* vowel in **release** *Roderick*, since the acoustics of this particular vowel would be strongly affected by the pronunciation of the funny *r* before it. Moreover, the degree of independence of the funny *r* from other features is not invariant. The funny *r* is now an accepted, although not majority, way of saying *r* in certain urban varieties of British English (Nolan and Oh 1996: 44, but see also Lindsey and Hirson 1999), and is a normal idiosyncrasy in New York (Wells 1982: 508). Thus it might no longer be correct to call the *r* idiosyncratic if the accent were otherwise recognisable as New York or urban British English; although it might be if the accent were recognisable as Australian, rural British, or General American. Either way, the probability of a match between two samples in funny *r* will be different, dependent on other features of the sample.

Given the importance of the notion of independence of evidence, it is reasonable to expect that a forensic-phonetic investigation should include some indication of the degree of independence of the parameters used.

Chapter summary

There are many different parameters that can be used to compare speech samples forensically, and the choice is ultimately determined by a linguistically informed analysis of the nature of the speech material. There is no ideal parameter, but the most important criterion is for a parameter to show a high ratio of between-speaker to within-speaker variation. There are compelling reasons for comparison to be both in acoustic and in auditory terms, and speakers are known to differ in linguistic as well as non-linguistic parameters. Linguistic knowledge of the language(s) in question, especially of factors conditioning the nature of between-speaker and within-speaker variation in the language, is necessary to evaluate differences between forensic-phonetic samples.

4

Expressing the outcome

It is important to be explicit about the intended outcome of a forensic-phonetic investigation, since that determines the conceptual framework for the whole approach. In Chapter 2 it was pointed out that the aim of a forensic-phonetic investigation was to determine the probabilities of the evidence, not the probabilities of the hypothesis. What this means is described in this chapter. It will be seen that the conceptual framework involves the use of basic principles of interpretation that derive from logic (Champod and Evett 2000: 239).

The outcome of a forensic-phonetic voice comparison should be an informed opinion that will aid the legal process. For example, it might help the court reach some kind of a decision, usually concerning the guilt or otherwise of the accused. Or it might help the police or prosecuting authority decide whether prosecution is sensible on the strength of the voice evidence. Or it might help the defence question the strength of the voice evidence against their client.

It might appear that the expression of this opinion is straightforward. Perhaps we expect a formulation involving degrees of probability that the voices are, or are not, from the same person, for example: 'Given the high degree of similarity between the two speech samples, there can be very little doubt that these two samples are from the same speaker', or 'It is highly likely, given the extensive differences between the speech samples, that they are from different people.' The conclusions of these statements are similar to the ones cited in Baldwin and French (1990: 10) and, in the author's experience, typical of the kind of statement expected by police, counsel, and judiciary when it comes to speaker identification. Other practitioners, e.g. Broeders (1999: 239) report similar expectations.

However, there are serious problems with this way of expressing the outcome of a forensic investigation, and here is the place to draw attention to them. Detailed general treatment can be found in the excellent introductory text *Evaluating Evidence* (Robertson and Vignaux 1995). Highly recommended shorter discussions can also be found in Evett's (1991) paper 'Interpretation – a Personal Odyssey', and in three papers that deal with forensic speaker recognition: 'Some observations on the use of probability scales in forensic investigation' (Broeders 1999); Champod and Evett's (2000) commentary on Broeders' paper, and 'The inference of identity in forensic speaker recognition' (Champod and Meuwly 2000).

The two problematic formulations exemplified above both express *the probability of a hypothesis given the evidence.* For example, 'It is likely (an estimate of probability) to be the same speaker (a hypothesis) given the high degree of similarity (the evidence)'. This is often abbreviated as $p(H \mid E)$, where p stands for probability, H for hypothesis, the vertical line | for given, and E for evidence. However, it is generally accepted that it is neither logically nor legally correct for the forensic expert to attempt to state the probability of a hypothesis given the evidence:

> It is very tempting when assessing evidence to try to determine a value for the probability of guilt of a suspect, or the value for the odds in favour of guilt and perhaps even reach a decision regarding the suspect's guilt. However, this is the role of the jury and/or judge. It is not the role of the forensic scientist or statistical expert witness to give an opinion on this.... It is permissible for the scientist to say that the evidence is 1000 times more likely, say, if the suspect is guilty than if he is innocent.
>
> Aitken (1995: 4)

Why? The main reason is a common sense, but also a logical one. An expert cannot make a sensible estimate of the probability of a hypothesis such as *these two samples were/were not spoken by the same speaker* on the basis of the scientific evidence alone. Access to all the evidence available to the court is needed for that.

This point can be easily demonstrated, adapting the example from Champod and Evett (2000: 8–9). Assume that a questioned speech sample has been compared in some detail with speech samples from a suspect, and found to be very similar – for argument's sake, let us say they agree in 100% of the many features considered. On the basis of the high degree of similarity (this is the evidence), the expert says that it is very likely (a probability judgement) that they come from the same speaker (the hypothesis). It transpires, however, that the questioned sample was obtained on a day when there is evidence to suggest that the suspect was not in the city where the crime was committed, and offender and suspect are subsequently also found to differ in blood type. In the light of this evidence, the probability that the two speech samples are from the same speaker is dramatically reversed – in fact it would hardly make sense any longer to entertain the possibility that the same speaker was involved.

Now assume that a questioned speech sample has been compared with a speech sample from the accused, has been found to differ in many respects, and the expert concludes that there is low probability of the samples coming from the same speaker. Independent evidence from two eye-witnesses puts the suspect at the scene of the crime, however, and rings similar to those the offender was seen wearing by the victim of the hold-up are found in the suspect's home. Now the probability of provenance from the same speaker is considerably higher than would appear from the scientific (i.e. forensic-phonetic) evidence alone.

Logically, then, in order to be able to quote the probability of the hypothesis given the evidence, the forensic scientist would have to possess knowledge of *all* the relevant evidence. Since this is almost never the case (Champod and Evett 2000: 239), it cannot

The likelihood ratio

First an instructive example from Robertson and Vignaux (1995: 13). Suppose it is known that 80% of children who have been sexually abused bite their nails. A child is suspected of having been sexually abused, and she does indeed bite her nails. The court wants to know the probability of sexual abuse. In order to evaluate the nail-biting evidence, however, it is logically clear that another piece of information is needed: what percentage of children bite their nails who have *not* been sexually abused? If this percentage is smaller than that of sexually abused children, then the nail-biting evidence will offer some support to the sexual abuse hypothesis; if it is greater, then the nail-biting evidence is actually a counter-indication.

Now a forensic-phonetic analogue. Suppose that it is known that 80% of speech samples from the same speaker are 'very similar' in feature x. You, the forensic phonetician, are given some speech samples that are indeed 'very similar' in feature x. The court wants to know the probability that these speech samples come from the same speaker. In order to help the court, the extra piece of information needed is: what is the proportion of speech samples 'very similar' in feature x that come from *different* speakers?

Thus two items of information are required: the probability of the evidence (nail biting; high degree of similarity in feature x) assuming that the hypothesis (sexual abuse; same speaker) is true, and the probability of the evidence assuming that it is not. Instead of trying to state the probability of the hypothesis given the evidence, which is the job of the court, forensic experts must attempt to quantify the probability of the evidence given the two hypotheses. In forensic phonetics, this typically equates to estimating how much more likely the degree of similarity between questioned and suspect speech samples is if they were spoken by the same speaker than by different speakers.

The evidence here, then, is the quantified degree of similarity between the speech samples, and two hypotheses are involved. The first, often called the prosecution hypothesis (H_p), is that the speech samples were spoken by the same speaker. The second, the defence or alternative hypothesis (H_d), can take many forms, but will usually be a version of the assertion that the speech samples were spoken by different speakers.

This allows the strength of the forensic evidence to be evaluated in a logical, and also commonsense, way. In order to understand this, let us return to the hypothetical forensic-phonetic example above.

We assumed that a high degree of similarity was observed between questioned and accused speech samples. This high degree of similarity constitutes the evidence. We assumed further that 80% of paired speech samples with this high degree of similarity have been shown to be from the same speaker. Thus the probability of observing the evidence assuming the samples are from the same speaker $p(E \mid H_p)$ is 80%. Now, in order to determine the strength of the evidence, we also need to take into account the percentage of paired speech samples with this high degree of similarity that have been shown to come from different speakers. Let us assume it is 10%. The probability of observing the evidence assuming the samples come from different speakers $p(E \mid H_d)$ is thus 10%.

The strength of the evidence is now given by the ratio of two probabilities: the probability of the evidence given the hypothesis that the two samples are from the same speaker, and the probability of the evidence assuming that the samples are from different speakers. This ratio is called the Likelihood Ratio (LR), and can be expressed in the formula at (4.1) where H_p is the prosecution hypothesis (the two speech samples come from the same speaker), and H_d is the defence hypothesis (the samples come from different speakers). Another way of understanding the LR is that the numerator represents the degree of *similarity* between the questioned and suspect samples, and the denominator represents how *typical* they are: the probability that you would find measurements like the questioned and suspect samples by chance in the relevant population (Evett 1991: 12).

$$\text{LR} = \frac{p(E \mid H_p)}{p(E \mid H_d)} \qquad (4.1)$$

Formula for likelihood ratio

Note that both numerator and denominator are of the form probability of evidence given hypothesis, or $p(E \mid H)$. In our hypothetical example, the former probability is 80%, the latter 10%, so the Likelihood Ratio for the evidence is 80:10 = 8. In words, the correct evaluation of the high degree of similarity between the speech samples is this: that the degree of similarity is eight times more likely to be observed were the samples from the same speaker than from different speakers. The reader might recall that this kind of reasoning was already used in the monodimensional discrimination example connected with Figure 2.2 earlier, with its equal error rate of 40%, and associated LR of (60%/40% =) 1.5.

The value of the LR expresses the degree to which the evidence supports the prosecution hypothesis. The greater the difference between questioned and suspect samples, the smaller becomes the numerator, and the more similar they are, the greater the numerator. The more likely you are to find values of the questioned and suspect samples in the relevant population the greater the denominator, and the less typical they are of the population, the smaller the denominator. LR values greater than unity give support to the prosecution hypothesis; LR values less than unity give support to the defence hypothesis. The amount of support in each case is proportional to the size of the LR.

How this works is shown in Table 4.1. In the top half of Table 4.1 the numerator – the similarity between questioned and suspect samples – is held at 80%, indicating a high similarity. The denominator, indicating the probability of observing the difference between questioned and suspect samples by chance in the relevant population, is then increased from a low 10% to 80%. It can be seen that if the similarity between the questioned and suspect samples is large (e.g. 80%), and the probability of observing that similarity by chance in the population is small (e.g. 10%), the Likelihood Ratio is

Expressing the outcome

Table 4.1 Simple illustration of how the likelihood ratio works

p(E | same speaker)	80%	80%	80%	80%	80%
p(E | different speakers)	10%	20%	40%	60%	80%
LR	8	4	2	1.333	1
p(E | same speaker)	10%	20%	40%	60%	80%
p(E | different speakers)	80%	80%	80%	80%	80%
LR	0.125	0.25	0.5	0.75	1

large (80/10 = 8). As the chance of observing the similarity between the questioned and suspect samples at random in the population increases from 10 to 80%, the LR decreases. When the LR reaches a value of 1, that indicates that you are just as likely to observe the difference between the questioned and suspect samples if they come from the same speaker as if they are chosen at random from two different speakers in the population, and the evidence is of no use.

In the bottom part of Table 4.1 the denominator is held constant at a high 80%, indicating a high probability of finding the difference between questioned and suspect samples by chance. The value of the numerator is then increased against this, indicating an increasing similarity between questioned and suspect samples. It can be seen that as the numerator increases, so does the value of the LR. The LR goes from a value of 0.125, indicating support for the defence hypothesis that different speakers are involved, to a value of 1, indicating once more that the differences between questioned and suspect samples are just as likely to be observed if they came from the same speaker or different speakers.

The interpretation of a LR less than 1, as in Table 4.1, is best handled in terms of its reciprocal. For example, given a LR of 0.125, as in Table 4.1, the odds in favour of the defence are 8 to 1, since the reciprocal of 0.125 is (1/0.125 =) 8. You are eight times more likely to observe the difference between questioned and suspect samples if they came from different speakers. This is, of course, the same value for the LR as the 8 in the upper half of Table 4.1. The only difference is that the evidence is now in favour of the defence.

As just illustrated, a LR value of unity is equivocal: it implies that one is equally as likely to observe the evidence if the accused is guilty as if they are not. It therefore does not distinguish between defence and prosecution hypotheses and has no probative value.

In Chapter 2 it was explained how the dimension size (i.e. the number of observations in a sample, see Figure 2.5) affects confidence limits. Dimension size has a different, but intuitively correct, effect from a Likelihood Ratio perspective. Intuitively, the evidence cannot be so strong if it is based on samples containing few observations as if it is based on samples made up of many observations. And this is indeed what happens. Dimension size correlates with the magnitude of the LR, such that the smaller the dimension size of the two samples being compared, the more the LR tends towards unity. Thus the number of observations in a sample translates nicely into the strength of evidence.

Combination of likelihood ratios

One of the important features of the LR approach is that it allows evidence from different sources to be combined in a principled way (Robertson and Vignaux 1995: 69–72). The different sources could be from different forensic investigations (e.g. gunshot residue, speaker identification) or – what we are concerned with here – from different aspects of the same investigation. For example, it was explained above that voices can and must be compared forensically with respect to many features. A LR can be estimated for each of these features, and an overall forensic-phonetic Likelihood Ratio ($OLR_{f\text{-}p}$) derived from the LRs for the individual features.

This is symbolised at (4.2), where $OLR_{f\text{-}p}$ stands for overall forensic-phonetic Likelihood Ratio; LR $E_{f\text{-}p}1$ stands for the likelihood ratio for the first piece of forensic-phonetic evidence $E_{f\text{-}p}1$; LR $E_{f\text{-}p}2$ for the LR from the second item of evidence, and so on. As explained above, each of the LR terms stands for the ratio of the probability of observing the difference in a particular phonetic feature assuming the prosecution hypothesis (that samples have come from the same speaker) to the probability of observing the difference assuming the defence hypothesis. Thus, for example, the likelihood ratio for the first piece of forensic-phonetic evidence (LR $E_{f\text{-}p}1$) is $p(E_{f\text{-}p}1 | H_p)/p(E_{f\text{-}p}1 | H_d)$.

$$OLR_{f\text{-}p} = f(LR\ E_{f\text{-}p}1,\ LR\ E_{f\text{-}p}2, \ldots,\ LR\ E_{f\text{-}p}N) \tag{4.2}$$

Combination of Likelihood Ratios from different forensic-phonetic features to determine an overall forensic-phonetic Likelihood Ratio

The *f* after the equality sign in the expression at (4.2) stands for *is a function of*. This means that there is a method of combining the individual LR values on the right-hand side of the equation such that a unique value for the overall forensic-phonetic LR on the left can be determined. This is like writing $y = f(x)$ (y is a function of x) to stand for the equation $y = x^2$.

Assuming that it is possible to calculate LRs for several pieces of forensic-phonetic evidence, how are they to be combined to yield an overall LR for the forensic-phonetic evidence? This depends on the degree to which the items of evidence on which they are based can be considered mutually independent. The combination of the LRs of different pieces of evidence is straightforward if they are independent: the combined LR is the product of their associated LRs. Thus, given a LR of 3 from one piece of forensic-phonetic evidence, and a LR of 4.2 from a second piece of evidence independent of the first, the combined LR is (3 × 4.2 =) 12.6.

If the different items of evidence are not independent, combination can become very much more complicated. For example, the combined LR for two items of evidence E1 and E2, where E2 is dependent on E1, is the product of two LRs. The first LR is for the first item of evidence E1, viz.: $p(E1 | A)/p(E1 | \sim A)$. (The '| ~A' part of the denominator means 'assuming that A is not true'.) The second LR is for the second piece of evidence E2 taking into account *both* the first piece of evidence E1 and the assumption A, viz.: $p(E2 | E1\ and\ A)/p(E2 | E1\ and \sim A)$.

It has already been pointed out in conjunction with requirements on forensic-phonetic parameters that it is not a straightforward matter to assess the degree of interdependence of forensic-phonetic data in the first place. Since the forensic-phonetic data that will be used in Chapter 11 to demonstrate the application of the approach show an overall rather high degree of independence, the problem of combination of LRs from dependent items of evidence will not be discussed further.

Whichever way LRs from different items of forensic-phonetic evidence are to be combined, it can be appreciated that, even though LRs from individual forensic-phonetic features are often not big, a large number of small-valued LRs greater than unity can build up into a high-valued combined LR (Robertson and Vignaux 1995: 73–76). Suppose, for example, that two speech samples were compared with respect to four independent features, yielding low LRs of 3, 2, 2.5 and 3, respectively. The combined LR for these values is $(3 \times 2 \times 2.5 \times 3 =)$ 45, which is much bigger than any of the individual LRs. By the same token, a single LR with a value much less than 1 can reverse the positive evidence from a set of otherwise highly valued LRs. An additional LR of 0.01 to our example above will mean that the combined LR now becomes $(45 \times 0.01 =)$ 0.45. A LR of 0.45 means that one is now $(1/0.45 =)$ about 2 times more likely to observe the differences between samples if they came from *different* speakers.

Verbal scales for the likelihood ratio

The value of the LR thus quantifies the strength of the evidence. Since the numerical form of a LR may not be readily interpretable to the court, translations into verbal scales have been proposed. One proposal, used at the Forensic Science Service, is shown in Table 4.2 (Champod and Evett 2000: 240).

Thus the LR of 8 for the above forensic-phonetic comparison (the difference between the speech samples is eight times more likely to be observed if the samples came from the same speaker) might be translated into 'There is limited evidence to support the prosecution hypothesis'. Likewise, a LR of 0.03 might translate into 'There is moderate evidence to support the defence hypothesis'.

There is some debate as to the merit of converting the numerical values of the LR into a verbal scale for the benefit of the court. On the one hand, juries are not

Table 4.2 Some proposed verbal equivalents for likelihood ratios

Likelihood ratio	Proposed verbal equivalent	
>10 000	Very strong evidence to support ...	
1000 to 10 000	Strong evidence to support ...	
100 to 1000	Moderately strong evidence to support ...	
10 to 100	Moderate evidence to support ...	
1 to 10	Limited evidence to support ...	Prosecution hypothesis
1 to 0.1	Limited evidence against ...	
0.1 to 0.01	Moderate evidence against ...	
0.01 to 0.001	Moderately strong evidence against ...	
0.001 to 0.0001	Strong evidence against ...	
<0.0001	Very strong evidence against	

supposed to be able to handle numerical data well and there is also said to be a danger that they give more weight to numerical as opposed to non-numerical evidence. On the other hand, verbal scales are counter-indicated by the fact that the meaning of words can vary both between and within the several interested groups of jury, expert witnesses, counsel and judiciary (Robertson and Vignaux 1995: 55–57). For example, another scale with slightly differing values and verbal interpretations was proposed, for DNA evidence, by Evett (1991: 19).

Log scaling of likelihood ratio

The strength of evidence is often construed metaphorically in terms of weight: we talk of the *weight* of evidence, the *scales* of justice. Weight is furthermore usually understood as resulting from the process of adding. For this reason, some, e.g. Good (1991: 89–90), have advocated the use of logarithms to combine LRs from independent pieces of evidence. Another reason is that logarithms help 'to comprehend the magnitude of large numbers' (Champod and Evett 2000: 241–2).

By first taking the (common) logarithm of the LRs, the LRs can then be added together, rather than multiplied. (The common logarithm of a number is the power to which 10 must be raised to get that number; thus the common log of 1000 is 3 because 1000 is $10 \times 10 \times 10$, or 10 to the power three (or 10^3). Multiplying a LR of 1000 by a LR of 100 to get an overall LR of 100 000 (or 10^5) is the same as adding the common logs of the LRs (since $\log 10^5 = \log 10^3 + \log 10^2$).

Some log equivalents of the likelihood ratios in Table 4.2 are given in Table 4.3.

Another reason why log scaling might be preferable is that, as explained above, a likelihood ratio of 1 indicates that the evidence is worthless. The common log of 1 is 0, and this might be better understandable as a numerical equivalent of worthlessness. However, it is also possible that the concept of 'very strong evidence' is better reflected in a number like 10 000, than 4. Again, there is no consensus concerning the use of log scaling of LRs.

Table 4.3 Logarithmic equivalents of some likelihood ratios

Likelihood Ratio	Log equivalent	Possible verbal equivalent	
>10 000	>4	Very strong . . .	
1000 to 10 000	3 to 4	Strong . . .	support for the
100 to 1000	2 to 3	Moderately strong . . .	prosecution
10 to 100	1 to 2	Moderate . . .	hypothesis
1 to 10	0 to 1	Limited . . .	
1 to 0.1	0 to −1	Limited . . .	
0.1 to 0.01	−1 to −2	Moderate . . .	support for the
0.01 to 0.001	−2 to −3	Moderately strong . . .	defence
0.001 to 0.0001	−3 to −4	Strong . . .	hypothesis
<0.0001	>−4	Very strong	

Prior odds

An extremely important role in the evaluation of forensic identification evidence is played by the concept of prior odds. In order to assess the amount of support for the prosecution hypothesis that both questioned and suspect voice samples came from the same speaker, the overall, i.e. combined, LR ratio from the voice evidence has to be evaluated in the light of the amount of belief in that assertion before the voice evidence was considered.

This belief is called the prior odds for the assertion, and can be formulated as at (4.3). This shows that the prior odds for the prosecution assertion A_p that both samples come from the same speaker (symbolised POA_p) is simply the ratio of the probability of the assertion being true, $p(A_p)$ (i.e. that the samples come from the same speaker) to that of it being false, $p(\sim A_p)$.

Suppose, for example, an incriminating telephone call was intercepted from a house in which five men, including the suspect, were known to be at the time of the call. In the absence of the voice evidence, the probability that the suspect made the call is 1/5 or 0.2. Odds are derived from probability by dividing the probability by one minus the probability, so the prior odds for believing that the suspect made the call are: $p(A_p)/[1 - p(\sim A_p)] = 0.2/[1 - 0.2] = 0.2/0.8 = 1/4$, or 4 to 1 against him making the call.

$$POA_p = \frac{p(A_p)}{p(\sim A_p)} \quad (4.3)$$

Prior odds for the prosecution assertion A_p that two samples come from the same voice.

Now suppose that the suspect's voice (but not those of the others) is available for analysis and compared with that of the incriminating call, and an overall LR of 20 estimated on the basis of the forensic-phonetic evidence. This means that one would be 20 times more likely to observe the degree of difference between the suspect's voice and the voice of the incriminating call if they came from the same speaker than if they were spoken by different speakers. The amount of support for the assertion that both calls were made by the same speaker is then the product of the overall forensic-phonetic LR and the prior odds, or $(20 \times 1/4 =) 5$.

This value, 5, basing on the prior odds and the voice evidence, becomes what is called the posterior odds for believing the assertion, given the initial conditions and the forensic-phonetic evidence. This is formulated at (4.4):

$$\frac{p(A_p \mid E)}{p(\sim A_p \mid E)} = \frac{p(A_p)}{p(\sim A_p)} \times \frac{p(E \mid A_p)}{p(E \mid \sim A_p)} \quad (4.4)$$

posterior   prior   LR for
odds for   =   odds   ×   evidence
assertion

Determining posterior odds for the assertion by combining prior odds and Likelihood Ratio for evidence where the term before the equality sign is the posterior odds, i.e. the odds in favour of the prosecution assertion A_p that the same speaker is involved given the voice evidence E, and the prior odds. The term immediately after the equality sign in (4.4) is the prior odds, i.e. the odds in favour of the prosecution assertion before the evidence E is presented, and the rightmost term is the Likelihood Ratio for the evidence.

It will be appreciated that the prior odds can have an enormous potential effect on the strength of the scientific evidence, since with a different prior you get a different posterior (Roberson and Vignaux 1995: 18). Consider the case of the telephone call just described. If only two men were known to be present in the house instead of five, the probability that the suspect made the call would now be 1/2, or 0.5. The prior odds for the suspect making the call would then be: $p(A_p)/[1 - p(\sim A_p)] = 0.5/[1 - 0.5] = 0.5/0.5 = 1/1$, or evens: the suspect is just as likely to have made the call as not. Taking the voice evidence into account makes the posterior odds $(1 \times LR = 1 \times 20 =)$ 20, and gives stronger support to the assertion that the suspect is the speaker than in the previous example with five occupants, *even though the LR for the voice evidence itself has not changed*.

It can also now be understood that it is the non-accessibility of the prior odds for the forensic expert that makes quoting the probability of the hypothesis given the evidence, $p(H | E)$, logically impossible. It can also be seen that it was in fact failure to take into account the prior odds that gave the nonsense results in the two straw-men examples quoted at the beginning of this section to demonstrate the inadvisability of quoting the probability of a hypothesis in the absence of all the evidence.

Despite the importance of the prior odds, the forensic phonetician is usually not provided with all the necessary information to estimate them for the voice evidence. Some information may be available, for example whether the incriminating telephone calls were made from the accused's house, or their mobile, or whether the accused has been identified by a police officer familiar with his voice. However, as will be explained below, there are also good reasons why the phonetician should insist on not being told such data and should concentrate on assessing the LR for the voice evidence on its own. Obviously, it will be part of the responsibility of legal counsel to ensure that the prior odds are taken into consideration at some point in the case.

Alternative hypothesis

The denominator of the likelihood ratio represents the probability of the alternative, or defence hypothesis H_d: the probability that values like those found in the questioned and suspect samples can be found at random in the population. In the discussion above, it was pointed out that H_d can take many forms, but is usually a version of the claim that the questioned and suspect samples come from different speakers. There are some important points to be made about the alternative hypothesis.

Firstly, it is important to understand that the choice of the alternative hypothesis can substantially alter the value of the Likelihood Ratio, and hence the strength of the evidence (Robertson and Vignaux 1995: 33–50). For example, a plausible narrowing in the defence account of the voice evidence will often be from that 'it was the voice of someone (an Australian male) other than the accused' to that 'it was the voice of someone (an Australian male) who *sounds like* the accused'. (The reason for bracketing the Australian male will be explained below.) Under the first hypothesis, the population to be considered will be all other male Australian speakers; under the second, it will be all male Australian speakers who sound like the accused.

Such a change in the defence hypothesis can have a considerable effect on the LR. For example, the probability of observing a high degree of similarity between two voice samples is going to be very high under the prosecution hypothesis that they come from the same speaker. However, the probability of observing a high degree of similarity is also likely to be high under the defence hypothesis that the samples come from two speakers who sound similar – certainly higher than if the defence were simply that it was just someone else speaking. Thus the value for the LR $[p(E \mid H_p)/p(E \mid H_d)]$ will tend towards unity under the similar-sounding speaker defence, and will consequently contribute less support for the assertion that the same speaker was involved.

This does not mean that the similar-sounding speaker hypothesis will automatically work in favour of the defence, however, for two reasons. Firstly, changing the hypothesis changes the prior odds. If the (Australian male) offender's voice belonged to someone else in Canberra, say, the prior odds would be ca. 144 000 to 1 against it being the accused (assuming 360 000 Canberrans, two-fifths of whom are adult males). With a LR for the voice evidence of, say, 300, the odds in favour of another speaker being the offender would still be large: still (1/144 000 × 300 =) 480 times more likely to be someone else than the accused.

These odds would shorten considerably if they reflected only the number of adult males in Canberra who actually sound like the accused. For argument's sake, let us assume there are 100 such male speakers who would be readily mistaken for the accused, and let us scale down the LR accordingly to 200. Now the odds in favour of the accused being the speaker are (1/100 × 200 =) 2 to 1 against: it is now 2 times more likely that the accused is the offender than not. Consider what would happen if the alternative claim were narrowed to 'It's the accused's brother'. We scale down the LR some more to 50 to reflect an assumed closeness between the two brothers' voices. Now the prior odds are evens, and the value of the posterior odds shoots up to (1/2 × 50 =) 25: the offender is now 25 times more likely to be the accused than not.

The second reason why the similar-sounding speaker hypothesis will not automatically work in favour of the defence is a rather important one that will be explained in greater detail below in the chapter asking *What is a voice?* It is that just because two voices sound similar to a lay ear does not necessarily mean that they are similar in all their phonetic characteristics that a trained ear will register. Thus it is not necessarily the case that the LR will be considerably higher for sound-alikes than for non-sound-alikes and thus the scaling-down of the LRs in the previous example was not necessarily warranted. The reader should insert the original putative LR of 300 in the 'similar speaker' and 'his brother' alternatives to see what the effect is on the posterior odds in favour of guilt.

We can now return to the bracketed Australian male of the second paragraph. The point here is a simple one: just saying 'different speaker' in the alternative hypothesis is never specific enough. Reference needs to be made to other features like sex and language.

Bayesian inference

Expressing the probability of the evidence given the competing defence and prosecution hypotheses is part of what is termed a Bayesian approach, after Thomas Bayes, an eighteenth-century English Presbyterian minister and mathematician. The essence of a Bayesian approach has been demonstrated above: it allows the updating of the odds in favour of a hypothesis (e.g. that both voice samples were from the same speaker) from new evidence and prior odds. If, for example, there is reason prior to submission of forensic-phonetic evidence for the court to believe that there is only a one-in-a-hundred chance for the accused to be guilty, a LR of 10 from the phonetic evidence will reduce these odds to a posterior odds of $(0.01 \times 10 =)$ one-in-ten. The use of Bayesian, as opposed to conventional, statistics will be totally new to some, novel to others, and controversial to yet others. In this section, therefore, the important arguments for and against Bayesian inference are rehearsed, both in forensics and in general, and this is done against the relevant historical background.

Historical background

Bayes was one of a group of scholars, called the classical probabilists, who were concerned with the problem of drawing inferences from evidence to causes; to truth; to hypotheses. In 1763 he proved the inverse of Bernoulli's theorem, and thus provided the answer to the question underlying the model of the scientific method itself: given the observed data, what is the probability of a hypothesis purporting to explain those data being true? (Gigerenzer *et al.* 1989: 29–31). This answer is now known as Bayes' theorem, or Bayes' law. It can be found expressed in terms either of probabilities or of odds (Robertson and Vignaux 1995: 226). The formula at (4.4), saying that the probability of a hypothesis is actually the prior odds times the likelihood ratio, is actually the odds version of Bayes' theorem.

From about the mid-nineteenth century a different understanding of probability became dominant, and classical probability, along with Bayesian inference, became the object of strong critique and ridicule. The classical construal of probability is as a measure of 'strength of belief' (Robertson and Vignaux 1995: 114), or of 'degree of uncertainty felt by the person making the inference' (Lindley 1982: 199). It is something predicated, in an individual case, of a hypothesis, of evidence, or of judgement. This classical concept yielded to the notion of probability, in a general case, as a property of *data*, as calculated from the 'frequency of an event in a long-run series of trials' (Robertson and Vignaux 1995: 114). For example, the probability that it will rain tomorrow is understood as calculated from the frequency of rainy days in the past.

This particular understanding of probability was common to two theoretically different schools of statistical inference that emerged in the early twentieth century, one developed by R. A. Fisher, and one by J. Neyman and E. Pearson. Although differing theoretically, both approaches have been institutionally amalgamated into a hybrid by mathematical statisticians, statistical textbook writers and experimenters, and it is this hybrid that is now established and taught as the dominant paradigm, to the virtual exclusion of Bayesian methods, in most of the social sciences (Gigerenzer *et al.* 1989: Ch. 3, esp. 106–9). Because of its concept of probability, the hybrid approach has come to be known as 'frequentist'.

Expressing the outcome

Since the beginning of the twentieth century, however, Bayesian inference slowly began to recover respectability and ground, especially in legal applications and decisions outside the experimenter's laboratory (Gigerenzer *et al.* 1989: 91, 233, 237). At the beginning of the 1990s, Bayesians and frequentists found themselves in a kind of complementary distribution, each predominant in different fields of application: Bayes in law, economics and modelling of human rationality, and frequentist in the experimental and social sciences (Gigerenzer *et al.* 1989: 232–3).

In roughly the last decade, Bayesian inference has again become widely accepted in statistics, and this is primarily due to a phenomenal increase in computer power. It is now possible to use computational brute force to solve Bayesian problems for which analytical solutions were previously not possible (Malakoff 1999: 1460–2).

Interestingly, no mention will be found of Bayes in the most recent reference work on the phonetic sciences (Hardcastle and Laver 1997), either specifically in its chapter devoted to experimental design and statistics in speech science, or in general. This is because experimental phonetics falls clearly under the rubric of the social sciences, and as such is presumably still in thrall to the frequentist paradigm. This may be one of the reasons why forensic-phonetic evidence will sometimes, perhaps often, be presented in frequentist terms.

Current acceptance

How accepted is a Bayesian approach nowadays? In order to answer this question, it is useful to examine increasingly smaller fields, going from general areas to forensic science to forensic phonetics; and to distinguish at each level theoretical acceptance from practical real-world use.

The Bayesian comeback was not easy, and there is no grand unified theory reconciling the philosophical differences between Bayesian and frequentist approaches. However, there is now a general acceptance that both Bayesian and frequentist approaches are appropriate, depending on the nature of the problem to be solved, and they now both 'peacefully coexist, at least in the context of application', even within a given discipline (Gigerenzer *et al.* 1989: 272). Generally, therefore, Bayesian approaches can now be encountered in a wide variety of areas where humans are trying to make inferences and predictions from data, and update their belief in hypotheses when new data come in (Malakoff 1999). These areas range from the more purely scientific, like astrophysics, statistical theory, or historical linguistics (Kumar and Rose 2000: 247–9) to the more real-world considerations of improving the resolution of magnetic resonance imaging devices; drug testing; determining limits on fishing catch; public policy; commercial speech and speaker recognition; insurance; and that infuriating anthropomorphism the Microsoft animated paperclip.

Of specific relevance to this book, Bayesian inference is now becoming more and more accepted as the appropriate theoretical framework for evaluating evidence in forensic science (Robertson and Vignaux 1995). Some, e.g. Gonzalez-Rodriguez *et al.* (2001: 1), would even say it has become firmly established. Champod and Meuwly (2000: 201) cite, for example, paternity cases, document examination, toolmarks, fingerprints, trace evidence, and 'forensic evidence in general' as areas where a Bayesian approach is received. To these they add forensic speaker recognition, arguing (pp. 195–6) that competing interpretative frameworks like verification, recognition, and error counts are inadequate for forensic purposes. Bite marks is another area for which Bayesian analysis is claimed to be particularly suited (Kieser *et al.* 2001). It is doubtful, however, whether it is anywhere near being received in fingerprinting or handwriting analysis.

The use of Bayes' theorem in forensic speaker identification was proposed as early as 1984 (Meuwley and Drygajlo 2001: 146), but the idea obviously did not catch on. In 1990, for example, the first author of Baldwin and French's *Forensic Phonetics* (1990: 10) presented scales for quoting the probability of the hypothesis ('. . . that they are the same person; . . . that they are different people') given the evidence. Three years later, a questionnaire from the International Association of Forensic Phonetics was also seeking to determine scales of opinion in expressing the probability of the hypothesis. In the latter half of the 1990s, however, papers advocating Bayesian methods in forensic speaker identification (e.g. Rose 1998a: 220, 1999a: 3; Koolwaaij and Boves 1999: 248; Champod and Meuwley 2000) began to appear. Three of the four papers on forensic speaker identification at the 2001 conference on automatic speaker identification were either explicitly Bayesian or used likelihood ratios.

The key word in this discussion of the degree of current acceptance of Bayes' theorem, both in forensic science in general and forensic speaker identification in particular, is *theoretical*. Although it is clear that there is much support in theory for the approach, it is very difficult to determine to what extent a Bayesian approach has been adopted *de facto* in forensic science, either by practitioners or in the courts.

According to Broeders (1999: 239), 'Most practising experts do not think this [indicating the degree of support for the hypothesis with a likelihood ratio] is a good idea'. (It is probably otiose to say that the author is an exception.) Champod and Evett (2000: 243) echo this: 'The great obstacle to progress from the current state is that most forensic scientists are unfamiliar with the theory and principles of logical interpretation'. They do, however, indicate that the Forensic Science Service is now instructing entrants in the fundamentals of probability theory and the principles of interpretation, and are confident that things will change rapidly.

It is not at all clear that the courts, either, are rushing to embrace the approach, although here it is probably sensible to draw a distinction between the judiciary and legal counsel on the one hand and the jury on the other. Whether jurors should be instructed in Bayesian inference has been discussed among judges, lawyers and scholars since 1970 (Gigerenzer *et al.* 1989: 264). Recent decisions make it pretty clear that the court believes that the reasoning of the jury should not be interfered with.

Although instances of the successful practical application of Bayes' theorem can be cited, in the author's experience very few members of the Australian judiciary are aware of, or really understand, its basics, despite the existence since 1995 of a clear introduction to the topic: Robertson and Vignaux's *Interpreting Evidence*. This is consistent with the findings of Sjerps and Biesheuvel (1999: 225), who report that in a small experiment most Dutch criminal lawyers and judges (most of them trainees, however) prefer the logically incorrect Dutch scale of judgements to Bayesian formulations. One reason for this is undoubtedly the inherent conservatism of the Law. Another reason, suggested to the author from the head of forensic services in Canberra, is that Bayesian inference is generally felt to be too complex.

All the more important, therefore, that the pros and cons of a Bayesian approach should be made explicit. We turn to this now.

Pros and cons

What important arguments can be adduced in favour of, and against, a Bayesian approach? Many of the arguments for a Bayesian approach are couched, naturally, in terms of how it stacks up against the competing frequentist position, also variously called the 'classical' or 'coincidence' or 'two stage' approach, widely practised in courts of law – see for example Robertson and Vignaux (1995: 113–33) and Evett (1991: 10–11). The extent to which frequentist analyses cannot handle forensic data, and there are many examples (Lindley 1982: 199), is then seen by default (rather illogically but, in the general absence of non-frequentist competing paradigms, naturally) as a point in favour of Bayes. There are also arguments for a Bayesian position that do not derive their cogency from reference to any competing paradigms. The most important of these is that Bayesian analysis has been shown to work.

Much of the critique of Bayesian statistics is in opposition to the application of Bayesian inference *per se*, rather than its specific application in forensic identification. Indeed it is often conceded that it is the proper way to evaluate scientific evidence (Gigerenzer *et al.* 1989: 265). So it is important to examine and rehearse some of the main arguments for and against a Bayesian approach both in forensics and in general.

Pro-Bayesian arguments

The main advantages of a Bayesian approach are that it has been shown to work with forensically realistic material; that it makes combining separate pieces and types of evidence easy; that its explicitness helps avoid common fallacies in inference; that it pays due attention to the individual nature of the evidence; and that frequentist analyses are demonstrably inferior in many important respects. These are addressed in turn below.

Empirical confirmation of theoretical predictions

Many advantages for the Bayesian approach – for example its applicability to all inference problems, its mathematical coherence, its explicitness – are, naturally, claimed by its adherents. However the most important is a practical one: it works (Lindley 1982: 198, 1990: 56). This is obviously such an important point that it deserves some discussion.

As shown above, Bayes' law predicts that same-subject data should be resolved with LRs greater than 1, and different-subject data should have LRs smaller than 1. This therefore provides a test for the method. The extent to which known different-subject data are resolved with LRs smaller than 1, and known same-subject data are resolved with LRs larger than 1, reflects how well the method works.

That this is indeed the case has already been demonstrated with three types of forensically common evidence: DNA, glass fragments, and speech. Evett *et al.* (1993: 503), for example, demonstrated that repeat DNA samples from 152 subjects were, as predicted, resolved with LRs greater than 1, whereas ca. 1.2 million pairs of DNA samples from different subjects were, again as predicted, associated with LRs of less than 1 in the vast majority of cases (only eight in a million comparisons of DNA from different subjects yielded an LR greater than 1). In a very small experiment, Brown (1996) compared glass fragments from broken windows with respect to their trace elements (such as calcium and iron), and found that most comparisons involving fragments from the same window had LRs greater than 1, and most comparisons with fragments from different windows had LRs smaller than 1 (p. 23).

As far as speech is concerned, recent forensic speaker identification work using Bayesian methods on forensically realistic speech material also demonstrates the correctness of the method. For example, Kinoshita (2001) has shown with Japanese speakers how, even with very few acoustic parameters, 90% of same-speaker pairs are resolved with LRs greater than 1, and 97% of different-speaker pairs have LRs smaller. Meuwly and Drygajlo (2001: 150), with Swiss-French speakers, obtained values of about 86% for same-speaker pairs with LRs greater than 1, and 86% for different-speaker pairs with LRs less than 1. Gonzalez-Rodriguez *et al.* (2001) for Spanish, and Nakasone and Beck (2001) with American English also show similar results.

Because same-subject data are resolved with LRs greater than 1, and different-subject data are resolved with LRs smaller, the approach can be used to successfully discriminate same-subject from different-subject pairs: if a pair has LR > 1 it is judged to be same-subject, if its LR < 1 it is evaluated as different-subject. This discrimination was actually the main goal of the speech papers quoted above.

The fact that the Bayesian approach works can thus be construed both from a theoretical and a practical point of view. Firstly, the theory can be said to work because its theoretical predictions have been empirically confirmed in several forensically important areas. Secondly, the theory can be said to work because it can be practically used to discriminate same-subject from different-subject data.

Ease of combining evidence

Another very important feature of the approach is that it makes combining evidence from different sources straightforward (Robertson and Vignaux 1995: 118). Thus if two speech samples are compared with respect to two different features, and found to differ significantly in one feature but not in the other, it is not immediately obvious how to interpret this. Within a Bayesian approach, however, it has been shown above that the evidence from separate sources is combinable by multiplying their respective LRs to derive a LR for the combined evidence. The importance of this solution cannot be overstressed, and it will be demonstrated later.

Explicitness

It is often pointed out that the explicitness of the Bayesian approach helps to understand and recognise the so-called prosecution and defence fallacies. These fallacies consist in the double-whammy of interpreting only one part of the LR and transposing the conditional.

For example, imagine that the probability of the evidence in a forensic speaker identification was 80% assuming that the questioned and suspect samples came from the same speaker, and 10% assuming provenance from different speakers. Thus the LR for the forensic-phonetic evidence is $p(E \mid H_p)/p(E \mid H_d) = 0.8/0.1 = 8$. The evidence is eight times more likely assuming that the suspect is guilty. One version of the prosecution fallacy with this example would be to focus on the 80% numerator and claim, by inverting the conditional, that because the probability of the evidence assuming guilt is 80%, the accused is 80% likely to be guilty.

This is called transposing the conditional because, in making the probability of the evidence given the prosecution hypothesis [$p(E\,|\,H_p)$] into the probability of the prosecution hypothesis given the evidence [$p(H_p\,|\,E)$], it is swapping the conditions of the probabilities. These terms are not interchangeable, however, as can be seen from the fallacy in the following standard zoological demonstration. What is the probability of it having four legs if it's a cow? Barring unusual genetic and road accidents, the probability for the four-legged evidence assuming the cow hypothesis is 1 [p(quadruped | cow) = 1]. Now transpose the conditional: what is the probability of it being a cow if it has four legs? [p(cow | quadruped) = ?] Certainly not 1: *QED*.

The other version of the prosecution hypothesis would be to focus on the denominator of 10% and invert the conditional to say that, since the probability of a match by chance is so low, it is very probable – 90% probable given the 10% match by chance – that the accused is guilty.

The same arguments can be used *mutatis mutandis* by the defence. Thus one defence fallacy would consist in putting a different spin on the transposed conditional of the denominator: the 10% match by chance means that it could be anyone of 10% of the relevant population. If the relevant population is 5000, that means anyone of 500 individuals! These examples show that both defence and prosecution fallacies give very different, extreme, interpretations to the evidence, the correct strength of which as quantified by the LR of 8 is moderate at best.

These errors in thinking are by no means uncommon (Evett 1991: 11; Robertson and Vignaux 1995: 91–4). Sjerps and Biesheuvel (1999: 225), for example, report that in a small experiment most Dutch criminal lawyers and judges (most of them trainees) were unaware that the current Dutch scale, with its transposed conditionals, is logically incorrect, and prefer it to Bayesian formulations. Errors are not confined to legal counsel: doctors have been shown to transpose the conditional in medical diagnosis as well as ignoring the prior (Gigerenzer *et al.* 1989: 257–8). The existence of an explicit model for forensic inference is therefore crucial.

Inferiority of analyses based on statistical significance

As explained above, hybrid frequentist analyses constitute the mainstay of modern descriptive statistics, and also 'a high proportion of situations involving the formal objective presentation of statistical evidence uses the frequentist approach with tests of significance' (Aitken 1995: 3). Simple examples are parametric tests like Student's t, or non-parametric tests like Chi-square. Student's t-test reports the probability of the difference between the mean values of two samples being significant. *Significant* here means that there is a high probability that the difference between the means of the two samples is not due to chance, and is therefore due to some other factor, usually assumed to be that the two samples have come from different sources, or populations. So, looking at the data in Figure 2.2 for example, one could t-test whether speaker RS has a significantly lower mean value for dimension 1 than DM. He does, which means that the difference between their means is most likely not due to chance; perhaps it is because they are different speakers.

However, the frequentist notion of statistical significance does not relate in any useful way to the origin of differences between speech samples. This is for the simple reason that statistically significant differences occur between two speech samples from the same speaker (Rose 1999b: 15–23), and non-significant differences occur between speech samples from different speakers (Rose 1999a). Thus, to look once again at the data in Figure 2.2, although there is a high probability that RS has a significantly lower mean value for dimension 1 than DM, a *t*-test also shows that DM's mean value for dimension 1 does not differ significantly from MD's. In other words, *just because a difference between two speech samples is statistically significantly different this does not mean that they were spoken by different speakers.* Likewise, *a non-significant difference does not necessarily imply same-speaker origin.*

Inferiority of approaches based on matching

A common approach to forensic analysis usually involves first deciding if there is a 'match' between suspect and questioned samples. If a 'match' is declared, then the investigation is continued, and it is calculated how probable it would be to find such a match by chance in the relevant population. If not, the suspect is excluded. This type of analysis is therefore often called a 'two-stage' approach (Evett *et al.* 1993: 498–9). It also falls under the heading of frequentist approaches because the criteria for a match are often couched in frequentist-distributional terms. For example, the suspect sample might be considered a match if it falls within a certain frequentist-distributionally defined distance of the questioned sample.

There are two main things wrong with the matching part of this approach. Firstly, there is an intuitively unsatisfactory all-or-nothing decision involved. Thus a suspect will be excluded if their values fall on one side of the cut-off for a match, but not if they fall on the other, even though the difference in the values may be minimal. This is often called the 'fall-off-the-cliff' effect (Robertson and Vignaux 1995: 118): a little bit more than the cut-off value you are still standing there; a little bit less you are gone. A Bayesian approach avoids this problem completely, since the mathematical functions involved are continuous, not discrete.

Secondly, matching has the undesirable consequence that it makes it certain that one would get the evidence assuming that the suspect is guilty, because a 'match' has been declared. This assumption will sometimes be correct, as for example when there is a match between the blood group of the suspect and that of a single bloodstain at the crime scene. Then, if the suspect is guilty, the probability of observing the same blood group at the crime scene as that of the suspect is 1. However, in many cases, and almost always where forensic speech samples are concerned, the probability of observing the evidence assuming that the samples have a common origin is not 1, and is sometimes very much less than 1. Therefore the assumption imposed by matching is false.

Again, this problem is avoided by a Bayesian approach, because it allows any value between and including 0 and 1, not just 1, for the numerator of the LR. In those cases where the probability of the evidence assuming common origin is 1, then the LR will be the reciprocal of the probability of finding the match in blood groups by chance in the relevant population. For example, if the probability of a match by chance is, say, 20%, then the LR will be 1/0.2 = 5. This result, where the numerator of the LR is 1, is the same as that derived by the two-stage approach. However, whenever any other value is required for the LR numerator, and this will be in the majority of cases, the two approaches will evaluate the evidence differently, and, as has been shown with DNA analysis, the evaluation of the two-stage matching approach will be incorrect (Evett *et al.* 1993: 502–4).

Frequentist analysis can also be shown to fail in more complex forensic situations, such as the so-called 'twin trace' problem. This is a scenario where the crime scene has two different bloodstains from two different perpetrators, and the suspect's blood group matches one of them. The correct strength of evidence cannot be assessed by a frequentist analysis in such a case (Evett 1991: 14). A Bayesian analysis, on the other hand, can correctly evaluate such data (Stoney 1991: 119, 122–3).

Focus on individuality

A trial is about evaluating evidence in a particular, individual case. 'Forensic scientists . . . must try to assess the value as evidence of single, possibly non-replicable items of information about specific hypotheses referring to an individual event' (Robertson and Vignaux 1995: 201). Frequentist approaches, however, emphasise the general at the expense of the individual: '. . . statistical [i.e. frequentist] justice is a contradiction in terms, for it replaces the individual with an average, the case at hand with a generality' (Gigerenzer *et al.* 1989: 260, see also p. 226). A Bayesian approach, although it makes use of general statements, is geared precisely to evaluate the individual case.

We shall return to this point later, when more theoretical ground has been covered, but the difference between the specific and the individual can still be appreciated by considering the examples of successful discrimination approvingly cited above. In these cases, it proved possible to successfully discriminate same-speaker from different-speaker speech samples on the basis of their associated LRs. But, as will be emphasised in due course, these are *general* data – they obtain over many different pairs of speakers. They do not contribute as much as they could to an *individual* case, where it is important to know the *actual value* of the LR – whether it is really big, moderately big, minuscule, or what.

Anti-Bayesian arguments

Undoubtedly the source of the greatest and continuing criticism of Bayesian inference centres on the notion of prior odds: that it is by nature indeterminate; and that it is incompatible with the presumption of innocence. Other objections concern the complexity of the approach, both mathematical and logical. These will now be briefly addressed.

Indeterminacy of prior odds

Within a Bayesian framework, a quantification of the prior odds is essential to determining the probability of the hypothesis, and, since it is the factor by which the LR is multiplied, it will have an enormous influence on the posterior odds. Most criticism concerns the fact that in many, perhaps most, real-world situations, the prior odds will not be known. Therefore, different investigators may have different estimates of the prior odds, and thus their estimates of the posterior odds in favour of the hypothesis will therefore also differ (Kotz *et al.* 1982: 205; Gigerenzer *et al.* 1989: 93, 227, 264; Lindley 1990: 45). This has also led to the epithet 'subjective' being applied to Bayesian probabilities, usually by frequentists.

An additional criticism related to the indeterminacy of the prior odds is this. The expert can legitimately be invited by the opposing side to calculate a whole lot of different prior odds corresponding to different possible assumptions. There is then a danger that the jury will react to the information overload and ignore the scientific evidence *in toto* (J. Robertson, personal communication).

The difficulty of estimating priors, or the fact that different experts may differ in their estimation, does not of course mean that it is not important and can be ignored. Suppose a patient has a positive mammogram. The probability that she actually has breast cancer will be very different depending on whether she has already presented with a lump in her breast, or whether she is asymptotic (Gigerenzer *et al.* 1989: 256).

There is, however, a sense in which the prior odds controversy is a red herring, at least for the forensic phonetician. The part of Bayes' theorem that has relevance for the forensic scientist is the likelihood ratio, not the prior odds. The forensic scientist is primarily concerned with assessing how probable the difference between questioned and suspect samples is assuming that they come (1) from the same, and (2) from different speakers. In the majority of cases the forensic phonetician does not even know the prior odds. Moreover, *pace* Champod and Meuwley (2000: 199), there are good grounds for insisting that they normally should *not* be given information relating to the prior odds, since, as will be explained in Chapter 9, this can lead to difficult and justified questions concerning the expectation effect. It is interesting to note that most papers making use of a Bayesian approach to discriminate same-subject from different-subject data simply use the LR in the absence of any prior (e.g. Nakasone and Beck 2001: 6).

Notwithstanding all this, there is one prior that can legitimately be asked of the expert forensic phonetician: what is the past performance, either of themselves, or the automatic system they use? (Koolwaaij and Boves 1999: 248ff).

It must be emphasised that all this does not make the problem go away: it needs to be reiterated here that the prior odds are vitally important; they must be addressed sometime in the legal process; and the assumptions behind them must be made explicit but kept simple.

Prior odds and the presumption of innocence

Prior odds are also implicated in the criticism that they are incompatible with the presumption of innocence (Gigerenzer *et al.* 1989: 264). The argument runs like this. Presumption of innocence implies a zero probability of guilt as prior. This in turn would make it impossible to apply Bayes' theorem, since any LR multiplied by a prior odds of zero is zero.

As explained above, Bayesians understand probability as a degree of belief. Therefore, the validity of this argument turns on the implicature between the meaning of the epistemic verbs *presume* and *believe*. (Epistemic verbs are 'verbs of thinking' with complex meanings involving semantic primes KNOW, TRUE, SAY, and perhaps other elements, as well as THINK (Goddard 2001).) If these two verbs are incompatible, then the meaning, or even part of the meaning of one must contradict the meaning, or even part of the meaning of the other. In other words, does *I presume Y is innocent* contradict *I believe Y is innocent*?

Since this is clearly a semantic problem, it is properly solved by semantic analysis, which is part of the discipline of linguistics. The appropriate analysis tool is well known: Natural Semantic Metalanguage (Wierzbicka 1996). Natural Semantic Metalanguage (NSM) expresses the meaning of words as combinations of basic units called semantic primes, and has been very successful in various applications, including legal semantics. Langford (2000), for example, uses NSM to describe the difference between the legal and lay meanings of *murder*, *manslaughter* and *homicide* (see also Goddard 1996).

According to NSM, the meaning of *I presume that Y is innocent* is specified by the following semantic components, adapted from A. Wierzbicka (personal communication):

I can't say: 'I know that Y is innocent'
I want to think for the time being: 'I know it'

On the other hand, *I believe that Y is innocent* involves the following components, adapted from Goddard (2001):

The prosecution says Y is not innocent
When others hear this, they can think it is true
I don't want to think that it is true
I think that Y is innocent

This breakdown of the semantic components allows us to see that there is no contradiction between the meanings of *presume* and *believe*. The crucial components are *I think that Y is innocent* (for *believe*), and *I can't say: 'I know that Y is innocent'* (for *presume*). These components are not contradictory, since the statement 'I think it is so but I don't know' entails no contradiction. Thus it can be shown by NSM analysis that incompatibility of prior odds with presumption of innocence is not a valid criticism of the legal use of Bayes' theorem.

Ignorance of defence hypotheses

In order to compute a LR, one must have knowledge of the alternative, or defence hypothesis. It was pointed out above that, although it will usually be an assertion that a different speaker was involved, the alternative hypothesis can, in fact, take many forms: it was not the accused, but someone else/it was not the accused, but his brother/ it was not the accused but someone sounding similar, and so on. The probabilities associated with different alternative hypotheses will be different, and therefore so will the resulting LR. In jurisdictions where the defence is not obliged to disclose its line of defence before the trial, the calculation of a LR by an expert appearing for the prosecution is therefore theoretically not possible.

It can be seen that this is not a critique of the Bayesian approach *per se*, but rather a practical problem (cf. Robertson and Vignaux 1995: 210–11). One solution would be for prosecution to calculate a LR assuming simply a defence hypothesis of different speakers, but to be prepared to recalculate in response to a different hypothesis.

Complexity

A frequently encountered criticism concerns the logical and/or mathematical complexity of Bayesian inference, which make it too difficult to explain in court (Evett 1991: 14; Gigerenzer *et al.* 1989: 265). It is true that the mathematics underlying Bayesian inference can be intimidating, involving horrendously complicated integrals. The court, however, is not expected to understand this – that is, after all, one of the reasons why an expert is required. Moreover, to paraphrase Evett (1991: 14), the maths is complicated because the situation is complicated, and if this is the correct way to evaluate forensic evidence, then that is how it must be.

As far as the *logic* behind Bayes is concerned, the author is not persuaded that it is excessively difficult to understand. In fact it could be argued that a formulation like 'You are ten times more likely to observe the difference between these speech samples if they came from the same person than if they came from two different people' seems to be conceptually rather simple, especially when compared with explanations of the logic underlying frequentist significance testing (Malakoff 1999: 1464).

A reality check

So much for the ideas. Now for the reality. Three things currently stymie the full application of a likelihood ratio approach in forensic phonetics. Firstly, there is the degree to which it is not received practice, either by the experts themselves, or in the courts. Broeders (1999: 239) thinks that the reasons for the former are that practitioners 'do not wish to stop giving what are in fact (disguised) categorical judgements' about the hypothesis, and that this is reinforced by the court's willingness to hear such categorical statements (e.g. it was/was not the speaker) qualified by a statement about the degree of the expert's uncertainty (e.g. I am pretty sure).

Now, it may seem that a statement like 'I am 80% sure that it isn't the same speaker' is equivalent to the LR based statement 'You are four times more likely to observe the difference assuming that the speakers are different than if they are the same'. (A probability of 80% is equivalent to odds of $(0.8/(1 - 0.8) =)$ 4 to 1.) But, as should be clear from the exposition in this chapter, there is a world of difference. It is simply not logically possible to estimate a probability like '80% sure' unless you know all the background information, i.e. the prior odds.

Clearly, it must be said that the current state of affairs with respect to the appropriate formulation of the forensic expert's conclusions presents a potentially very serious problem. Moreover, it is one not confined to forensic-phonetic evidence, since the appropriate formulation of conclusions has come under the critical spotlight in many other areas of forensic science too (Broeders 2001: 9). What happens, for example, if the court is not aware of, or does not understand, the logically correct approach to estimating the strength of the scientific evidence, and expects, or insists, on a statement of the probability of the hypothesis given the evidence? *One way of avoiding this is for the expert to ensure that the approach informing their forensic analysis is totally clear from the outset to all interested parties.* They will need to explain that they will not be, indeed cannot be, trying to say whether the same speaker is involved or not. They will need to explain that they will rather be trying to assess the strength of the evidence by estimating how much more likely it is to observe the differences between questioned and suspect samples under the competing prosecution and defence hypotheses. The other way of avoiding the problem of course is for interested parties to read this book! In any case the onus is on both the expert and the institutions they serve. Forewarned is forearmed.

Broeders (2001: 9) says that the use of likelihood ratios in forensic speaker identification remains unrealistic because experts are generally unable to adequately quantify their findings, and this is the second factor inhibiting the approach. In order to estimate the denominator of the LR, we need to know what is typical. For example, how many speakers in the population have a funny *r* sound? How many speakers differ to the extent observed in their *ee* vowel acoustics? In other words, it is necessary to have an idea of the statistical distribution of the relevant parameters in the relevant populations.

It is certainly true that we still possess only restricted knowledge of such things for forensically realistic, natural, unconstrained speech, even for the commonly measured parameters. This is a direct consequence of two things. Obtaining such data requires an enormous amount of time, and there is a pervasive lack of funding support, in Australia at least, for the kind of research necessary to determine them. However, *pace* Broeders, it is vitally important to realise that this scarcity of quantitative data does not prevent a realistic implementation of the approach. Indeed, this is probably the area where phoneticians come into their own and have the most positive role to play in providing information that will be of use to the court.

An expert phonetician is firstly able to note the inevitable similarities and differences between speech samples, and is able to identify from among these the forensically important instances. These will be the similarities and differences that occur in comparable environments and that are not a function of other, linguistic or situational, variables (Nolan 1997: 761). To do all this they rely on their linguistic knowledge of the phonological variation of the language(s) in question and their empirical knowledge from extensive exposure to similar data. They are then able to describe, quantify and evaluate the important differences and similarities between speech samples. Perhaps two samples will show many differences that are not attributable to linguistic or situational variables; perhaps two samples might show many similarities, but the similarities are such as could be expected to be found at random in the population; perhaps they might show a degree of similarity that one would only rarely expect to encounter in different speakers.

It will be possible to evaluate some of the differences and similarities between the samples, especially the acoustic ones, in terms of relatively 'hard' LRs. This is where information, if not on the appropriate reference distribution at least on a comparable one, is either available or can be compiled by the expert. Other similarities and differences will have to be evaluated in terms of 'soft', or intuition-based LRs. However, providing there is enough comparable speech material, there are usually so many points of comparison that a phonetician can extract that it will be possible to arrive at a likelihood ratio estimate.

Thus typical statements might be: 'Both the questioned and suspect samples show an abnormal lateral fricative allophone for /s/, together with many features of a lower-class

Glaswegian accent with the same admixture of Australian English phonology. In addition, using the distributions in the Bernard (1967) corpus for Australian English, the difference between the samples in vowel formant frequencies can be estimated to be between four and ten times more likely to be observed from the same speaker than from different speakers. The incidence of a lateralised /s/ in the Australian male population is softly estimated at between one in 200 and one in 300, which gives a LR of at least (1/0.005 =) 200. Ignoring the dialectal features, the agreement found between questioned and suspect samples in these particular features is therefore at least (4 × 200 =) 800 times more likely were they to have come from the same speaker than different speakers. This, prior odds pending, lends considerable support to the prosecution hypothesis that they were spoken by the same speaker.' Or 'The similarity in formant frequencies between the questioned and suspect samples is great, but according to the Bernard (1967) corpus, it is such that it could be equally expected to occur by chance between many males with this accent. Therefore the LR is probably not very different from one and, pending prior odds, there is little if any degree of support for the prosecution hypothesis.'

Of course, the phonetician is never absolved from being explicit about the basis of both their 'hard' and 'soft' LR estimates, and the contribution of these estimates to the overall LR. One thing, independent of Bayesian or frequentist formulations, is clear: *any statement of probability proffered in a strict sense must refer to population data*. Unless it is known, or estimated, how a feature distributes in the population, nothing can be said about the probability of observing it, at random, from that population. A forensic-phonetic expert expressing a probability statement, or giving a LR based on probability statements, must therefore be prepared to clarify its status as population-based or intuition-based, or both.

A third reason why the LR approach cannot yet be fully implemented for forensic speech evidence is that the appropriate statistical models have not yet been adequately refined. This is because of the high degree of complexity inherent in variation between natural speech samples. Both of these problems – ignorance of parametrical distributions and unrefined statistical modelling – will be demonstrated in some examples of the calculation of LRs from real acoustic data given in Chapter 11. This shortcoming is not as important as the previous two factors, because it can be shown that good results can be obtained with current statistical models.

Whatever the nature of the observation data, and the remaining imperfections of the statistical models, it can hardly be denied that one point of major significance of the LR approach lies in 'providing a conceptual framework within which our thinking about the identification process can take place' (Broeders 1999: 240).

Chapter summary

The aim of a forensic-phonetic identification should be to express its outcome in terms of a Bayesian likelihood ratio. This is the logically correct way of quantifying the strength of forensic identification evidence, and should constitute the conceptual framework for all forensic-phonetic comparisons. The LR should state how much more likely it is to observe the differences between questioned and suspect voice samples assuming that they have come from the same speaker than if they have come from different speakers. The LR must be combined with the appropriate prior odds to estimate the posterior odds in favour of guilt. The nature of the defence, or alternative, hypothesis can affect the LR and the prior odds must be considered with care.

Initial studies with forensically realistic speech samples have shown that the LR approach makes the correct theoretical predictions and discriminates well between same-speaker and different-speaker pairs. Scarcity of hard data and lack of statistical knowledge, both of the distribution of important parameters in the population and of how to adequately model those distributions, currently constitute impediments to full implementation of the approach.

Even with only partially quantified data, however, the LR approach still provides the correct framework within which forensic phoneticians can conceptualise their activity, and express their conclusions based on extensive linguistic and empirical knowledge. In this way, it is possible for them to provide speaker identification evidence that is both extremely useful and highly interpretable.

5

Characterising forensic speaker identification

It might seem that recognising someone by their voice is a simple notion, but there are many different circumstances under which this can happen. Your voice could be recognised by your friend when you ring them up, for example, or it might need to be automatically recognised by a computer in order for you to have access to your bank statement over the phone. There are thus many different types of speaker recognition, of which forensic speaker identification is one. There are also different types of forensic speaker identification, and it is important to be able to classify the type of forensic speaker identification evidence involved in a case because its proper evaluation depends on the type. The aim of this chapter therefore is to describe the different types of forensic speaker identification and to explain how forensic speaker identification relates to other types of speaker recognition. The chapter draws heavily on the survey in Chapter 1 of Nolan's (1983) *The Phonetic Bases of Speaker Recognition*, which improves on an earlier (1976) classification by Bricker and Pruzansky.

Starting with the notion of speaker recognition, the first important distinction addressed is that between the two main classes of speaker recognition task: *identification* and *verification*. It is then examined how forensic speaker identification relates to this distinction. Cross-cutting the difference between identification and verification is the second basic distinction, between *naive speaker recognition* and *technical speaker recognition*. This is explained next, and again the relationship between them and forensic speaker identification is clarified. Since earwitness evidence is of obvious forensic relevance, the section on naive speaker recognition considers the different variables affecting the accuracy of voice identification by naive subjects. Finally, an approach to forensic speaker recognition is described that is important but difficult to classify – voiceprint (or aural-spectrographic) identification.

Speaker recognition

Forensic speaker identification can often be found classified as a kind of speaker recognition (e.g. Nolan 1997: 744–5). Speaker recognition has been defined as 'any decision-making process that uses some features of the speech signal to determine if a particular person is the speaker of a given utterance' (Atal 1976: 460 fn.1).

Given the discussion on the likelihood ratio in the preceding chapter it can be appreciated that, since in forensic speaker identification the decision as to whether or not an utterance was spoken by a particular speaker is properly the domain of the court, this characterisation is not entirely appropriate. Another aspect wherein Atal's characterisation is not totally correct for forensic speaker identification is that it strongly suggests that an unambiguous, categorical outcome is expected: the person is either determined to be or determined not to be the speaker of a given utterance. In the forensic case the outcome should be a ratio of probabilities. Despite these shortcomings, it is clearly still helpful to persevere with the idea of forensic speaker identification as a kind of speaker recognition.

Speaker identification and verification

There are two main classes of speaker recognition task, called *identification* and *verification* (Furui 1994: 1; Nolan 1997: 745). The distinction between them rests firstly on the type of question that is asked and secondly on the nature of the decision-making task involved to answer that question. Comparing the characteristics of speaker identification and verification will help clarify the nature of the task in typical forensic speaker identification.

Speaker identification

The aim of speaker identification is, not surprisingly, identification: 'to identify an unknown voice as one or none of a set of known voices' (Naik 1994: 31). One has a speech sample from an unknown speaker, and a set of speech samples from different speakers the identity of whom is known. The task is to compare the sample from the unknown speaker with the known set of samples, and determine whether it was produced by any of the known speakers (Nolan 1983: 9).

This is shown schematically in Figure 5.1, where a simple speaker identification experiment is represented with a reference set of 50 known-speaker samples. In Figure 5.1, the unknown sample on the left is compared with that from known speaker 1 (John), then known speaker 2 (Bruce), and so on. The question mark represents the question: are these two speech samples from the same speaker? If it is decided that the unknown sample is the same as one of the known speakers, say known speaker 4, then that identifies the speaker of the unknown sample as Roderick.

Speaker verification

Speaker verification is the other common task in speaker recognition. This is where 'an identity claim from an individual is accepted or rejected by comparing a sample of his speech against a stored reference sample by the individual whose identity he is claiming' (Nolan 1983: 8). This situation is represented in Figure 5.2. This time Roderick wants access to some highly sensitive material. He phones the verification system up requesting access, and gives his name and identification number. The system has

Characterising forensic speaker identification

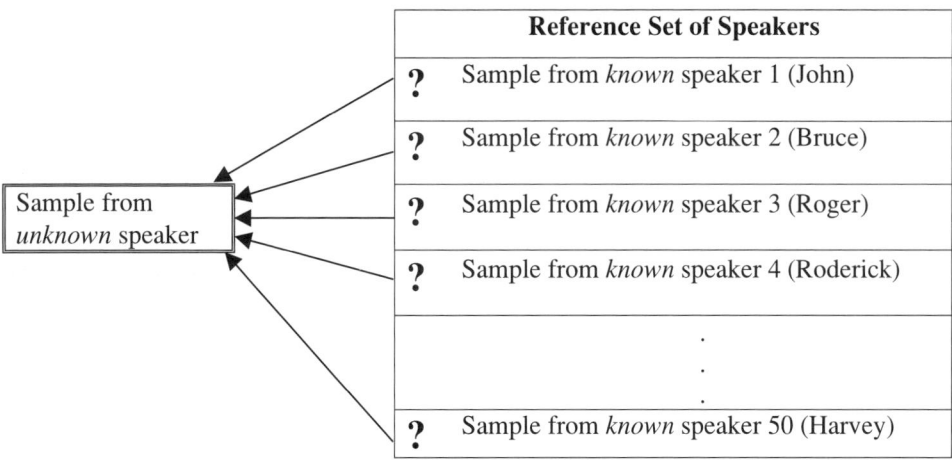

Figure 5.1 Schematic representation of speaker identification

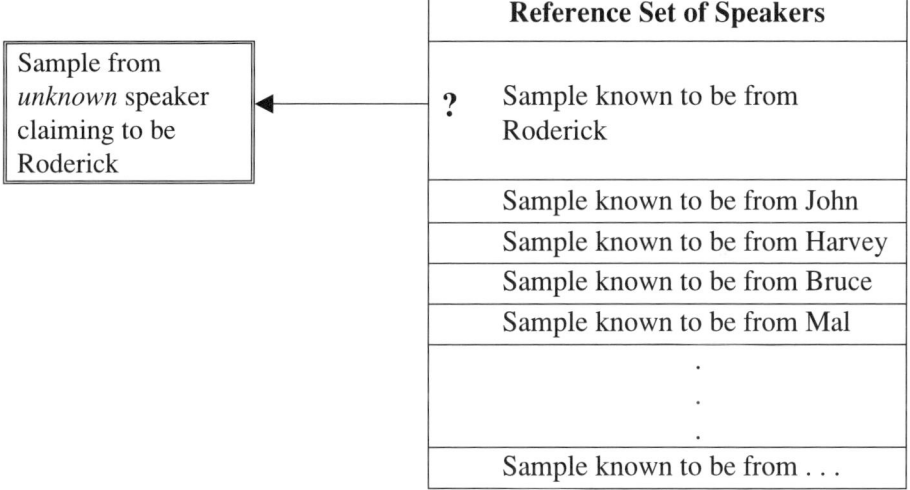

Figure 5.2 Schematic representation of speaker verification

examples of Roderick's voice in storage, which it automatically retrieves and compares with that of the sample tendered by Roderick. If the two voice samples are judged similar enough, Roderick's claim is verified and he is given access.

Comparison between identification and verification

It is clear that both identification and verification have in common a discriminant comparison between a known and an unknown speech sample in order to answer the question of whether the samples come from the same speaker or not. Nevertheless, apart from the difference in the applications for which they are intended, there are several other differences between them. The most important distinction concerns the properties of the reference set of speakers involved, and has to do with the distinction firstly between *open* and *closed* sets, and secondly between *known* and *unknown* sets.

Open and closed set identification

In speaker identification, the reference set of known speakers can be of two types: closed or open. This distinction refers to whether the set is known to contain a sample of the unknown voice or not. A *closed* reference set means that it is known that the owner of the unknown voice is one of the known speakers. An *open* set means that it is not known whether the owner of the unknown voice is present in the reference set or not.

This is a very important distinction. Closed set identification is usually a much easier task than open set identification. Since it is known that the unknown speaker is one of the reference set, the closed set identification task lies in (1) estimating the distance between the unknown speaker and each of the known reference speakers, and (2) picking the known speaker that is separated by the smallest distance from the unknown speaker. The pair of samples separated by the smallest distance is then assumed to be from the same speaker (Nolan 1983: 9). Because the nearest known speaker is automatically selected in a closed set identification, no threshold is needed.

In an open set identification, on the other hand, one cannot assume that the pair of samples separated by the smallest distance is automatically the same speaker. One has to have a pre-existing threshold, such that samples separated by distances smaller than the threshold will be correctly deemed to be from the same speaker.

Both closed and open sets can occur in forensic case-work, although the latter, where you do not know if the putative offender is among the suspects or not, is usually far more common. Since the task usually becomes very much simpler with a closed set, the distinction between open and closed set tasks is an important one in forensic speaker identification, and it is important for law enforcement officers to be aware of the distinction.

Known and unknown set membership

The distinction between closed and open set does not have any meaning for verification: for the process to make any sense, it must be assumed that the individual whose identity is being claimed is among the reference set of speakers. However, another property of the reference set of speakers does: whether it is *known* or not. In the typical verification scenario, the reference set of speakers is known. It will usually be the set of a firm's employees; the set of a bank's clientele; a set of parolees, the identity of all of whom will need to be verified from time to time. But the set does not have to be known, and this is typical of forensic speaker identification, where the questioned voice is usually one of a set whose membership is not known.

Number of comparisons

Identification and discrimination differ in the amount of paired comparisons that need to be undertaken. In speaker verification, as shown in Figure 5.2 the discriminant comparison occurs only once, between the unknown sample that is claimed to be from a given individual and the known sample from that individual. In identification, the comparison occurs iteratively, between the unknown sample and each of the known samples, as in Figure 5.1 (Nolan 1983: 9). This difference has several important consequences. In particular it determines the number and type of decisions that can be made; the effect on performance of the size of the speaker pool involved; and the evaluation of the performance of verification and identification strategies. These are addressed below.

Type of decision

In identification, only two types of decision are possible. Either the unknown sample is correctly identified or it is not. Verification is more complicated, with four types of decision. The decision can be correct in two ways: the speaker is correctly identified as being who they say they are, or not being who they say they are. And it can be incorrect in two ways: the identity claim of the speaker can be incorrectly rejected (the speaker is who they say they are but is rejected), or incorrectly accepted (the speaker is an impostor but is nevertheless accepted).

Evaluation of performance

For identification, a single figure will characterise its performance: this is usually the number of errors expressed as a percentage. For example, a speaker identification system can be characterised as having a 2% error. This means that in the long run – in a large number of trials using different questioned voices and different sets of reference speakers – it makes an identification error 2% of the time. Types of incorrect identification will depend on whether an open set is involved or not. With a closed set, an incorrect identification will be a false identification – the questioned sample is said to be from speaker x when it is not. With an open set, an incorrect identification can either be a false identification or it can be a missed hit, where a match is present between the questioned sample and one of the speakers but is missed by the system.

Verification, once again, is more complicated. In order to characterise verification performance, two figures are needed. One figure quantifies the number of false acceptances, also known as *false identifications*, or *missed detections*. This is the percentage of times the system says that two samples are from the same speaker when they are not. The other figure quantifies the number of false rejections, also known as *false eliminations* or *false alarms*, which is the percentage of times the system has failed to detect the presence of the same speaker.

The figures can be plotted against each other for different conditions, the resulting graph being known as a *receiver operating characteristic* (ROC), e.g. Bolt *et al.* (1979: 90, 91); Gruber and Poza (1995: section 97), or *detection error trade-off* (DET), e.g. Nakasone and Beck (2001). Such a curve makes explicit how well the system performs in discriminating same-subject from different-subject data.

Imagine that a system has been developed to verify, say, 100 clients. In Chapter 2 it was pointed out how in a verification system the number of false eliminations and false rejections are in a trade-off relationship that can be altered by changing the threshold of the system for deciding when two samples are similar enough to be

Forensic Speaker Identification

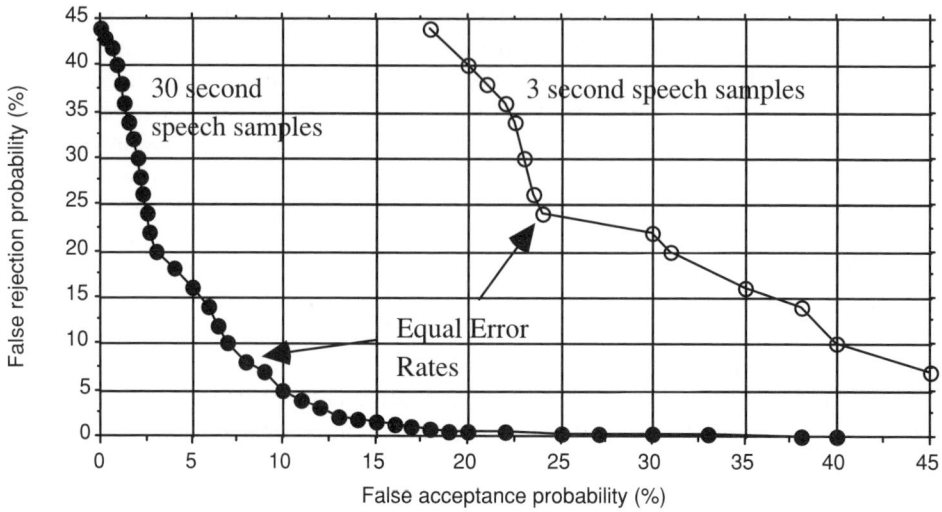

Figure 5.3 Hypothetical detection error trade-off curves showing the performance for a speaker verification system on speech samples of differing duration

considered the same. The system is tested to see how well it can verify each of these 100 clients using many different thresholds. It is tested not only for how well it can verify the voice of a particular *bona fide* client, of course, but also for how well it can reject an impostor. For each of the different thresholds, a point is marked on a graph that represents the average performance of the system over all 100 speakers. Moreover, it is desired to find out whether the amount of speech required from each speaker affects the discrimination performance. Figure 5.3 shows some hypothetical, but realistic, DET curves for this speaker verification system (they are based on actual data from Nakasone and Beck (2001)).

Two curves are shown: one representing the performance of the system when it is presented with speech samples of 30 seconds duration, and one when it only has 3 seconds of speech to work on. The horizontal axis marks the percentage probability of false acceptances and the vertical axis shows the percentage probability of false rejections. It can be seen that, for the curve representing 30 seconds of speech, a value of 10% on the false acceptance horizontal axis corresponds to a value of 5% on the vertical false rejection axis. This means that when the system is set up so that in 10% of the cases it evaluates a different-speaker pair as the same (a false acceptance, an impostor is admitted), it will evaluate a same-speaker pair as different (a false rejection, a *bona fide* applicant is rejected) 5% of the time.

The trade-off in detection errors can be clearly seen. If it is desired to minimise the probability of a false acceptance (a clearly desirable feature in forensics, since we want to avoid incriminating the innocent – saying that suspect and offender are the same when they are not) then a corresponding increase must be reckoned with in the probability of rejecting *bona fide* applicants (in FSI, missing some guilty parties). Thus it can be seen that a false acceptance rate of 10% will be accompanied by a 5% probability of false rejection, but if we shift the threshold to lower the false acceptance rate to 5%, the rate of false rejections will increase to 16%.

As explained in Chapter 2, a verification system can be set to deliver the same percentage of false rejections as false acceptances – the so-called equal error rate (EER). It can be seen that the EER of this system on 30 seconds of speech is about 8%.

DET curves are useful for showing how a system performs under different conditions. In Figure 5.3 a second curve is shown for the system's performance on the much shorter amount of 3 seconds' speech. As can be seen, the position of the curve indicating the system's performance on 3 seconds of speech has been displaced upwards and to the right of the curve for the performance on the 30 second samples. The system's equal error rate for this condition is about 24%. This indicates that the system can perform better if it can base its verification decision on a greater amount of speech. (This is a reminder that the amount of speech available is an important consideration in speaker recognition.)

It can be appreciated that identification is more easily characterised than verification. Because of this it is the simplest paradigm for testing the performance of speaker identification parameters.

Relationship between size of speaker pool and performance

A speaker identification task requires $n + 1$ decisions for a speaker pool of n speakers (deciding that the unknown voice is one of the n known voices or none of them). Because of this, its performance degrades with increasing number of speakers in the pool. You can thus expect a speaker identification system to perform much worse if 100 speakers are in the pool rather than 20. Because of the greater number of possible decisions in verification, the evaluation of verification systems is more complex. However, since speaker verification requires a simple binary comparison (deciding whether the questioned voice is who it claims to be or not), its performance will approach a constant that is essentially independent of the number of speakers to be verified, providing that the distribution of the speakers' characteristics is not excessively biased (Naik 1994: 32; Furui 1994: 1).

Differential distortion in questioned voice

In both verification and identification, it is possible that the questioned voice has deliberately been distorted by its owner, but the distortion will be of different kinds. In speaker verification, mimicry must be reckoned with, as an impostor tries to gain unlawful access. In identification, on the other hand, disguise is a possibility.

Relationship between forensic speaker identification and speaker identification/verification

The question of how forensic speaker identification relates to the two major task paradigms of identification and verification has been discussed by several authors (Nolan 1983; Künzel 1994: 136; Broeders 1995: 155–8; Champod and Meuwly 2000). It is an important question, and worth addressing here, because it is clear that speaker identification and verification have served, and continue to serve, as models for frequentist, or non-Bayesian forensic speaker identification (that is, approaches that attempt to state the probability of the hypothesis that the same speaker is involved from the evidence of the difference between questioned and suspect speech samples.) It is clear that forensic speaker identification shares with both speaker identification and verification the comparison of unknown and known speech samples in order to derive information relating to the question of whether they have come from the same speaker or not.

It is also easy to think of speaker identification and verification scenarios that correspond to forensic situations. For example, it might be the case that one had a questioned voice sample, which would correspond to the unknown sample on the left in Figure 5.1. One might also have voice samples from several suspects: these would correspond to the known speaker samples on the right. The task would then be to find the strength of the evidence supporting identification of the speaker of the questioned sample as one of the known suspect samples. In the unlikely event that it was known that the questioned sample was from one of the suspects, the conditions for a closed set identification would exist. Normally, however, this is not known, and so the test would be an open one. Given these parallels, it is not surprising that forensic speaker identification has been likened to speaker identification (Künzel 1994: 136), and indeed the term identification appears in the title of this book.

A parallel between forensic identification and speaker verification might be the very common situation in which the police are claiming that the questioned sample comes from a single suspect. Indeed it is often pointed out (e.g. Nolan 1983: 9; Doddington 1985; Naik 1994: 31; Broeders 1995: 158) that, from the point of view of the nature of the task, the decision-making process involved in forensic-phonetic comparison of suspect and incriminating speech samples should be considered a kind of verification. (Then why is it so often called identification, as in the title to this book? Nolan (1983: 9, 10) suggests that this is because the circumstantial effects associated with identification – possibility of lack of cooperation from suspect, possibility of disguise attempts – are to be expected in the forensic case.)

It is therefore possible to draw parallels between forensic speaker identification on the one hand and both speaker identification and verification on the other in the general nature of the question asked. Neither does there seem, for example, to be much difference between seeing in forensic speaker identification an attempt to give the strength of the evidence either in favour of identifying the speaker of a questioned speech sample as a suspect (i.e. an identification task), or in favour of the police's claim that the questioned speech sample is from the suspect (i.e. a verification task). Despite the above parallels there remain important differences between forensic speaker identification and speaker verification and identification, and these now need to be discussed.

Knowledge of reference data

One major difference between automatic speaker verification/identification and forensic speaker identification is that *in verification and identification the set of speakers that constitutes the reference sample is known, and therefore the acoustic properties of their speech are known.*

It was demonstrated above that in order to discriminate between same-speaker and different-speaker samples, thresholds were needed, such that if the difference between the samples exceeded the threshold the sample was considered to be from different speakers, and if not, the sample was considered to be same-speaker. Now, because the properties of the speech of each of the members of the speaker pool in speaker verification and identification are known, threshold values for discrimination can be estimated directly from them, and can therefore be maximally efficient. Often a threshold is set by reference only to the distribution of between-speaker distances, and it can even be updated to keep track of any (known!) long- and short-term changes in the speech of the members of the speaker pool (Furui 1981: 258).

In forensic speaker identification, however, the reference set is not known, and consequently the acoustic properties of its speakers can only be estimated (Broeders 1995: 158). The constitution of this reference pool will also in fact differ depending on circumstances. For example, if the defence hypothesis is that the questioned voice is not that of the accused but of someone with a voice similar to the accused, the reference set will be all speakers with a voice similar to the accused (of a certain sex in a certain area at a certain time). It will be demonstrated in Chapter 8 on speech acoustics how adequate knowledge of the reference sample is essential to the correct evaluation of the difference between samples as between-speaker or within-speaker.

Bayesian approach

The second main difference between speaker identification and verification on the one hand and forensic speaker identification on the other arises from the Bayesian imperative for forensic science discussed in Chapter 4. Speaker identification and verification aim for a categorical answer to the hypothesis that the same speaker is involved, given the evidence of the difference between questioned and reference samples. Forensic speaker identification, on the other hand, as explained in Chapter 4, should aim to provide a likelihood ratio, which reflects the probability of the evidence assuming both same and different origin: how much more likely the magnitude of the difference between samples is if they came from the same than from different speakers. Two consequences of this difference require comment: the categorical nature of the decision, and the threshold that the categorical decision requires.

Categoricality

Speaker identification and verification ultimately require a categorical decision: are the questioned sample and the reference sample from the same speaker or not? Such a decision can be problematic in forensic speaker identification for reasons that can best be illustrated with closed set identification, but that can be extended to open sets too.

Recall that a closed set comparison is where it is known that the reference set contains the test item, or in other words it is known that the questioned voice, or the voice to be identified, is one of the set of suspect voices. For example, it may be that it is known that the questioned voice belongs to one of two identical twin brothers, but it is contested which.

Assume, then, that it is known that the questioned voice sample Q is either from suspect sample S_1 or suspect sample S_2, and that the three samples are compared with respect to a phonetic parameter P1. The difference in P1 between Q and S_1 is much smaller than the difference in P1 between Q and S_2, and in a closed set identification, this means that S_1 is automatically, and fairly uncontroversially, identified as the source of the questioned sample. Suppose now, however, that the difference in P1 between Q and S_1 was only marginally bigger than the P1 difference between Q and S_2. Now it would seem wrong simply to identify S_1 as the offender, given that both S_1 and S_2 are very similar to Q (Champod and Meuwly 2000: 196).

Contrast this with a Bayesian approach. The likelihood ratio associated with the parameter P1 will be

$$\frac{\text{the probability of observing the value for P1 in Q assuming that the speaker is } S_1}{\text{the probability of observing P1 in Q, assuming that the speaker is not } S_1}$$

Since, because it is a closed set, 'not S_1' implies S_2, this is the same as dividing by the probability of Q's P1 in S_2. If the probability of observing Q's P1 assuming the speaker is S_1 is 0.75, and the probability of observing Q's P1 assuming S_2 is 0.42, the LR for P1 is 0.75/0.42 = 1.8. One would be 1.8 times more likely to observe the value for P1 if S_1 was Q than S_2. Note that it is not categorically stated that S_1 and Q are the same; or that S_2 is not Q; only that Q is so many times more likely to be S_1 than S_2.

Threshold

As already explained, in order to arrive at a categorical decision, speaker identification and verification require thresholds. From one point of view it might be claimed that, although they may differ with respect to the nature of the outcome, the means by which it is arrived at in speaker verification/identification and forensic speaker identification is probably not too different, since a Bayesian analysis is actually one of the methods used in automatic speaker verification (Furui 1994: 3). The difference is that in verification the LR is treated as a threshold and converted into a categorical statement, such that a pair of samples differing by less than the threshold is considered a match, and a pair differing by more not a match. Champod and Meuwly (2000: 195–6) point out that the concept of identity embodied by the use of such a threshold 'does not correspond to the definition of forensic individualisation'. This is because in forensic speaker identification the probability of a random match between a questioned sample and one of the reference samples is usually greater than zero, and so it is not possible to conclude from a match, as is done in speaker identification, that the suspect is identified, or verified.

Control over samples

Another difference between speaker verification/identification and forensic speaker identification is in *the degree of control that can be exercised over the samples to be compared*. As pointed out above, a high degree of control means a high degree of comparability, which is conducive to efficient recognition. In speaker verification, for example, there is total control over the reference sample, which is stored and retrievable as templates in the verification system. There is also a high degree of control over the test sample, since the individual wishing to be correctly identified can be expected to be compliant with requests to repeat, and re-repeat if necessary, the text of any desired reference templates. Control even extends to randomisation of test samples to guard against the use by impostors of recorded samples of the *bona fide* applicant's voice.

An additional consideration is that the reference templates can be deliberately constructed to contain material with high individual-identifying potential (Broeders 1995: 156) – material, that is, containing parameters which allow *a priori* high discrimination rates.

In forensic speaker identification, on the other hand, very little control is typically possible over the questioned sample: it may be an incriminating telephone call; a voice recorded during an armed robbery; or an obviously disguised voice (it is reported that a large percentage of cases are disguised). The degree of control over the suspect sample, and hence the concomitant degree of comparability with the questioned sample, varies, but it is also often small enough to create serious difficulties. For example, a common type of language data supplied for forensic speaker identification is the recording of a suspect's voice during police interview. It can be appreciated that this can differ substantially from, say, the offender's threatening voice recorded during an armed robbery. Even getting the suspect to repeat the incriminating text, which might be construed as a means of obtaining comparability, is fraught with difficulties that only a linguist can be in the position to evaluate.

The upshot of this lack of control is that it is necessary, if forensic speaker identification is to be possible, for experts to know how voices differ as a function of typically encountered circumstances in order to correctly assess the comparability between samples. The second decision to be made in the initial assessment of forensic samples is whether they are comparable in this sense (the first is whether the quality is good enough.)

Summary: Verification and identification

Forensic speaker identification has been compared to both the two major speaker recognition tasks of identification and verification, but in spite of some similarities it differs from them in several major respects. These include the non-categorical, Bayesian nature of forensic speaker identification, the unknown and variable nature of the reference sample involved, and general lack of control over both questioned and suspect samples.

We can now turn to other classifications within speaker recognition. From the point of view of forensic speaker identification there are several distinctions to be addressed, and it will help to have a figure (Figure 5.4) that illustrates how these interrelate.

Figure 5.4 shows, firstly, the field of Speaker Recognition containing its sub-types. The first important thing that Figure 5.4 shows is the division of the field of Speaker Recognition into the two sub-fields of Naive and Technical Speaker Recognition, and this is the distinction that will be discussed first. The technical speaker recognition field contains computerised analysis, of which both acoustic forensic analysis and automatic speaker recognition are parts. Finally, a box is shown for Forensic Speaker Identification, which intersects the fields of both Naive and Technical Speaker Recognition, giving boxes with the two main areas of Auditory and Acoustic Forensic Analysis.

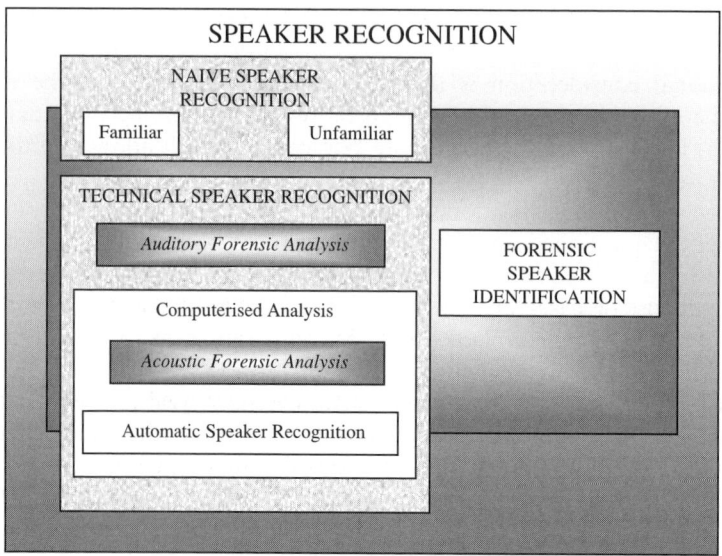

Figure 5.4 The interrelationship of different types of speaker recognition

Naive and technical speaker recognition

Probably everyone has experienced recognising speakers by their voice (Rose and Duncan 1995: 1), and often this recognition is used in evidence. For example, a rape victim might claim to have recognised her assailant by his voice in a voice line-up. However, the fact that this kind of speaker recognition is 'performed by untrained observers in real-life (or experimentally simulated real-life) conditions' (Nolan 1983: 7) sets it apart from recognition performed by experts, where analytic techniques are used.

On the basis of this distinction, then, speaker recognition can be classified into the two main types shown in Figure 5.4: Naive Speaker Recognition, where 'normal everyday abilities' are used, and Technical Speaker Recognition, also called Speaker Identification by Expert (Broeders 2001: 6), which uses specialised techniques (Nolan 1983: 7, 1997: 744–5). It is the latter, expert-technical, type that is of interest in this book, although the extent to which naive listeners can identify voices is of obvious forensic importance and will also be discussed.

Technical speaker recognition

Auditory forensic analysis

The nature of the specialised techniques referred to is the basis for a further classification of technical speaker recognition. There is, firstly, the use, by phoneticians, of specialised auditory techniques to describe and compare speech samples with respect to both their phonetic quality and voice quality. These techniques have already been alluded to in Chapter 3 describing the different types of parameters used in forensic speaker identification. This approach has been termed *technical speaker recognition by listening* (Nolan 1983: 7, 8), but since its main application is in forensics, it can be called *auditory forensic analysis*, which is also somewhat less of a mouthful. The auditory forensic analysis box is shown in Figure 5.4 as part of technical speaker recognition.

Computerised analysis

Nowadays, of course, the image more readily associated with technical speaker recognition is that of a voice being recognised by computer. This is part of an approach that can be called *computerised analysis*. For the purposes of this book, it is necessary to point out two different ways in which computers are used in speaker recognition. The first way is that in the popular consciousness, with a computer programmed to implement, without human intervention, powerful signal processing statistical techniques on speech acoustics to recognise a voice. All the decision making is done by the program, so a high level of objectivity is claimed for the procedure. All the human does (after they have written the program of course) is give the appropriate speech data to the computer and start the program. Actually, however, the human has to be involved in aspects of the procedure that can crucially affect its performance, such as deciding on the many parameter settings for the program (Braun and Künzel 1998: 15), so it is not correct to say that automatic procedures are totally free from human decision.

The approach just described has been called *commercial speaker recognition* (Künzel 1994: 135, 1995: 68), since that is where it finds its major application. Telephone banking and high-security access are two examples. It is also commonly referred to as *automatic speaker* (or *voice*) *recognition* and this is the term used here. Automatic speaker recognition is indeed an impressive reality, although not in forensic phonetics, and only under well-controlled conditions.

The second way in which a computer is used in speaker recognition is that typically encountered in forensics, where it is used interactively, to automatically extract comparable acoustic data from forensic speech samples after they have been selected by the investigator. In the absence of a common term for this, it can be called *acoustic forensic analysis*, to contrast with, and complement, the term *auditory forensic analysis* used above. Both acoustic forensic analysis and automatic speaker recognition can be seen as sub-parts of the computerised analysis box in Figure 5.4.

It is probably fair to say that it is automatic speaker recognition that most people have in mind when they think of forensic speaker identification. Künzel's (1995: 68) caricature, 'And why can't you use one of those computerised devices they have in detective films which can verify voices in just a couple of seconds?', is not far off the mark. Certainly this is the picture one gets from episodes in popular television dramas like *Law and Order* and *Crime Scene Investigation*, and the Harrison Ford film *Clear and Present Danger*.

It is therefore vital to emphasise that *automatic speaker recognition and acoustic forensic analysis are very different*, and especially the role of computers in both is different (van der Giet 1987). It was repeatedly claimed during the last two decades that a fully automated system for forensic speaker identification was not possible:

Forensic Speaker Identification

> [Es] hat sich gezeigt, daß der derzeitige Stand der beteiligten Wissenschaften nicht ausreicht, um ein vollautomatisches System zur forensichen Sprechererkennung zu konstruieren, ja daß sogar aus gründsatzlichen Erwagungen ein solches System auf absehbarer Zeit auch nicht konstruierrt werden kann.
>
> [[It] has been demonstrated that the current state of the relevant sciences is inadequate to construct a completely automatic system for forensic speaker identification. Indeed, basic principles say that, for as far as we can predict, it cannot *be* constructed.]
>
> van der Giet (1987: 122)

> ... although computer-based tools for acoustic analysis ... aid the expert in drawing his conclusions, a fully automatic voice identification device is certainly not in sight, due to the vast number of imponderables and distortions of the speech signal which are typical to the forensic situation.
>
> Künzel (1994: 135)

> ... it is unrealistic to anticipate a fully automatic procedure that will be able to extract a sufficiently comprehensive speaker profile from a questioned speech sample, given the variety of speech styles encountered in forensic conditions.
>
> Broeders (1995: 161)

Although the basic message remains that automatic speaker identification systems still cannot be generally implemented in the forensic context, it is possible now to see a very slight shift in acceptance:

> Although the development of state-of-the-art speaker recognition systems has shown considerable progress in the last decade, performance levels of these systems do not as yet seem to warrant large-scale introduction in anything other than relatively low-risk applications.
>
> Broeders (2001: 3)

Automatic speaker recognition and acoustic forensic analysis differ primarily in two respects: (1) the degree to which some kind of human decision is involved, and (2) the type of acoustic analysis implemented. Quintessential automatic speaker recognition is just that: automatic, with minimum human involvement and human decisions excluded. In reality, however, types of automatic speaker recognition can differ to the extent to which some kind of human decision is involved – perhaps in selecting what part of the speech signal is to be analysed automatically (e.g. Wolf 1972). In forensic acoustic analysis, however, there has to be a very much greater amount of human involvement. This is especially to decide whether samples are of good enough quality for analysis; to select comparable parts of speech samples for computerised acoustic analysis; and to evaluate the results that the computer provides.

Text dependency

One distinction in automatic speaker recognition that might be taken to reflect a degree of involvement of the investigator is commonly referred to as text-dependent vs. text-independent analysis. In the former, the same text, i.e. the same words or utterances, is used for both training the recognition device and testing it. In the latter, speaker recognition is attempted on utterances that are not controlled for lexical content (e.g. Furui 1994: 1).

Text-dependent speaker recognition performs better than text-independent speaker recognition (Nakasone and Beck 2001). This is because some of the overall difference between speakers will be contributed by different speech sounds. If you compare speaker A saying *bee* with speaker B saying *car*, probably a large part of the overall difference between the two samples will be a function of the difference between the sounds in the two words. If differences between sounds are controlled for, for example by using the same word, so that speaker A's *car* is compared with speaker B's *car*, then observed acoustic differences will more directly reflect difference between speakers, and can be used to recognise them.

This distinction, in text dependency, is important because it is obvious that the forensic situation resembles more closely the text-independent task, since it cannot be assumed that questioned and suspect speech samples will contain all the same words. (This is one consequence of the lack of control over samples mentioned above.) However, neither is it the case that questioned and suspect samples will never contain lexically comparable material (Broeders 1995: 157), and in the author's experience quite often the same word, or even phrase, will recur in both questioned and suspect samples with useful frequency. Nevertheless, the crucial question of comparability remains. It is obviously necessary to know what decrease in recognition can be expected with increasing loss of control over speech sounds in the samples to be compared.

As mentioned, the acoustic parameters used to recognise speakers also tend to differ considerably between the two approaches. The well-controlled nature of automatic speaker recognition allows the use of mathematically and statistically very sophisticated and very powerful, and very abstract, so-called *automatic signal processing parameters* and techniques that have been developed mainly by speech engineers (Furui 1994). Acoustic forensic analysis, on the other hand, most often uses traditional parameters that have originated in linguistic-phonetic research and that relate more directly to audible phonetic features. 'Traditional' and 'automatic' parameters were contrasted in Chapter 3, where it was emphasised that there was room for both in forensic acoustic analysis, depending on the individual case circumstances.

Performance of automatic systems

Under ideal, or very favourable, conditions, automatic speaker recognition can be highly efficient. One leading forensic phonetician already noted in 1994 that it was 'not much of a problem either scientifically or technologically' (Künzel 1994: 135), and in addition noted manufacturers' claims that the performance of such systems (telephone banking, telemonitoring of individuals on parole) comes close to 1% equal error rate. (That is, in only 1% of cases is the correct speaker not correctly identified, and in only 1% of cases a bogus speaker is identified as who they claim to be; see Figure 5.3.) One currently advertised system, that of Voice ID systems Inc., cites an even lower (presumably equal) error rate of 0.48% (VoiceID 2001: 2). This is for performance on the TIMIT research data base of 630 speakers from eight American dialect regions. However, automatic systems can differ considerably in the demands they have to meet: the heterogeneity and number of their client base, for example, or the amount of within-client variation that has to be accommodated. It is therefore not possible to find a single figure that will characterise them all.

Gish and Schmidt (1994) give an idea of the capability of an automatic system under forensically quite realistic conditions. They report a suite of speaker identification experiments using the conversational speech of 45 speakers recorded from different phone numbers over different long-distance telephone lines from speakers representing all regions of the United States and including crosstalk and noise. The error rates range from 6% to 29%, depending on the scoring method used (p. 30).

More recently, Nakasone and Beck (2001) have reported on the FBI's forensic automatic speaker recognition algorithm FASR. On forensically realistic data – that is non-contemporaneous spontaneous speech samples recorded from different outside telephones – they were able to verify 50 different speakers, using 40 impostors per speaker, with a best EER verification performance of 13.8%, which figure is bracketed by those just quoted of Gish and Schmidt (1994). Performances on other, forensically unrealistic data were between about 13% and 15%. Nakasone and Beck say that the FASR system has an overall recognition capability comparable to that of the highest-performing commercial system the FBI evaluated. It is interesting to note that despite this performance, they still do not consider their automatic approach ready for court evidence, and envisage that 'it is possible that the FASR can become a useful investigative tool pending further studies and improvement'.

The generally excellent performance of automatic speaker recognition systems is extremely important, for two reasons. Firstly it demonstrates the logically prior consideration for FSI that the *task of identifying and verifying speakers by their voice is not only possible, but can be done very well*. Secondly, automatic speaker recognition constitutes the obvious benchmark for comparing performance in all types of forensic speaker identification, where conditions are usually very far from ideal. Some of the limitations on FSI that arise from the non-ideal real-world conditions have already been mentioned in Chapter 2.

Conditions on forensic-phonetic speaker identification experiments

Given the drastic effect of real-world conditions on forensic-phonetic identification, experiments to test the efficacy of forensic-phonetic parameters should obviously attempt to simulate real-world conditions as closely as possible, like that of Nakasone and Beck (2001). Such experiments should conform at least to the following three main conditions.

- Natural conversation should be used. If it is necessary to elicit individual forms containing the parameters to be tested, for example high vowels in stressed syllables, the desired degree of control can be exercised by using a version of the map task (e.g. Elliott 2001: 38).
- Within-speaker comparison should involve non-contemporaneous, i.e. different-session, data (Rose 1999b).

- As many as possible similar-sounding, or at least not too different-sounding, subjects should be used. This means controlling at least for sex and accent.

Ideally, in addition, the appropriate recognition model should be discrimination of same-speaker from different-speaker pairs, and performance should also be tested using different transmission channels, for example over different telephone lines. Many of these conditions are also applicable to naive speaker identification experiments, to which we now turn.

Naive speaker recognition

Common to all the above approaches is the application of an analytic technique, informed either by Phonetic and Linguistic knowledge or by signal processing statistics and signal detection theory. As mentioned above, a term that has been used to refer to this activity is *Technical* (or *Expert*) *Speaker Identification*. The implicit contrast is with recognition by listeners untrained in such techniques: a useful term for this is *Naive Speaker Recognition*. The distinction between technical and naive speaker recognition is represented in Figure 5.4.

Common experience suggests that untrained human listeners can and do make successful judgements about voices, and the natural human ability to recognise and identify voices has been accepted in courts for several centuries (Gruber and Poza 1995: section 99). Nolan (1996) points out that earwitness evidence becomes important when a voice has been witnessed in the commission of a crime; when visual identity was disguised, e.g. by masks; or when only a voice sample of a suspect, and not the suspect themselves, is available. An earwitness may claim to recognise a voice previously heard, or subsequently identify the suspect as the offender.

Two important things to realise about naive speaker recognition from a forensic standpoint, however, are that it is not as good as assumed, and it is also extremely complicated; as a result, earwitness testimony is extremely difficult to evaluate. This is because many different factors are known to influence the ability of naive listeners to identify or discriminate between voices, and little is known of the way in which these factors interact. In addition to the familiarity of the voice claiming to be recognised, other variables are known to be operative in naive speaker recognition. Some have to do with the listener: for example, how generally good (or bad) they naturally are in recognising voices; how familiar they are with the language in which they claim to recognise the speaker; whether they expect to hear a particular voice or not; how old they are. Some variables have to do with the speaker. They include, for example, how distinctive their voice is; how well they command a second language; or if they have disguised their voice. Some variables have to do with non-linguistic properties of the target sample: how long it lasts; how good its quality is. The most important of these are briefly addressed below.

Familiarity

Probably the most important variable contributing to naive speaker recognition is the extent to which the listener is familiar with the voice. Thus naive speaker recognition can be categorised as *familiar* and *unfamiliar*. This difference is represented in the naive speaker recognition box in Figure 5.4.

Familiar naive recognition

In familiar naive recognition the listener is familiar with the voice to be identified because the listener has been exposed to what they assume is the same voice on sufficient prior occasions for their brain to have constructed a reliable template of the voice's features. A familiar voice can usually be identified because the listener actually knows its owner. This happens when you can identify members of your family, or friends, or media personalities, by their voice alone. An example of familiar forensic recognition might be when a close friend or relative of the suspect is asked whether they can identify the questioned voice as coming from the suspect. It is also possible, however, for a familiar voice to be recognised without being identified – where the listener can say 'I have heard that voice before, but don't know whose it is'. (This is one occasion where it is necessary to distinguish between the senses of *recognition* and *identification*.) It has been claimed (Emmorey *et al.* 1984: 121) that it may even be necessary to distinguish between familiar-intimate and familiar-famous voice recognition.

In unfamiliar naive recognition a listener is able to recognise a voice with minimal prior exposure. This might occur in a test where listeners may be exposed to one voice and then be required to recognise it from a set of unknown voices. This is a common paradigm in many speaker recognition experiments. It is also a typical scenario in a voice line-up, for example. A bank robber has given instructions to the teller during the robbery and the teller is later tested in a voice line-up as to whether they recognise the voice of the robber from a group including (or not) the suspect. Experiments with both normal and brain-damaged subjects (van Lancker and Krelman 1986) have indicated that the recognition of familiar voices makes use of different cognitive strategies, in very different brain locations, from the discrimination of unfamiliar voices. In particular, recognition of familiar voices appears to be the result of pattern-matching activity specialised in the right hemisphere.

The reason the distinction between familiar and unfamiliar recognition is important is that unfamiliar recognition has been shown in many studies to be much worse than familiar, and therefore, *ceteris paribus*, familiar recognition evidence must be accorded greater strength than unfamiliar (which, it will be shown below, has to be treated with extreme caution). For example, Rose and Duncan (1995: 12, 13) showed that in tests to see whether pairs of same-speaker *hello*s could be discriminated from pairs of different-speaker *hello*s, listeners who were familiar with the speakers had an overall error rate of 26% compared to one of 55% for listeners who did not know the speakers. The same test with longer utterances showed the same familiarity advantage, with overall error rates of 14% and 35% respectively. Both these tests show familiar recognition to be on the average about twice as good as unfamiliar recognition.

Although it shows greater recognition scores than unfamiliar recognition, it is important to realise that the reliability of familiar speaker identification also tends to be overestimated. There is one well-known case (Ladefoged and Ladefoged 1980: 49) in which one of the world's pre-eminent phoneticians points out that he was not able to identify his mother's voice from a single word, or single sentence, in a controlled experiment designed to investigate familiar recognition. Less anecdotally, in experiments carried out by the author and others (Rose and Duncan (1995); Foulkes and Barron (2000); McClelland (2000)) familiar listeners have been shown to mis-identify foils as friends or close relatives; mis-identify friends as other friends; and fail to be able to identify close friends. Rose and Duncan (1995), and Ladefoged and Ladefoged (1980), for example, quote error rates of between 5% and 15%, and 17% and 69% respectively, depending on the amount of data used (short words vs. a sentence vs. a longish text).

Foulkes and Barron (2000) used members of a close social network to gain an idea of the kind of error rate to be expected for familiar recognition over the telephone. Each of their nine male subjects was required to identify a shortish (about 10 seconds) stretch of speech over the telephone from 12 different voices: nine of the voices were of their mates, two were from foils, and they were also played their own voice. The test was open set, i.e. listeners did not know who to expect.

Foulkes and Barron report a mean rate of 67.8% correct identification of ten familiar voices by nine listeners, and use this to point out that familiar recognition is not as reliable as is assumed. However, the most important thing in their results is obviously the strength of the evidence associated with a claim of a naive witness that they identify a familiar voice, or in other words the likelihood ratio:

$$\frac{\text{probability of the identification claim assuming the familiar listener is correct}}{\text{probability of the claim assuming they are not (i.e. false identification)}}$$

This can be calculated from their results by estimating the percentage correct identification of familiar voices and the rate of incorrect identification of any voice as a familiar voice. Thus for example one of their respondents (Bill) was played 10 familiar voice samples (including himself) and correctly recognised 6 of them. However, out of the total of 12 samples played to him – that is the 10 familiar samples and the 2 foils – he incorrectly identified 4 as familiar voices (p. 186). This gives a likelihood ratio of ([6/10]/[4/12] =) 1.8. Bill is on average just under twice as likely to be correct as incorrect if he claims to recognise a familiar voice over the phone. In the whole experiment, there were 61 out of 90, or 67.8% correct identifications of familiar voices, and 12 out of 108, or 11.1% false identifications. This gives an overall likelihood ratio for the group of (0.678/0.111 =) 6.1. The best estimate of the strength of evidence *based on these data* is therefore that a witness will be on average six times more likely to be correct as incorrect if they claim to recognise a familiar voice over the phone.

It hardly needs to be emphasised that these figures are valid for the circumstances of this experiment only. Thus, since it will be shown below that recognition improves with amount of speech available, the results would almost certainly have been better if the listeners had been exposed to more than the ca. 10 seconds involved.

Unfamiliar naive identification

The relatively poor performance of unfamiliar recognition was mentioned above. Bull and Clifford (1999) provide a very useful summary of the results of experiments published since 1980 testing the reliability of unfamiliar earwitness testimony, and also describe the many factors known to affect accuracy of identification. They make three important points. Firstly, they emphasise (p. 216) that the experimental evidence continues to reinforce the earlier conclusion that 'long-term speaker identification for unfamiliar voices must be treated with caution'. On the basis of the accumulated experimental evidence, they furthermore advise against the sole use of earwitness identification, echoing their earlier judgement (p. 220): that 'prosecutions based solely on a witness' identification of a suspect's voice ought not to proceed, or if they do proceed they should fail'. Similar admonitions can be found in other sources. Thus Deffenbacher *et al.* (1989: 118): '. . . earwitnessing is so error prone as to suggest that no case should be prosecuted solely on identification evidence involving an unfamiliar voice'. Finally, (p. 220) they endorse the opinion that 'it would be wrong to exclude earwitness identification from the fact finding process so long as the identification procedures were conducted properly and, presumably, some other type(s) of quality evidence were available'. (By 'properly conducted procedures' is presumably meant the use of voice line-ups, since these can clearly have probative value. Otherwise it is difficult to see the advantage of admitting earwitness evidence in the presence of other evidence if it is not to be admitted on its own.)

Variation in listener ability

A very important factor in the discussion of naive response is the inherent between-subject variation in recognition ability. There is ample evidence from naive speaker recognition studies that not all humans are equally good at recognising familiar (or indeed bad at recognising unfamiliar) voices (Ladefoged and Ladefoged 1980: 45; Hollien 1995: 15; Foulkes and Barron 2000: 182). In an experimental study with 110 American subjects, for example, the individual subject's success rate in recognising 25 famous voices ranged all the way from totally correct (100%) to chance (46.7%) (van Lancker and Kreiman 1986: 54).

A forensically more relevant study showing the between-subject inherent variation in recognition ability is the already mentioned experiment by Foulkes and Barron, where between-subject differences are emphasised. (It is more relevant because it deals with recognition over the telephone and incorporates misidentifications.) The worst overall performance in their admittedly very small group of nine subjects was 58.3% (i.e. 7 out of 12 voices correctly responded to), and the best was 91.7% (11 out of 12 voices.).

In LR terms for recognition of familiar voices, Bill had the lowest LR of 1.8, and Hugh the highest with 10.8. Probably because of the smallness of the test, some subjects made no false identifications and so their LR has a denominator of zero and is undefined. If these subjects' data are smoothed by assuming an error of 1 in 24, or 4%, likelihood ratios of up to 17.5 are observed. It is also interesting to note that a value of 15 can be derived for Ladefoged's performance in Ladefoged and Ladefoged (1980: 49).[1] This range, from 1.8 to 17.5, effectively extends over an order of magnitude. Its large size therefore presents obvious problems for interpretation in court,

---

1 Ladefoged correctly identified 24 out of a pool of 29 familiar speakers on the basis of a 30-second passage, and misidentified 3 out of 53 speakers as familiar. This gives an approximate LR of 83%/6% = 15.

since it means that the strength of evidence from an individual witness in a specific case can range from negligible to moderate. The 95% confidence limits for the LRs for these data are ±4.4. That is, one can be 95% confident that they will lie between ±4.4 of the mean LR of 9.5. *On the basis of these figures*, then, we could be 95% confident that an earwitness's performance could be associated with LRs of between 5 and 14.

Of course, some between-subject variability might be accounted for in terms of yet other variables. Age is one, at least for adults. Older listeners (over ca. 40) are not as good on average as younger listeners (Kreiman and Papçun 1985; van Lancker and Kreiman 1986: 55, 56; Bull and Clifford 1999: 217). However, at least some of the residual variability is presumably not referable to another obvious factor and must count as random. Sex does not seem to correlate with naive recognition ability, and the effect of sightedness is unclear (Bull and Clifford 1999: 217).

Three final variables related to the listener need to be mentioned. Accuracy improves considerably if the listener participates actively in a conversation with the target speaker rather than just listens passively to their voice; and it also improves if the witness is under stress when exposed to a target voice (Nolan 1996). What about the listener's confidence in their identification: is that an indication of accuracy? Some experiments, e.g. Rose and Duncan (1995: 14), have shown positive correlations between listeners' certainty and their accuracy. However, Bull and Clifford (1999: 218) point out that the relationship might only hold for recognition tasks that are easy, and suggest for the present that 'earwitness confidence should not be taken to indicate that one witness is more likely to be correct than another witness'.

Variable distinctiveness of speaker's voice

Another important and well-documented factor is that some voices are identified better than others (Papçun *et al.* 1989; Rose and Duncan 1995: 12, 16), and it can therefore be assumed that some voices carry more individual-identifying content than others. There is a small amount of evidence to suggest that this is because the more auditorily distinctive voices have parametric values that lie farther away from the average (Foulkes and Barron 2000: 189–94), but there may be more to it than that. It may also be the case that a voice is badly identified because it has a wide range of variation that takes it into the ranges of other voices (Rose and Duncan 1995: 16). Indicative of this is the fact that, since different tokens of the same word from the same speaker are differentially recognised, even different tokens of the same utterance from the same speaker can differ in their individual-identifying content (Rose and Duncan 1995: 12). This is one reason why it is important to have a lot of data from the target voice.

Very few recognition experiments have been conducted where the distinctiveness of the voice has been taken into account, so the contribution of this effect is not well known. Bull and Clifford (1999: 218) surmise that distinctiveness can correlate with how badly voices are recognised as a function of delay. It can also be noted here that voice distinctiveness is a very important consideration in the composition of earwitness line-ups. It is important that the voices of a group of speakers in a line-up all have the same chance of being chosen at random and none of them is distinctive for any reason (Broeders and Rietveld 1995: 33, 35–6).

Non-linguistic properties of stimulus

There are several other commonsense features relating to the stimulus that influence recognition. These are *exposure* (comprising *quantity* and *elapsed time*), and *quality*. Thus voices are more reliably identified or recognised by naive familiar listeners if the listeners are exposed to a lot of speech. Ladefoged and Ladefoged (1980: 49) report, for example, that if the speech consisted of just *hello*, the error rate was 69%, compared to 34% for a single sentence, and 17% for 30 seconds of speech. Bull and Clifford (1999: 217) mention the problem that in reality it is often not clear how much speech an earwitness has heard, because earwitnesses have been shown to considerably overestimate this amount. They also observe that it may not be the amount of speech *per se* that affects reliability, so much as the probability that greater amounts of speech give a listener greater exposure to the within-speaker variation involved, and thus enhance recognition. They say that it is important for more accurate recognition that the earwitness needs to have been exposed to the voice on more than one occasion (which will presumably increase the voice's familiarity).

A very important factor is delay. Performance diminishes with the time elapsed between hearing a reference voice and being required to recognise/identify it (although Nolan (1996) notes a 'curious plateau or even slight improvement in the 1–3 week interval'). Broeders and Rietveld (1995: 34) point out that results from experiments investigating the effect show rather different magnitudes, and they attribute this to differences between experiments in other variables. Some rates from actual experiments are 50% after 10 minutes, 43% after one day, 39% after a week, and 32% after a fortnight (Clifford *et al.* 1981); 83.7% after one week declining to 79.9% after three weeks (Rietveld and Broeders 1991); 83% after two days declining to 13% after five months (McGehee 1937).

In view of the variability of results presumably arising from differences in the conditions of experiments designed to investigate delay, Bull and Clifford (1999: 220) are only able to surmise that 'earwitness performance may well be poorer with delays of months but may not necessarily be poorer with delays of a few days'. Nolan (1996) makes perhaps the more useful practical point that 'it is important to carry out [earwitness] identifications as soon as possible after the crime'.

Not surprisingly, recognition is also better with good-quality recordings than with poor – for this reason it is usually worse over the telephone than over a hifi system (Köster *et al.* 1995; Köster and Schiller 1997: 24), although that also depends on the differential effect of the channel degradation on the acoustic structure of the stimulus (Schiller *et al.* 1997: 15–16).

Language

Most people know that some countries, such as Canada and Switzerland, have more than one official language. However, even in countries with a single official language, like the United States, France, Australia or Great Britain, there exist large groups of speakers that use other languages. In many other countries with a high level of dialectal diversity, like China, Thailand and Vietnam, there is a more complex interaction between the Standard Language and the dialects. Speakers switch between local, or regional, forms of speech and the standard depending on the details of the social interaction, or use amalgams from both regional and standard (Lehman 1975: 11–13;

Diller 1987: 150–5). In fact most countries in the world are practically, if not constitutionally, multilingual, and 'multilingualism is the natural way of life for hundreds of millions all over the world' (Crystal 1987: 360). Because of this there are many naturally bilingual or even multilingual (or bi- or multidialectal) speakers (Rose 1994, 1998b). Thus it is not uncommon for forensic speaker identification to involve more than one language or dialect. Moreover, the different languages or dialects can be involved in a variety of ways.

Various researchers (e.g. Ladefoged and Ladefoged 1980: 45) have commented on the role of the language in speaker recognition, and several studies have researched aspects of this role. From the point of view of speaker recognition, it is useful to think in terms of two variables: the listener, and the language of the sample they hear (the 'target' language). In the case of the listener, they can be monolingual, bilingual, or multilingual (Köster and Schiller 1997: 18). In the case of the target language, it can be the same as that of the listener; a different language from that/those spoken by the listener; or the same as the listener's language but spoken with an accent. For example, a monolingual witness might be asked to recognise a suspect speaking in a different language, or claim that they can recognise a speaker speaking with a heavy accent in their own language.

Target language as variable

Several studies have shown that recognition is better in the listener's native language than in a language they are not familiar with. That is, someone trying to recognise a voice in a language they do not know does not perform as well as someone trying to recognise a voice in a language they do. For example, Goggin *et al.* (1991) showed that monolingual English listeners recognised English-speaking voices better than German-speaking voices, and monolingual German listeners did the same, *mutatis mutandis*. Goggin *et al.* (1991: 456) also quantified the magnitude of the effect: voice identification improves by a factor of 2 when the listener understands the language relative to when the message is in a foreign language.

It has been hypothesised that the native language advantage in studies like these comes from the fact that native listeners are able to make use of linguistic as well as non-linguistic cues. Support for this view comes from experiments by Schiller *et al.* (1997), who showed that when some of the linguistic cues were suppressed, the native language disadvantage disappeared.

Bilingualism, of course, is not a simple phenomenon: people can have many different types and degrees of command of a second language (Crystal 1987: 362). Like many immigrants to Australia and the United States from non-English-speaking countries, their competence in English may range from a knowledge of only a few words to idiomatic use of syntax and vocabulary. Or, like their children's knowledge of the parent's language(s), they may understand it to different degrees, but not speak it. One way in which differing degrees of bilingual competence manifests itself is in pronunciation (called *accent* in linguistics): very few bilingual speakers have a perfect native accent in both their languages, and the pronunciation of one of their languages will show through in the other. Since it can show through to different degrees with different speakers, this constitutes a potentially useful set of parameters in forensic speaker recognition, but is more an aspect of technical speaker recognition.

Forensic Speaker Identification

The effect of trying to recognise speakers speaking the listener's native language with an accent has been investigated in several studies. These have shown that it is more difficult to recognise a speaker with an accent than one without. Goldstein *et al.* (1981), for example, showed that, when the amount of data is reduced, voices speaking English with a Chinese or Black American accent are not as well recognised by speakers of General American English as voices with a General American English accent. Thompson (1987) demonstrated the same thing with Spanish-accented English, and showed furthermore that accented voices could still be recognised better than foreign voices: the Spanish-accented voices were still recognised better by English listeners than voices speaking Spanish.

Listener's linguistic competence

Turning now to the variable of the listener, the next obvious question is whether there is any effect on speaker recognition from whether the listener is in any sense bilingual. Results of experiments to investigate this have shown that familiarity with a non-native target language does generally confer some advantage (Schiller *et al.* 1997: 4, 5; Köster and Schiller (1997: 25). Goggin *et al.* (1991), for example, showed that, although monolingual English listeners had increasing difficulty recognising voices in native English, Spanish-accented English and Spanish, bilingual Spanish–English listeners recognised all three with equal facility. In Köster *et al.*'s (1995) experiment, English listeners with a knowledge of German outperformed English speakers without a knowledge of German in recognising a native German speaker speaking German. More complicating factors may be present, however, and it is still not clear to what extent the *degree* of competence in a foreign target language affects recognition (Köster and Schiller 1997: 25).

Differential similarity between listener's language and target language

It is a sensible question whether the linguistic distance between the listener's language and the target language has any effect. Is it easier for a listener to recognise someone speaking in a foreign language that is in some sense more similar to their own? Köster and Schiller (1997) found no evidence that this was so. They found, to be sure, that English listeners outperformed Spanish listeners in recognising German speakers, and this would be predicted from the relative typological closeness involved. However, both English and Spanish listeners were bettered by Chinese listeners, which clearly runs counter to the hypothesis, since Chinese is linguistically considerably further away from German than either English or Spanish.

Expectation effect

We tend to hear who we expect to hear: 'One is far more likely to identify a voice as a given person's if one is expecting to hear that person's voice' (Ladefoged 1978; cf. also Ladefoged and Ladefoged 1980: 47; Broeders 1995: 155). The role of expectation in naive speaker recognition is well known. Thus voices of unfamiliar foils in recognition experiments are regularly confused with those that the subjects expect, for whatever reason, to hear (Rose and Duncan 1995: 10, 11). The expectation effect is discussed further in Chapter 9 on speech perception.

Naive speaker recognition: Discussion

It is clear from the preceding sections that, contrary to what we may believe from personal experience, naive speaker recognition is anything but straightforward. Since a bewildering number of different factors can affect the accuracy of naive speaker recognition, not all earwitness statements will be equally reliable. It is therefore necessary to realise that when a witness is claiming to naively recognise a voice, careful attention must be given to the individual circumstances. In particular, an attempt should be made to classify the earwitness claim with respect to whether it can best be represented as familiar or unfamiliar recognition; whether the listener is familiar with the target language(s) or not; the quality and quantity of the speech material; the delay involved; and whether or not there is an expectation effect.

It is one thing to list these factors, as I have done above, and to demonstrate their effect. It is quite another to know how to evaluate them in specific circumstances. I have illustrated above how the LR associated with one of these identification tasks can be calculated, but there are enormous complexities in applying the method in specific cases.

Firstly, as demonstrated above, it is clear that a wide range of LRs can be expected for a given circumstance because of considerable between-subject differences in ability. This difficulty is then compounded by the fact that, because some voices can be recognised more easily than others, different voices will be associated with different LRs. (What is the probability of recognising this particular voice assuming that it is indeed this voice, and assuming that it is not?) So it is possible that the evidence given by someone with a moderate ability to identify voices will have the same strength as that of someone who is really good, if the voice that the former witness is claiming to identify is really distinctive, while the latter is claiming to identify a non-distinctive voice. The contribution of voice distinctiveness has not as yet been systematically studied, so there is little quantified data available to use to estimate LRs. On top of this come problems with integrating the other known variables. It may be the case that ceiling effects are present, for example, such that the effect of one factor may be so great as to make the effect from another inconsequential (Schiller *et al.* 1997: 15).

What is to be done, therefore? It may be that an individual case is sufficiently close to experimental conditions in the variables listed above, in which case some kind of LR estimate can be attempted by an expert. Often, however, real world case-work continues to confront us with messier circumstances than are envisaged in controlled experiments. For example, the author was recently involved in a case where both the speaker and the listener were operating in two non-native languages in which they both had good competence. A native speaker of Cantonese was claiming to be able to identify a voice of a native speaker of Shanghainese speaking English in one sample and Mandarin in another! There simply is no data on the reliability of naive speaker recognition under this degree of linguistic complexity.

If it is not possible to arrive at an estimate of the strength of the earwitness evidence, some courts may still be happy with an unquantified weighting. In this case it is best to follow the advice of Köster and Schiller (1997: 26) that more weight should be given to recognition if the earwitness is a native speaker, or has some knowledge of the target language, but that results must be treated with care if an earwitness does not understand the target language. To this of course can be added that familiar recognition can usually be given more weight than unfamiliar, and that in the case of unfamiliar recognition the warnings advised above should be heeded concerning its general unreliability and unsuitability as a sole source of evidence in building or prosecuting a case.

The other option is to counter the unreliability of earwitness identification by testing its accuracy in a voice line-up (Broeders 2001: 5–6). Properly constituted, the result of a voice line-up can substantially increase the evidentiary value of an earwitness identification (Broeders and Rietveld 1995: 29). However, 'The requirements that need to be met in administering such an auditory confrontation are quite formidable' (Broeders and Rietveld 1995: 25), and a cost–benefit analysis must inevitably be considered.

The many procedural considerations crucial to conducting a voice line-up that will have probative value are outside the scope of this book, but are covered in the following sources, and their reference bibliographies: *Speaker Identification by Earwitnesses* (Broeders and Rietveld 1995); *Consideration of guidelines for earwitness lineups* (Hollien 1995); *Criteria for earwitness lineups* (Hollien et al. 1995); *Preparing a voice lineup* (Nolan and Grabe 1996); *A study of techniques for voice line-ups* (Nolan and Kenneally 1996); and *Getting the voice line-up right: analysis of a multiple auditory confrontation* (Butcher 1996).

Familiarisation in auditory forensic analysis

Although the difference between familiar and unfamiliar recognition is an important distinction in naive speaker recognition, it is important to realise that familiarisation also plays an important role in technical speaker recognition, specifically in auditory forensic analysis (see Figure 5.4). In naive voice recognition it has been shown that familiarity confers superiority. It is also well known from forensic content determination studies that the more one becomes familiar with the voice, the more one is able to hear of the (degraded) content of a recording. It follows therefore that becoming familiar with a voice may very well aid all sorts of judgements about it.

How this might come about is not clear. One of the main messages of this book is that a voice has many different components, and that aspects of a sound can sometimes be part of one component, for example a speaker's voice quality, and sometimes part of another, for example their phonetic quality (phonetic and voice quality are discussed in detail below). Becoming familiar with a voice may help in correctly assigning sound features to the appropriate component, thereby bringing both components into sharper focus.

Humans are simply programmed to pay attention to and process voices, whether we are phonetically and linguistically trained forensic phoneticians or not. How we do this is not known, although, as mentioned above, it is probable that the cognitive strategies for recognising a familiar voice are different from those for recognising an unfamiliar voice.

Phoneticians are also humans. Most forensic phoneticians would agree that it is important to make use of the natural human ability to cognitively process voices. Indeed, it is not clear how one can override it. Because of this, *familiarisation with the voice(s) by repeated listening is also a valid and extremely important part of the forensic-phonetic approach*. (This will happen as a matter of course as part of the necessary preliminary familiarisation with the content of the samples that is necessary to select items for auditory-phonetic and acoustic analysis.) Another reason for recognising the value of familiarisation comes from the extremely restricted amount of material that can usually be quantified acoustically. In comparison to this, brains presumably can take in and process the *totality* of the material in forming an impression as to overall similarity or difference in the different voice components.

Aural-spectrographic (voiceprint) identification

One further approach to speaker recognition, difficult to classify under the previous headings, must be addressed: *aural-spectrographic voice identification*. This was originally termed *voiceprint identification* and refers to a method of speaker identification developed and then commercialised in the mid-1960s by L. G. Kersta (Kersta 1962).

The method was first based only on *visual* comparison of spectrograms. Spectrograms are very useful graphic representations of the acoustics of speech, and will be described in detail in Chapter 8. Figure 5.5 shows a spectrogram of the word *hello* spoken by a male speaker of Australian English. Later, during the 1970s, a combination of visual examination of spectrograms and auditory response was introduced: the examiner both looks at the spectrograms and listens to the speech samples (Hollien 1990: 215). Because of the inclusion of the listening part, and also because of problems associated with the term *voiceprint*, the name of the method (voiceprint identification) was changed by some practitioners. A commonly found term (e.g. in McDermott *et al.* 1996) is the 'aural-spectrographic' method, and this will be used below.

Aural-spectrographic voice identification evidence can still be found in investigative, corroborative and substantive use today. For example, the FBI is using it for investigative purposes (Nakasone and Beck 2001) as are also the Japanese police (Osanai *et al.* 1995; Osanai, personal communication, 2001). It is still admitted as evidence in US courts (Broeders 2001: 7), and at the time of writing the author is aware of a case involving voiceprint evidence in Australia. Its use is also documented

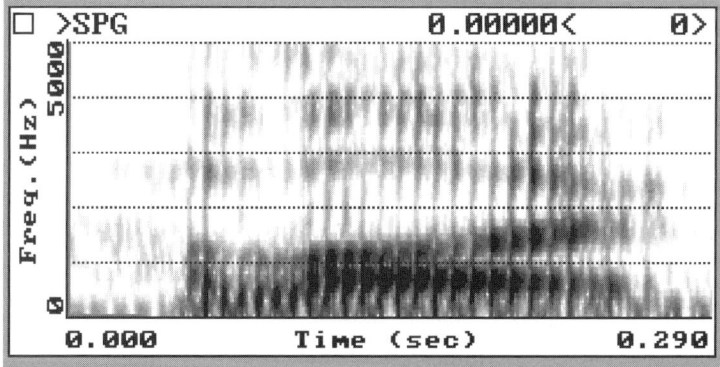

Figure 5.5 Spectrogram of Australian English *hello*

in Israel, Italy, Spain and Colombia to at least 1994, although by this date it had been abandoned in Germany and the Netherlands (Künzel 1994: 138).

In spite of its fairly widespread use, aural-spectrographic recognition in forensics has been, and in some areas – for example the United States – continues to be, the centre of considerable controversy on both scientific and legal fronts. One of the world's leading forensic speech scientists, for example, has testified that he believes voiceprinting 'to be a fraud being perpetrated upon the American public and the Courts of the United States' (Hollien 1990: 210). Another equally prominent forensic phonetician has stated that 'it can only be deplored that some countries not only tolerate but have even recently introduced the technique' (Künzel 1995: 78–9).

Within the relevant scientific community, phoneticians and speech scientists on the one hand argue against voiceprinters on the other. The method's merits are debated within the academic legal community; and in legal practice the controversy is reflected in the fluctuations of trends in admissibility judgements. It has also generated an enormous amount of animus, obloquy and almost *ad hominem* argument. The controversy surrounding the method thus requires discussion in some detail.

There are several useful summaries of the voiceprint controversy. Nolan (1983: 18–25) covers up to 1983. Hollien (1990, Chapter 10) takes it to 1990, and Künzel (1994: 138–9) four years further. None of these authors is well disposed towards the method. Gruber and Poza (1995) is currently the most up to date and, running to ca. 200 pages, the most comprehensive treatment of the use of voiceprints in forensic speaker identification. It is obligatory reading for anyone involved in a case with voiceprints, and also contains much of use for forensic speaker identification in general. The second author was technical advisor for an important monograph commissioned by the FBI in 1976 to evaluate the method (Bolt *et al.* 1979), and also completed a two-week course under its originator (section 55, fn. 79),[2] so it is claimed (section 38) that the paper provides a unique insider perspective, and 'a fair, and evenhanded, assessment of the scientific and legal merits of voicegram [i.e. aural-spectrographic/voiceprint] evidence' (section 2). Despite this, it remains severely critical of the method.

An example

As already made clear, the aural-spectrographic method involves comparing questioned and suspect voices both by how they sound and how they look on spectrograms. It will be useful at this point to follow the example of the technique's originator (Kersta n.d.) and get the reader to attempt a visual voiceprint comparison. This will not only provide an example of what visually comparing voiceprints is like, but will also serve to reinforce some crucial questions for forensic speaker identification in general.

Figure 5.6 presents six spectrograms of the word *hello* each spoken by a different Australian male: consider these six as suspects. Figure 5.5, presented above, also of *hello*, is the voice of someone who has committed a crime. *You do not, of course, know whether the speaker of the questioned hello is among the suspects.* You have to compare

---

2 References to this work in this book specify the section in the paper where the relevant material is to be found, not the page. This is because readers may have to have recourse to the paper in downloaded, unpaginated format.

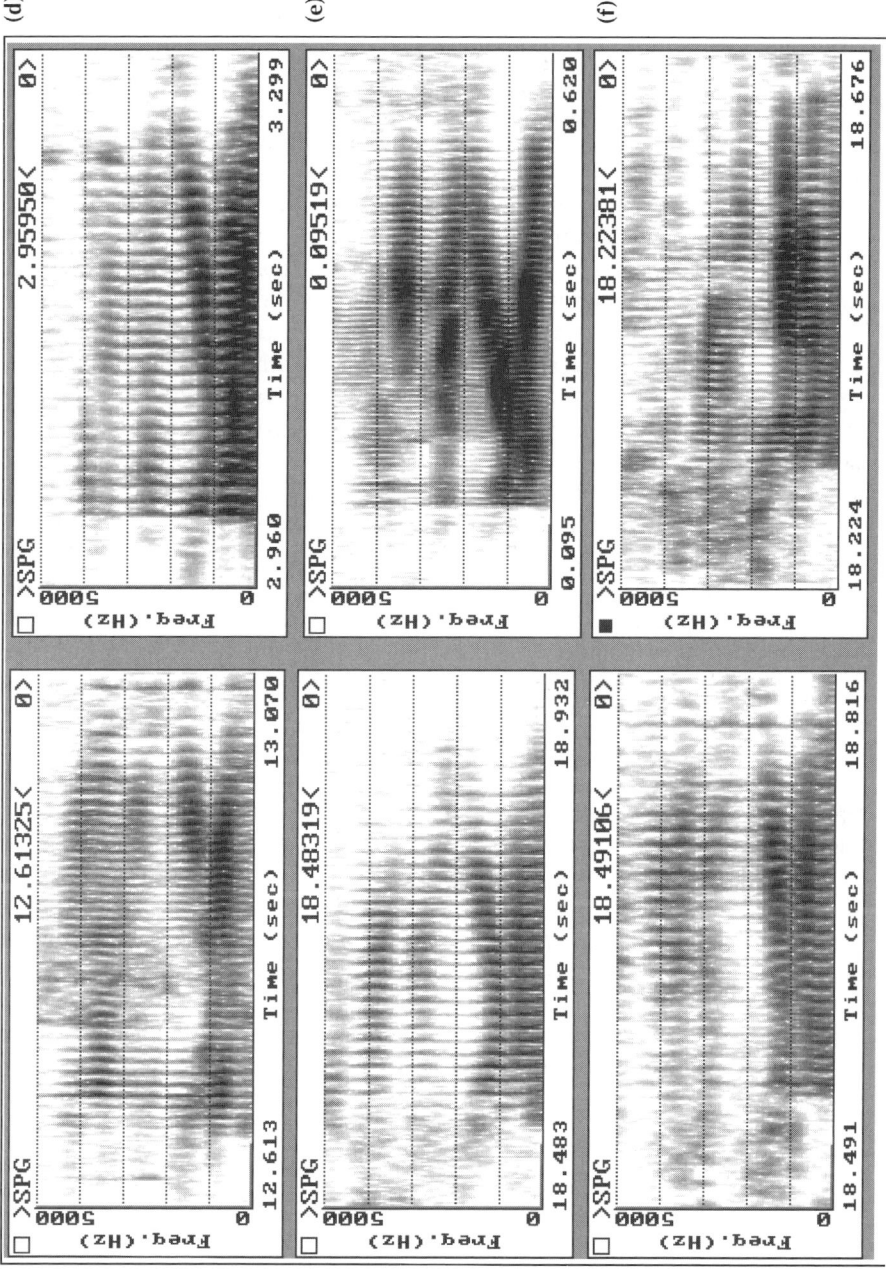

Figure 5.6 Suspects' data: spectrograms of six different male speakers of Australian English saying *hello*

the spectrogram in Figure 5.5 with the six spectrograms in Figure 5.6. You then have to decide whether the spectrogram in Figure 5.5 is sufficiently similar to one of the spectrograms in Figure 5.6. In that case you identify one of the suspects as the offender. Or you may decide that it is not sufficiently similar, and accordingly eliminate all the suspects (in reality, several more options involving degrees of confidence would be available to you.)

If I were to ask you to now take a few minutes to make your decision, I expect that, like me, you might eyeball the data quickly, and appreciate that it is not at all easy. You will also certainly object that it is unreasonable to be asked to make a decision because of one or more of the following reasons. (1) You do not know what to look for – which bits signal speaker identity – in these complex acoustic patterns; (2) you have not been told when two spectrograms are sufficiently similar to be considered coming from the same speaker, or (3) when two spectrograms are sufficiently dissimilar to have come from different speakers; (4) you have not been told how representative the spectrograms are of the speakers' voices; (5) you have not been told how to appropriately weight your decision criteria – whether for example you are to avoid false identification at any cost, and thus inevitably have to accept some false eliminations.

There is of course also a more important objection. In terms of the conceptual framework presented in the previous chapter, one should actually not be attempting to 'match' samples/spectrograms in the first place, but rather be attempting to give a likelihood ratio for the evidence. That is, one should be attempting to determine the probability of observing the difference between two spectrograms under the competing assumptions that they come from the same speaker and that they come from different speakers. From this point of view objections (2) and (3) above are more appropriately expressed as: you do not know the probability of observing the difference between two spectrograms of *hello* assuming that they have come from the same speaker and assuming that they come from different speakers. However, although this is an important criticism from the point of view of forensic speaker identification methodology, it is perhaps unfair from the point of view of illustrating a voiceprint comparison, and I will persevere with the assumption that one is trying to determine a match.

Although objections (2) and (3) are more appropriately expressed in Bayesian terms, it should be clear from previous discussion that the questions implied by the remaining objections (1) and (4) are not confined to voiceprint comparison, but apply to forensic speaker identification in general. Thus in all forensic speaker identification we need to ask: (1) which bits of the samples do we compare? and (4) how representative are the samples of the speakers' voices? Both these questions have already been touched upon in previous chapters.

In view of all these objections, let us make the task at least a little easier by revealing that the offender is actually among the suspects. The nature of the task has therefore now been changed from a difficult, but forensically realistic, open set identification to an easier, but forensically unrealistic, closed set identification. Other ways in which the data are both realistic and unrealistic can also be revealed at this point. Unrealistic aspects of the data are that they are controlled for intonation, all *hello*s being said with the same pitch features; and that they were recorded with the same equipment. In many other respects, the data remain forensically realistic. They are from naturally produced speech; the two *hello*s said by the offender are non-contemporaneous, being from recordings separated by one year; and all six speakers are sound-alikes, having been experimentally shown by Rose and Duncan (1995) to be misidentified by close members of their respective families in open tests. In any case, the guilty suspect is now the one whose spectrogram differs the least from the offender's spectrogram.

If I ask you to make your decision now, you may still object that knowing the closed set nature of the test does not help very much because objection (1) still obtains: you do not know which bits to compare. From the point of view of a voiceprint comparison, however, it is interesting to note that this does not really constitute a valid objection. This is because, as will be discussed below, you are supposed to compare everything, and linguistic knowledge is not a prerequisite.

You should now take a few minutes to identify the guilty suspect. The answer is given below.[3]

Having given an example of voiceprint comparison, I will now canvass the two major sources, one legal and one linguistic, of the controversy surrounding the aural-spectrographic method. The linguistically related problems, which will be dealt with first, concern primarily the methodology, and claims made about it. The legal side of things concerns the admissibility of evidence based on the method.

Linguistic controversy

Methodology

The first problem is that there is in fact no single agreed-upon methodology, although Gruber and Poza (1995: sections 54–71), who devote a whole section to the discussion of aural-spectrographic methodology, do mention two protocols. One has been developed by the Voice Identification and Acoustic Analysis Subcommittee (VIAAS) of the International Association for Identification (IAI), and has been published by the IAI in its journal (VCS 1991). The other protocol is that of the FBI (Koenig 1986: 2089–90). Since the FBI helped in developing the VIAAS/IAI protocol, both are very similar, and are obviously not independent.

The essential features of the method appear to be that the expert obtains samples from the suspect that are in some sense the same as those in the questioned sample. Then the suspect's sample(s) are compared aurally and spectrographically with the questioned sample(s), and the examiner reaches a conclusion as to whether or not they come from the same speaker.

Very many problems, both general and specific, have been associated with the method, and many aspects of it have been criticised in the literature. Probably the most serious general one is that it is simply not made clear what exactly the comparison of voices involves (Hollien 1990: 215). This is serious, because it prevents any proper assessment of the method and, more importantly, replication by the scientific community at large. Moreover, what is revealed about the method is not very

---

3 The offender is speaker C.

reassuring from the point of view of relevant received knowledge, since descriptions reveal a very poor understanding of the way speech works. There is no point in providing an exhaustive list of criticisms of the method here; these can be found in Gruber and Poza (1995). It is important to address the main specific ones, however, and this is done below.

Duplication of questioned sample

One important aspect of the method is that it is only claimed to work reliably when highly similar material, for example the same words, is available for comparison. Requiring data that are in some sense comparable is fair enough: this would be seen by most forensic phoneticians as a desideratum. One of the most worrying aspects of the method, however, is that, in order to obtain 'the same' data, the IAI protocol makes clear that the suspect is ideally not only to repeat the questioned material, but to try to imitate it in nearly all respects:

> Ideally, the exemplar should be spoken [by the suspect] in a manner that replicates the unknown talker, to include speech rate, accent, (whether real or feigned), hoarseness, or any abnormal vocal effect. . . . In general, the suspect is instructed to talk at his or her natural speaking rate: if this is markedly different from the unknown sample, efforts should be made through recitation to appropriately adjust the speech rate of the exemplar. . . . Spoken accents or dialects, both real and feigned should be emulated by the known speaker, ... If any other unique aural or spectrally displayable speech characteristics are present in the questioned voice, then attempts should be made to include them in the exemplars.
> VCS (1991: 373–9)

Essentially the same approach can be found in one current apologia for aural-spectrographic identification, taken from the website of a practitioner:

> The examiner can only work with speech samples which are the same as the text of the unknown recording. Under the best of circumstances the suspects will repeat, several times, the text of the recording of the unknown speaker and these words will be recorded in a similar manner to the recording of the unknown speaker. For example, if the recording of the unknown speaker was a bomb threat made to a recorded telephone line then each of the suspects would repeat the threat, word for word, to a recorded telephone line. This will provide the examiner with not only the same speech sounds for comparison but also with valuable information about the way each speech sound completes the transition to the next sound.
> McDermott *et al.* (1996: 2, 3)

There are serious flaws inherent in forcing such a mimicry. Firstly, there is the obvious consequence of the procedure. Requiring the suspect to try to make their voice as much like that of the questioned sample has the dangerous potential of placing an innocent person at risk of being misidentified (Gruber and Poza 1995: section 63). Note in particular the instruction for the suspect to mimic 'the spoken accents or dialects' of the questioned sample. Within many languages, dialectal differences are potentially powerful individual identifiers and discriminators. Aspects of speaking tempo, too, are promising speaker-specific parameters (Künzel 1997: 48), and it seems an outrageous suggestion that the method should actually attempt to deliberately exclude them. One possible justification for this might be that it would certainly count as a cogent argument for exclusion to be able to say that, even though a suspect mimicked the questioned speech, unaccountable differences persisted. It is not clear that this would be worth the risk.

Furthermore, three specific areas of ignorance in the process mean that it cannot be guaranteed that an innocent suspect will not incriminate themselves thereby. Firstly, the examiner cannot know how far 'the suspect had to diverge from his or her natural speaking style in order to achieve the degree of precision in replicating the unknown sample' (Gruber and Poza 1995: section 63). Secondly, given that people differ in their ability to mimic, it might be very difficult to tell to what extent an individual's failure to mimic the questioned sample was a consequence of their natural inability to mimic, or a reluctance motivated by guilt. Thirdly, there is no instruction how to judge when an appropriate degree of replication has been reached. This is not surprising, since it can only be done with a proper analytical model of the voice that distinguishes phonetic quality from voice quality, and this is not part of the approach. Finally there is the problem whether a suspect can be legally required to give a voice sample, and if so, whether they can then be legally required actually to mimic the questioned exemplar.

Comparison of samples

The protocols do not make clear precisely how, once obtained, the samples are to be compared. Two important things require clarification here. The first is: what aspect of the samples is to be compared? This corresponds to the forensic-phonetic parameters discussed in the previous chapters. The second is: how are they to be compared?

As parameters for comparison, the IAI protocol lists 'general formant shaping, and positioning, pitch striations, energy distribution, word length, coupling (how the first and second formants are tied to each other) and a number of other features such as plosives, fricatives and inter formant features' (Gruber and Poza 1995: section 59). The FBI protocol states that examiners make 'a spectral pattern comparison between the two voice samples by comparing beginning, mean [sic] and end formant frequency, formant shaping, pitch timing etc., of each individual word'. It does not state what aspects are considered in the aural examination. Further examples of the visual and aural features compared can be taken from another practitioner's website:

> Visual comparison of spectrograms involves, in general, the examination of spectrograph [sic] features of like sounds as portrayed in spectrograms in terms of time, frequency and amplitude. . . . Aural cues . . . include resonance quality, pitch, temporal factors, inflection, dialect, articulation, syllable grouping, breath pattern disguise, pathologies and other peculiar speech characteristics.
>
> AFTI (n.d.: 2)

As Gruber and Poza (1995: section 59 fn11) point out, these features, although sounding impressively scientific, are not selective but exhaustive, and the protocol amounts

Forensic Speaker Identification

to instructions to 'examine all the characteristics that appear on a spectrogram'. Likewise, they point out that the instructions as to aural cues also amount to telling the examiner to take everything he hears into account. Finally, they point out that it is no use just to name the parameters of comparison: one has to be specific about what particular *aspect* of the feature it is that is to be compared.

Criteria for decision

After comparing the suspect and questioned samples, the examiner arrives at a decision: 'When the analysis is complete the examiner integrates his findings from both the aural and spectrographic analyses' (McDermott *et al.* 1996: 3). The IAI protocol allows for seven possible decisions. These are: identification, probable identification, possible identification, inconclusive, possible elimination, probable elimination, and elimination. The decisions are based on two main criteria: the amount of matching data available, and the degree of similarity in that data. For example, an identification is to be returned if there are 20 or more matching words with all three formants (an acoustic feature), 90% of which words are very similar aurally and spectrally. Indicative of probable elimination are: not fewer than 15 words with two or more usable formants, at least 80% of which words must be very dissimilar aurally and spectrographically (Gruber and Poza 1995: section 60).

The problem here is that although the amount of data required for a particular decision is carefully specified, it is never stated what the criteria for similarity are. As illustrated above with the simulated spectrogram comparison, one naturally wants to know how similar is similar and how dissimilar is dissimilar. Moreover, the assumption is clear that similarity suggests identification, and dissimilarity exclusion. The discussion on the likelihood ratio in Chapter 4 made it clear that *similarity – the numerator of the LR – is only half the story: typicality also needs to be taken into account*. Finally, the amount of data required for a decision is not so clear after all:

> In order to arrive at a positive identification the examiner must find a minimum of *twenty speech sounds* which possess sufficient aural and spectrographic similarities. There can be no difference either aural or spectrographic for which there can be no accounting.
> McDermott *et al.* (1996: 3, 4) [author's emphasis]

Obviously, a decision based on 20 speech sounds may not be so reliable as one based on 20 words, which might contain anything up to one hundred different speech sounds in different contexts.

Independence of modalities

In the initial stages of its development, under its originator L. Kersta, the method was intended to be purely visual, perhaps inspired by fingerprint methodology. The method as now described involves two perceptual modalities: visual and aural. Practitioners claim that the combination of both modalities yields better results than visual on its own. This claim may have a certain *a priori* plausibility, since visual and auditory processing at least take part in different hemispheres of the brain. Any task involving complex visual pattern recognition, e.g. recognising faces, is processed better by the right hemisphere, whereas the left hemisphere has long been known to be the location for speech and language processing (Blakemore 1977: 161–4). However, it really depends on whether the voices can be considered as familiar or not, since, as was mentioned above, it is claimed that familiar voices, as is typical in complex pattern matching activity, are also recognised in the right hemisphere.

In fact, however, there appears to be little scientific evidence for the superiority of a combined aural-visual approach (Gruber and Poza 1995: sections 74, 118). What does appear likely, however, is that ears are better identifiers than eyes, at least when it comes to voiceprints (Nolan 1983: 23–4). If it is the case that the set of voices that are correctly identified by visual methods alone is a subset of those correctly identified aurally, then it is hard to see how a combined approach can be argued for.

If it is not clear that the combined aural-visual method is superior, there is no logical necessity for the visual comparison of spectrograms: the examiner could be expected to be able to perform as well by just listening. Further, since – as will be pointed out below – the aural part of the method actually appears to involve no analytic skill, there is no evidence that the examiner is performing any differently from an experienced layperson listening to the voice samples. Thus it is not clear whether this kind of testimony is warranted at all, since it duplicates what is within the capabilities of the jury. As Nolan (1983: 24) observes:

> ... the voiceprint procedure can at best complement aural identification, perhaps by highlighting acoustic features to which the ear is insensitive; and at worst it is an artifice to give a spurious aura of 'scientific' authority to judgments which the layman is better able to make.

Lack of theoretical basis

All of the previous critiques essentially follow from one thing: the method completely lacks a theoretical basis (Gruber and Poza 1995: section 72). What is a scientific theory? Science is a methodology (or really, since different sciences have different methodologies, a set of methodologies) for trying to find out and explain things about the world. In order to do this it has to posit entities, with properties. A theory is a set of hypotheses that specifies these entities and their properties, and describes how they relate. For many, another important aspect of a scientific theory is that it should be capable of being shown to be wrong.

The rudimentary elements of a theory of forensic speaker identification have actually already been presented in Chapter 2. There it was shown how voices can be discriminated using *entities* (e.g. the dimensions like 'average pitch'), the entitites' *properties* (e.g. their quantified values like 120 'average pitch' units), and *how the entitites relate* (i.e. how they can be combined) to arrive at a decision whether the samples are from same or different speakers. To the extent that it fails to correctly discriminate same or different pairs of speakers, such a theory, or its sub-parts, can also obviously be shown to be wrong.

When applied to the aural-spectrographic method, the explicandum is *how voices can be discriminated by that method*. For it is clear that, under certain conditions and to a certain extent, they can be. The entities that need to be invoked for the explanation

115

are those aspects of voices that can be abstracted, quantified, and compared, such that a quantified comparison will result in successful discrimination. However, as stated above, although properties of voices for comparison are indeed listed, these are not selective but exhaustive. Moreover, it is not specified what properties of the entities are to be examined, and as a consequence there is no quantification:

> ... despite many attempts since Kersta's time to impart a scientific tone to current spectrographic pattern matching techniques, none of the explanations by forensic examiners, including those of scientists, professional forensic examiners, and law enforcement personnel, has described a comparison methodology that relies on even one specific feature that can be used consistently across speakers.
>
> The lack of specific characteristics upon which examiners can rely to provide information of particular discriminatory value ... in the spectrographic patterns ... makes it clear that ... there is no scientific underpinning for the comparison methodology itself. ...
>
> Gruber and Poza (1995: section 55)

One reason for the absence of a particular theoretical base is clear. The aural-spectrographic approach is not analytic but gestalt-based: decisions are based on overall impressions (Gruber and Poza 1995: section 72). Since it is not analytic in approach, it is hardly surprising that the method relegates or even deliberately eschews theories from the speech sciences that attempt to explain acoustic patterns of speech analytically, in terms of the way they are produced and their communicatory significance. Thus, according to the originator of the technique, the application of the method does not include a mastery of or even an understanding of the principles of speech science. The IAI considers a high school diploma a sufficient educational standard for a practitioner, with speech science knowledge desirable (Gruber and Poza 1995: section 124). The FBI, too, emphasises signal analysis and pattern recognition skills for conducting voice identification examinations over formal training in speech physiology, linguistics, phonetics, etc., although a basic knowledge of these fields is considered important. They require a minimum of a BSc but no specialised knowledge other than from a two-week course in spectrograms (Koenig 1986: 2090).

The absence of a theoretical model for the aural-spectrographic method is seen by Gruber and Poza (1995: section 72) to follow of necessity from a lack of thorough knowledge about how humans pattern-match, both visually and aurally. However, it is not clear that this assumption is correct, since theories certainly exist, albeit competing ones, on how humans match patterns, as for example in face recognition (Bower 2001).

Whatever the reason for the absence of theory, since one of the main functions of a scientific theory is to explain, its lack can have serious consequences. It will mean that a practitioner will have difficulty explaining to the finder of fact the reasoning process leading from his observations to his conclusions:

> To the degree that subjective, gestalt pattern matching is at the core of spectrographic comparisons ... it is questionable whether the examiner will ever be able to successfully complete this step [of explaining their reasoning].
>
> Gruber and Poza (1995: section 111)

Claims

The claims made by proponents of the method have also been the source of much controversy. They have been of two types: one concerning the nature of voiceprints, and one concerning the reliability of the method.

Firstly, the term *voiceprint*. The term was not coined by practitioners, but it is likely that its adoption by them was intended to strongly suggest that, like fingerprints, voices are unique and invariant, and can be identified forensically with equal ease (Gruber and Poza 1995: sections 9, 10). In 1962 for example it was stated:

> Voiceprint identification is a method by which people can be identified from a spectrographic examination of their voice. Closely analogous to fingerprint identification, which uses the unique features found in people's fingerprints, voiceprint identification uses the unique features found in their utterances.
>
> Kersta (1962: 1253)

> As each one of the ridges of your fingers or on the palm of your hand differ from each other, so do all of the other parts of your body. They are unique to you ... including your voice mechanisms.
>
> Nash, quoted in Hollien (1990: 224)

As made clear in the foregoing chapters, voices show complex within-speaker variability. In addition they are constrained by functionality, are only a very indirect record of anatomy, and cannot practically be uniquely identified. In these senses, voices just are not like fingerprints, and the term and the analogy are bad. It is for obvious reasons usually avoided by forensic phoneticians and speech scientists.

In response to objections from the wider scientific community, the term voiceprint was eschewed by some practitioners in favour of more neutral terms like *spectrogram*, as in 'aural-spectrographic identification'. It is interesting to note that something similar happened with the term *DNA fingerprinting*, which was changed to *DNA typing* or *profiling* by geneticists who wanted to avoid the implications of the unique identification of individuals associated with fingerprinting (Cole 2001: 290). However, neither the term voiceprint nor the associated claim have disappeared, as witness the following, taken from the Web in 2001.

> The fundamental theory for voice identification rests on the premise that every voice is individually characteristic enough to distinguish it from others through voiceprint analysis.
>
> AFTI (n.d.: 1)

Since the originator of the method believed, however misguidedly, in the ability of voices to be uniquely identified by voiceprints, he can hardly be blamed for using the term voiceprint. However, it is difficult to construe its continued use as other than mischievous.

Lack of validation

'Extremes must usually be regarded as untenable, even dangerous places on complex and subtle continua' (Gould 2000: 267). Another source of controversy centres around the claims made about the reliability of the method. Early on it was claimed that it was virtually infallible. As might be imagined, such claims do not resonate well within the scientific community, whose members generally would arrogate, on the basis of experimental experience, the modest claim of Shakespeare's soothsayer in *Antony and Cleopatra*:

> In nature's infinite book of secrecy
> A little I can read.

Subsequently experiments were carried out, both by proponents of the technique and by speech scientists, to test its reliability. A controversy then arose in part from a critique of the methods and inferences in the experiments carried out by the proponents of the method, and in part from the rather large discrepancy found to exist between the error rates in the two sets of experiments. The experiments carried out by its proponents to test the method are of great importance, since it is their results that are quoted in court in support of the proven reliability of the technique. Moreover, as has just been pointed out, since the technique lacks a theoretical basis, its actual performance constitutes the sole basis left to it as a claim to be considered as a serious method of voice identification. They thus need to be summarised.

Two sets of data are usually referred to as constituting strong support of the method's ability to identify voices. One is the results of Tosi *et al.*'s well-known (1972) experiment; the other is the report of a survey in Koenig (1986).

The Tosi study

Tosi *et al.* (1972) was the most extensive study conducted on the method, and its results had a significant positive impact on the admissibility of voiceprint evidence in American courts. About 35 000 identification trials were performed over a two-year period using groups of between 10 and 40 speakers taken from 250 randomly chosen male university students.

The trials were conducted under different conditions, many of which were not forensically realistic. However, some tests did investigate the effect of using non-contemporaneous data and open- and closed-set identification, which do most closely approximate the forensic model (Hollien 1990: 222). Judgement was by visual inspection of spectrograms only, not listening. For open-set non-contemporaneous comparisons – the type of comparison which best approximates forensic reality – the error rates were a little more than 6% false identification and almost 12% false elimination.

The first thing these results show is that the method clearly works. The strength of evidence associated with such rates can be appreciated with the likelihood ratio. With a mean false identification error rate of 6%, the probability of a voiceprint identification assuming the suspect was guilty is 94% and the probability assuming they are not is 6%. This gives a likelihood ratio of (94%/6% =) ca. 16. You are 16 times more likely to have the examiner claim an identity if the suspect and questioned voices are the same than if they are not. Likewise, the mean false elimination rate of 12% is associated with a likelihood ratio of (88%/12% =) 7. The examiner is seven times more likely to say that the two samples are from different voices if they indeed are. In terms of the verbal strength scales for the LR presented in the previous chapter, a LR of 7 would count as 'limited', and a LR of 16 would be at the low end of 'moderate'.

As many have pointed out, however, the main question is how well the results of this experiment can be extrapolated to the real world, given the fact that the study was a highly controlled laboratory investigation with a forensically unrealistic heterogeneous population (Hollien 1990: 220, 222; Gruber and Poza 1995: section 127). Tosi *et al.*, in the conclusion to their 1972 study, claimed that there were good reasons to believe that the results would actually be *better* under real-world conditions, and this assumption – sometimes called the Tosi extrapolation – has naturally been quoted in support of the method in court. Gruber and Poza (1995: sections 79–87) however, give compelling reasons why this cannot be so, and conclude:

> In summary a strong case can be made that, lacking any scientific substantiation, careful researchers [i.e. Tosi *et al.*] should not have made such a strong claim that the results from an experimental study performed under nearly ideal conditions could justify the extrapolation that examiner performance in real-world situations would be better than those of the experimental subjects.

The Koenig survey

A study that promised to give the answer to the expected rates under real-world conditions, and that is often cited in support of the voiceprint technique, is a letter to the editor of the prestigious *Journal of the Acoustical Society of America* (Koenig 1986). In this letter are reported the results of a survey of 2000 voice identifications made by the FBI over a period of 15 years. Out of the 2000 cases reviewed, examiners felt able to pronounce judgements in 696, or ca. 35%. In these 696 cases it was claimed that the FBI could demonstrate both impressively low false identification and false elimination error rates of much less than 1% (Koenig 1986: 2090).

These figures appear to refute the above-quoted judgement of Gruber and Poza (1995) concerning increase in error rates attendant upon real-world conditions, and confirm the predictions of Tosi *et al.* (1972). However, it has been correctly pointed out (Shipp *et al.* 1987; Hollien 1990: 223–4; Gruber and Poza 1995: section 52) that the criteria used for assessing the correctness of identification or elimination were inadequate. The FBI, namely, deemed the examiner's judgement correct if it was 'consistent with interviews and other evidence in the investigation' (Koenig 1986: 2089). However, such considerations clearly do not constitute independent proof of correctness, even conviction or exculpation do not do that. Moreover, as pointed out by Hollien (1990: 224), there remains the pragmatically curious situation that the FBI does not permit the use of aural-spectrographic evidence in court even though it claims that its discriminatory power is so high. The FBI's restriction of voiceprint evidence to investigative purposes is still operative at the time of writing (Nakasone and Beck 2001).

It was mentioned above that another source of controversy is the inexplicable discrepancy between the very low error rates reported in studies such as Tosi *et al.* (1972) and Koenig (1986), and the consistently very much higher rates obtained in studies conducted by speech scientists who are not proponents of the approach (Hollien 1990:

222–4). Hollien (1990: 224), for example, quotes error rates of between 20 and 78% for such studies, and notes that obvious potential explanations like differences in the degree of skill of examiners will not account for the discrepancy.

It appears, therefore, that there remains inadequate confirmation of the low error rates claimed by proponents of the technique. Gruber and Poza (1995) note that the situation regarding knowledge of error rates had not changed from 1979, when it was summarised as follows:

> Voice identification by aural-visual methods can be made under laboratory conditions with quite high accuracy, with error rates as low as 1 or 2% in a controlled non-forensic experiment. . . . At the same time, the Committee has seen a substantial lack of agreement among speech scientists concerning estimates of accuracy for voice identifications made under forensic conditions. The presently available experimental evidence about error rates consists of results from a relatively small number of separate, uncoordinated experiments. These results alone cannot provide estimates of error rates that are valid over the range of conditions usually met in practice.
>
> Bolt *et al.* (1979)

In view of this, it is probably best to treat such claims of low error rates with extreme caution.

Legal controversy

Admissibility of voiceprint evidence

Given the frailty of justification from error rates, proponents quite often argue for the robustness of voice identification from the position of acceptability in court. Not surprisingly, given its controversial status, evidence based on voiceprint/aural-spectrographic technique has sometimes been admitted and sometimes not. Gruber and Poza (1995: sections 99–121) devote a whole section to the question. A perusal of the list of admissibility judgements in the United States courts in McDermott *et al.* (1996: 4–10) shows that voiceprint/aural-spectrographic evidence has been admitted about half the time.

Gruber and Poza (1995: section 120) identify three historical patterns in the admissibility judgements. General inadmissibility characterised the period before the positive results of the Tosi *et al.* (1972) study; general admissibility the few years after it; followed again by general inadmissibility as increasing numbers of scientists voiced their concern about the Tosi study.

Typical of this last period is the decison, in 1985, of what Gruber and Poza (1995: section 50) call 'the last full scale inquiry into the scientific and legal reliability and validity of voicegram identification evidence'. The hearing lasted two weeks and the court concluded:

> . . . that the aural spectrographic analysis of the human voice for the purposes of forensic identification has failed to find acceptability and reliability in the relevant

scientific community, and that therefore, there exists no foundation for its admissibility into evidence in this hearing pursuant to the law of California.

Like the previous judgement, admissibility judgements in the United States were largely based either on the Frye or McCormick standards for scientific evidence (Hollien 1990: 21, Gruber and Poza 1995: sections 104, 105). The Frye test (Frye 1923) made criterial the general acceptability of the method in the relevant scientific community. Of course, this simply shifted the argument to one of who the relevant scientific community was deemed to be, and proponents of the aural-spectrographic method argued for the exclusion of members of the general speech-science community as incompetent (Nolan 1983: 205). However, this was not always successful, and, since it had been demonstrated that the method does not generally find acceptance in the general speech-science community, when the Frye test has been applied, courts have tended to exclude expert evidence based on spectrograms (Gruber and Poza 1995: section 104). The application of the somewhat more lax McCormick requirement that the expert's evidence should be 'relevant', however, has usually resulted in admittance (Gruber and Poza 1995: section 105).

The Frye general acceptance test was held to be superseded, at least in the United States supreme court, in a 1993 decision in Daubert v. Merrell Dow Pharmaceuticals Inc. (Foster *et al.* 1993: 1509; Robertson and Vignaux 1995: 204). Daubert (1993) proposed more rigorous criteria for determining the scientific validity of the underlying reasoning and methodology used by an expert witness. These criteria are:

1. Whether the theory or technique can be, and has been tested.
2. Whether the technique has been published or subjected to peer review.
3. Whether actual or potential error rates have been considered.
4. Whether standards exist and are maintained to control the operation of the technique.
5. Whether the technique is widely accepted within the relevant scientific community.

According to these criteria, the aural-spectrographic method does not shape up very well (Gruber and Poza 1995: section 106). As described above, it still cannot be said to have been adequately tested. Neither can it be said to have undergone peer review through publication, at least not to the degree that obtains for other types of speaker identification, on which a vast refereed journal literature now exists. Neither has it been widely accepted in the relevant scientific community, if the latter is taken to include speech scientists and phoneticians. Although some error rates exist, there is still controversy as to their validity, and even the favourable estimates are not particularly good. Finally, although there certainly appear to exist standards controlling its operation, e.g. those promulgated by the IAI, these standards are, as shown above, open to serious question.

Because Daubert is not binding on other jurisdictions, and because the aural-spectrographic evidence can be admitted under other, non-scientific criteria, it is not clear what effect it will have on admissibility. The latest data available suggest that in 1998 admissibility in the United States ran to six states and four federal courts; exclusion to eight states and one federal court (Broeders 2001: 7). Perhaps the only sure thing is that courts will continue to differ.

Summary: Aural-spectrographic method

Aural-spectrographic voice identification relies on a combination of visual overall-pattern matching of spectrograms and overall auditory impression. It is controversial and not well received in the speech science community. This is not because of the primary observation data – most forensic phoneticians would agree that it is vital to both listen to speech samples and look at spectrograms. However, they do it analytically. The bad reputation of the aural-spectrographic method is mostly because of its lack of an explicit theoretical base, the poor understanding of speech implied by some of its methods, and as yet untenable claims as to its accuracy. There are also doubts whether the method is any better than can be expected from the auditory performance of a competent layperson. Because of these and other factors it falls well short of the Daubert standards for scientific evidence. In addition it has been claimed that, given the existence of superior techniques for forensic speaker identification, its continued use is 'perverse and unethical' (Braun and Künzel 1998: 12). It is also non-Bayesian, and is therefore subject to the same criticisms as any approach that attempts to quote the probability of a hypothesis given the evidence.

In the midst of all this negative criticism it is important to remember that there is absolutely no doubt that voices can be identified by the aural-spectrographic method, although the extent to which it can be done under forensically realistic conditions remains unclear. It thus seems reasonable to assume that, if it confirms to all safeguards specified in the IAI protocol, and manages to circumvent the problem of mimicry, the testimony of an experienced practitioner will have some probative value, if only because it is the result of careful and systematic visual and auditory comparison of adequate amounts of comparable data. It will still have to be made clear that a theoretical scientific basis for the technique does not exist, however, and the problem of expressing the outcome of an aural-spectrographic comparison in Bayesian terms also remains. Perhaps these aspects will emerge from the answers to the important questions that should be put to any expert giving evidence on voice identification:

- Precisely what parameters were used to compare the samples?
- How can the parameters be justified?
- In what way were the parameters quantified?
- What decision procedures were used? What, for example, were the thresholds?
- How can these decision procedures be justified?
- What is the probability of observing the differences between samples assuming same speaker origin/different speaker origin?

Chapter summary

This chapter has described the important distinctions within speaker recognition, and the different types of speaker recognition that they give rise to. In so doing it has characterised forensic speaker identification as a kind of technical speaker recognition most similar in its nature to speaker verification, but with an unknown set of reference speakers, and governed by a non-categorical Bayesian approach.

In contrast to naive speaker recognition, forensic speaker identification is informed by analytical knowledge of phonetics and speech science. It involves both auditory and computerised analysis, and thus consists of the twin branches of auditory and acoustic forensic analysis.

Naive speaker recognition can also have forensic applications. Its accuracy, however, is known to be affected by many variables, which must therefore be taken into account in estimating the strength of earwitness evidence. This is difficult, and so, unless it can rest on the results of a properly constituted voice line-up, earwitness evidence must be treated with caution.

The controversial aural-spectrographic, or voiceprint identification method involves unquantified gestalt comparison of the way voices sound and look on spectrograms. It differs from auditory and acoustic forensic analysis in lacking a theoretical base and appearing to rely on naive linguistic and visual pattern-matching abilities.

This concludes the first part of the book, the part introducing the basic ideas. We now move to the specific topics within the question of what speech is like.

6

The human vocal tract and the production and description of speech sounds

It was explained in Chapter 3 that the forensic-phonetic comparison of speech samples must be in both auditory and acoustic terms, but that the auditory description is logically and operationally prior. This means that many parts of forensic-phonetic reports are usually couched in the vocabulary of the descriptive phonetician. For example, it might be noted that 'the voices in both questioned and suspect recordings have an idiosyncratic palatal affricate instead of fricative'; or that 'the speaker in the questioned sample realises his voiceless dental fricative phoneme as a flat postalveolar before a following /r/, whereas the suspect has the normal allophone'; or that 'both recordings show a diphthongal offglide to /ou/ that sounds fronter and rounder than normal'.

Whatever does all this mean? The terms in these descriptions are articulatory: they describe vowels and consonants by the way they are assumed to have been produced. Since the ways that vowels and consonants are produced are traditionally described by reference to the anatomy of the vocal tract (dental, nasal, etc.), an understanding of them presupposes some familiarity with it.

What this chapter will do, therefore, will be to introduce the reader to the parts of the vocal tract that are important for speech production, and give some examples of how speech sounds are produced with the different parts. In this way the chapter aims to give a fairly non-technical account of how speech sounds are produced and described, and thus gives an idea of the analytic framework for the description of speech sounds referred to in Chapter 3. Although all of this analytical framework is forensically important, making possible as it does the auditory comparison of forensic speech samples, at certain junctures points of especial forensic-phonetic relevance will be mentioned. In outlining the structure of the vocal tract, this chapter will also provide a foundation for the subsequent account in Chapter 8, of equal forensic-phonetic significance, of how a speaker's vocal tract anatomy relates to their acoustic output.

Vowels and consonants are not the only type of speech sound, however. Intonation, stress and tone must also be considered. Intonation, stress and tone belong to the category of speech sounds called *suprasegmentals*, and vowels and consonants belong to the category of *segmental* speech sounds. Intonation, stress and tone are discussed later in the chapter.

The final section of the chapter looks at how individuals can differ in how fast and how fluently they speak.

The vocal tract

The human vocal tract is a most remarkable sound-producing device. It is responsible, when driven by the brain, just as much for the clicks of a Kalahari Bushman as for the vowels of a Phnom Penh Cambodian; for Tyrolean yodelling as for Central Asian Tuvan throat singing; for the rasp of Rod Stewart as for the sublimeness of Kiri Te Kanawa.

From an evolutionary point of view, the individual parts of the vocal tract are primarily associated with vegetative functions – the tongue for mastication, the pharynx and epiglottis for swallowing, the vocal cords for effort closure in defecating or giving birth, etc. However, starting about 1.6 million years ago, these structures evolved in their sound-producing capability to enhance our ability to use language through its primary medium of speech. Moreover, the vocal tract has also evolved in parallel with the peripheral hearing mechanism, so that the latter is tuned to respond best to the range of sounds that are produced by the former (Lieberman 1992: 134–7). These evolutionary developments also went hand in hand with expansion in the neural circuitry for controlling muscles for speech and for making sense of the incoming acoustic signals, and it is through palaeoneurological evidence indicating increases in brain size and in the nerves necessary to control tongue movement and breathing that we can begin to date the emergence of speech.

The basic dichotomy

Many parts of the human anatomy can incidentally influence speech output. Only some of these parts, e.g. the muscles between the ribs, or the tongue, are structures that speakers deliberately manipulate to produce speech sounds. However, for the purposes of this survey, it is necessary to focus only on the two structures that are basic to the production of speech. These are the vocal cords and the supralaryngeal vocal tract (SLVT). They are basic in the sense that they constitute two independently functioning and controlled, but precisely aligned, modules in speech production.

A good idea of the independent contribution of these two modules to speech can be obtained simply by doing the following. First take a deep breath. Say an 'ee' vowel (as in *heat*) on a sustained, comfortable pitch, and then change to an 'ah' vowel (as in *heart*) while maintaining the pitch. Still maintaining the pitch, change to an 'oo' vowel (as in *hoot*), and back to 'ee'. What you have done is to change the contribution of the supralaryngeal vocal tract, to produce different vowels, while keeping constant the contribution of the vocal cords, to produce the same pitch. Now try changing the contribution of the vocal cords while keeping the contribution of the supralaryngeal vocal tract constant. Take a deep breath, and keep saying an 'ee' vowel while you change pitch by going up and down. The pitch can be changed on any other vowel too, of course.

Description of speech sounds

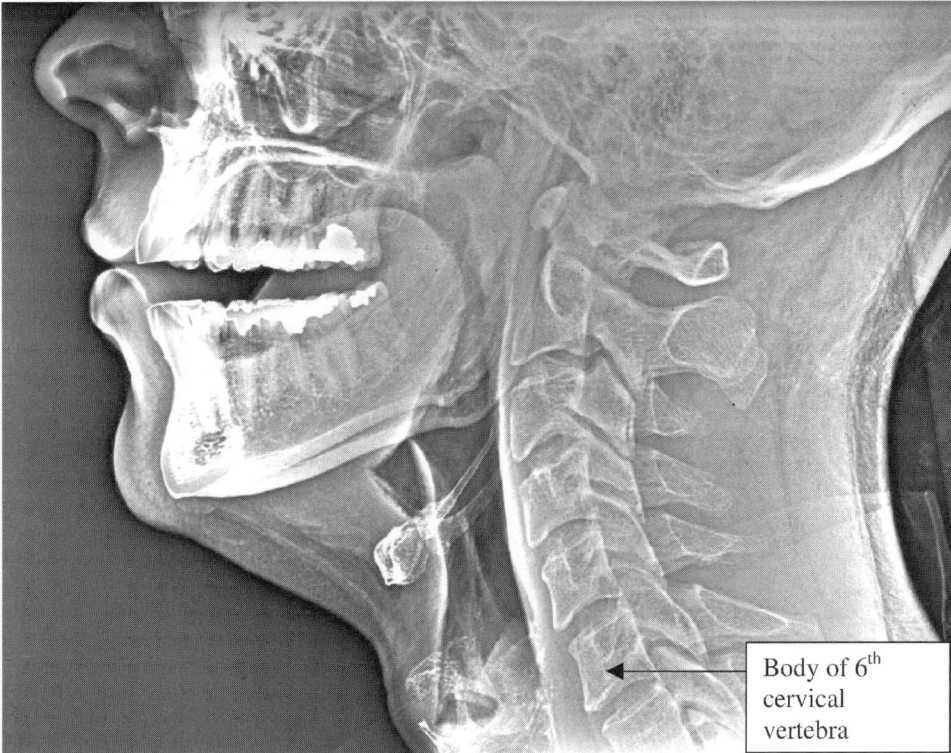

Figure 6.1 X-ray of vocal tract for [u]

The supralaryngeal vocal tract and the location of the vocal cords are shown in Figure 6.1, an X-ray of the author's vocal tract taken in the middle of a sustained 'oo' vowel similar to that in the (British) English word *who*. At the bottom of Figure 6.1 in the middle, in front of the body of the sixth cervical vertebra, can be seen the (top of the) larynx, or voice box. This is visible as the white structure causing the visible bulge in the speaker's throat called the adam's apple. In the middle of the larynx is a small dark triangular shape. This is a small ventricle immediately below which the vocal cords (which cannot be seen directly in the picture), are located.

The structures above the larynx in front of the vertebrae constitute the supralaryngeal vocal tract. These structures include for example the tongue, lips and teeth, which can easily be identified in Figure 6.1. Note how the lips are close together and slightly pursed, and the tongue appears bunched up towards the top and back of the mouth. These are typical supralaryngeal articulations involved in producing an 'oo' sound. It can also be noted that I was holding my larynx in what is probably an abnormally low position in this particular token of the vowel.

Vocal cords

The vocal cords are two small lips of elastic tissue with complex histological structure that are located in the larynx. They stretch from front to back across the top of the

Forensic Speaker Identification

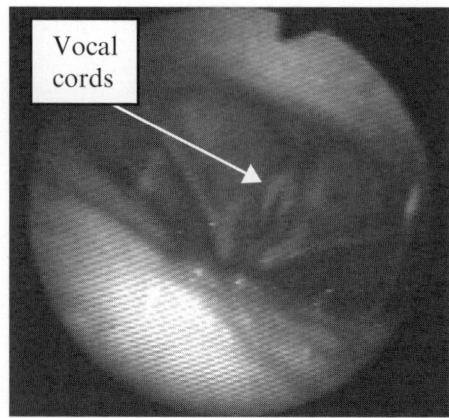

Figure 6.2 The vocal cords *in situ*

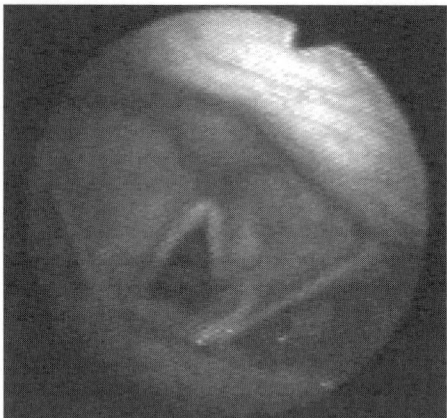

Figure 6.3 Vocal cords in abducted position

wind-pipe, or trachea. Figure 6.2 is a view down the author's throat, or pharynx, to show what the vocal cords look like when seen vertically, from a distance of about 7 cm. (The picture is a still captured from a video made with a fibreoptic endoscope inserted horizontally through the nose and then downwards into the top of the throat.) The front of my head is facing towards the top and slightly towards the right. The cords are visible as two tiny white strips at the bottom of the throat.

There are two extremely important ways in which the cords can be manipulated. They can be opened (abducted) or closed (adducted), and they can be tensed or relaxed. When the cords are held open, a small triangular gap is formed between them with its apex at the front. This gap between the cords is called the glottis (adj. glottal). The cords in abducted position are shown in Figure 6.3. The cords are viewed again from above, but from much closer, and in abducted position. The cords can be seen to have been pulled apart at the back, thus creating a triangular open glottis. Figure 6.4 shows the cords in adducted position. Now the cords can be seen to have been approximated at the back, so that they lie parallel.

Description of speech sounds

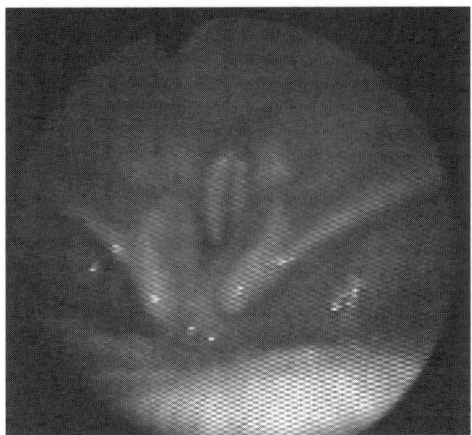

Figure 6.4 Vocal cords in adducted position

The cords can be tensed either actively, by contraction of the vocalis muscle of which they are in part composed, or passively, by contraction of a muscle that joins two important cartilages in the larynx: the crico-thyroid muscle.

These two manipulations – opening and closing of the glottis, and changing the tension of the cords – combine to participate in the production of speech sounds that are basic to all languages. When you are speaking, your vocal cords are continually being abducted and adducted, and at the same time being stretched and relaxed.

Voicing

Upon adduction, with the correct tension, the cords will vibrate when suitable pressure below the cords (called subglottal pressure) is supplied by the bellows action of the lungs. This vocal cord vibration is called voicing or phonation. Sounds thus produced are called 'voiced'. Vowels are usually voiced, as are consonants like 'm' [m], 'n' [n], 'l' [l] and 'r' [ɹ]. (For the time being, when referring to speech sounds, I shall now use both a representation in quotes to help the reader, followed by the proper phonetic symbol. As explained in Chapter 3, phonetic symbols are by convention written in square brackets.)

Voicing can be easily experienced with several simple experiments. Take a deep breath and say alternating long 'z' [z] and 's' [s] sounds (as at the beginning of *zoo* and *Sue*). The vibrations associated with voicing can be felt directly by simultaneously palpating the larynx. Alternatively, the vibrations of voicing transmitted through the tissues of the head can be felt by putting the palm on the top of the head. The sound of the voicing vibrations transmitted through the tissues of the head can also be amplified by putting both hands over the ears. By doing this while making [s] and [z], it can be appreciated that [z] is voiced. [s], on the other hand, lacks the vibrations and is consequently a voiceless speech sound. One of the important aspects of voicing can be seen in the fact that its absence or presence can signal the difference between speech sounds.

It can also be pointed out here that, apart from voicing, the speech sounds [s] and [z] are produced in the same way, that is, with the same supralaryngeal articulation. This anticipates the important notion of componentiality of speech sound structure – that speech sounds are composed of smaller units (one of which, for example, is voicing).

The reader should now try to determine which of the following sounds is voiced and which voiceless: 'v' [v], 'f' [f], 'th' [θ] as in *thing*, 'th' [ð] as in *this*. Take a deep breath, produce a sustained token of each sound, and see if you can tell whether your cords are vibrating.[1]

Pitch

By changing the tension in the cords when they are vibrating, the frequency of their vibration can be controlled. Increased tension results in higher frequencies, relaxation in lower. This is the mechanism underlying the changing pitch of speech, since high vocal cord vibration frequencies are perceived as high pitch and low frequencies as low.

Pitch is put to very many uses in speech. For example, in English it can indicate the difference between a neutral statement, e.g. *He's going*, and a question, e.g. *He's going?* In Tone Languages, differences in pitch signal difference in word identity. Say 'ma' [ma] on a high level pitch. That means *mother* in Standard Chinese. Now say 'ma' [ma] with a low, then rising pitch. That means *horse*.

It was pointed out above that when you are speaking, your vocal cords are continually being opened and closed, and at the same time being stretched and relaxed. It can now be appreciated that the opening and closing contributes to the production of voicing, and the stretching and relaxing to the production of pitch.

Phonation types

The vocal cords can not only vibrate, they can be made to vibrate in different ways. This gives rise to different sounds and constitutes the third dimension of vocal cord activity in addition to the two already described (voicing and pitch). Differing modes of vocal cord vibration are called phonation types, the most common of which is called modal phonation, or modal voice. (The term modal is thus used in its statistical sense, meaning most commonly occurring form.) Depending on the language, different phonation types are used to signal different types of information.

This can be best demonstrated with the phonation type called creak (also known as glottal fry, or vocal fry). Creak can be most easily produced deliberately in the following way. First say an 'ah' vowel (as in *far*) on a comfortably high pitch, and then gradually lower the pitch of the vowel, trying to see how low you can go. At a certain point, the phonation type will change to something that sounds like a series of individual pulses (or a very similar phonation type called creaky voice, which is a combination of normal vocal cord vibration and creak). Creak, or creaky voice, is transcribed with a diacritic tilde ('~') beneath the creaked segment, thus: [a̰]. You could try to produce a sustained 'ah' vowel, alternating, without stopping, between modal and creaky phonation: [a a̰ a a̰ a a̰ a . . .]

In some languages, phonation types are recruited for linguistic purposes. In a few languages they are systematically used to signal differences between words. For example,

---

1 [v] and [ð] are voiced; [f] and [θ] are voiceless.

in a variety of Thai called Thai Phake the only difference between the words *trade* and *to dance* is that the latter is said with a creaky phonation type: *trade* is [ka] and *to dance* is [ka̰] (Rose 1990). (Northern) Vietnamese is another language that is known to use creak(y voice) contrastively, that is, to contrast words.

Different phonation types are also used to signal paralinguistic information, for example the speaker's emotional attitude. In British English, for example, creak(y voice) affects a world-weariness or boredom on the part of the speaker. In the Mayan Language Tzeltal it is used to signal commiseration and complaint (Laver 1994: 196).

Creak(y voice) can also be extralinguistic, as a characteristic of an individual's voice. Laver (1994: 196) notes that it is not uncommon in American males, and older American females. One well-known, and often mimicked, example of this is the voice of the former Australian prime minister Bob Hawke.

Besides modal voice and creak(y voice) there are many other phonation types. Other well-known ones are breathy voice, whisper, and falsetto.

Forensic significance: Vocal cord activity

Parameters associated with vocal cord activity, especially pitch and phonation type, are considered very important in forensic phonetics. This is because, as illustrated above with phonation type, they can be used extralinguistically, as a characteristic of the speaker.

However, it is also forensically important to remember two things. Firstly, traditional parameters are never exclusively indicators of personal identity. The fact that, as just illustrated with phonation type, they can be *linguistic*, i.e. part of a language's sound system, or signal *paralinguistic* information like the speaker's temporary attitude, is often overlooked. The probability of observing similarities or differences in phonation type between samples assuming that they have come from the same speaker, and assuming they have come from different speakers, will thus be very different if the language under investigation is Northern Vietnamese than if it is American English.

Secondly, it is important to remember that although an individual's voice can be characterised, say, by the use of a non-modal phonation type, it is of obvious significance in the evaluation of forensic speech samples that, apart from pathological cases, the phonation types a speaker uses are largely under their control: they are deliberately creaky or breathy, for example. The fact that a speaker characteristically uses a particular articulatory setting does not mean that they always do.

Summary: Vocal cord activity

The sections above have shown that vocal cord activity can signal differences between sounds in three ways: (1) by the absence or presence of vibration; (2) by differences in rate of vibration; (3) by differences in mode of vibration.

The supralaryngeal vocal tract

The other main independent module of speech production is the supralaryngeal vocal tract (SLVT), which extends from just above the vocal cords to the lips. The important

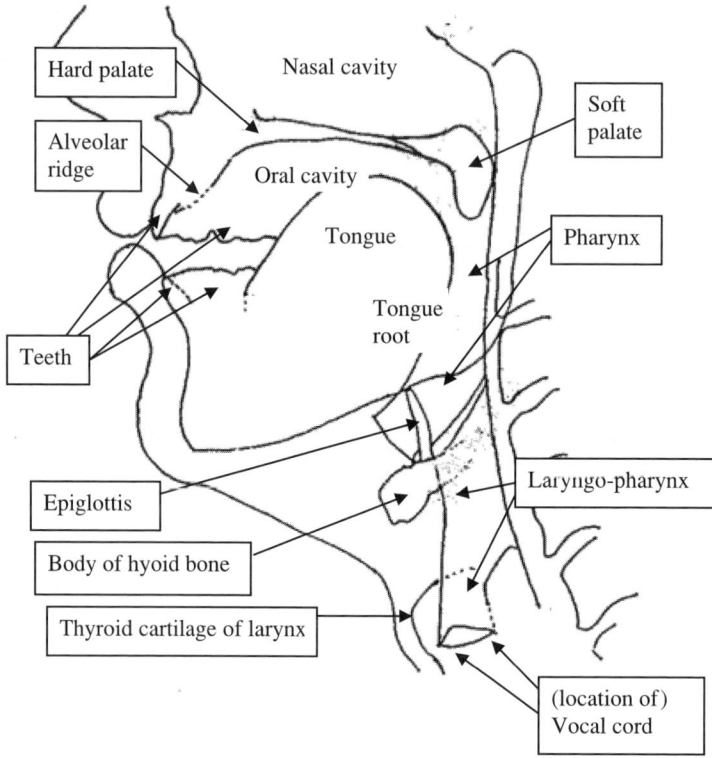

Figure 6.5 Outline of some important vocal tract structures

parts of the SLVT are labelled in Figure 6.5, which has been traced from the X-ray in Figure 6.1. The reader is encouraged to identify the structures in Figure 6.5 in the picture of the real thing.

Oral and pharyngeal cavities

The most important parts in the SLVT are three cavities: the oral cavity, the pharyngeal cavity (or pharynx), and the nasal cavity. These are marked in Figure 6.5. Most obvious are the horizontally oriented mouth (oral cavity), and, at right angles to it, the vertically oriented pharynx (pharyngeal cavity). The nasal cavity is separated from the oral cavity by the hard and soft palate.

The oral cavity is bounded at the front by the lips, at the bottom by the top of the tongue (dorsum), and at the sides by the cheeks. The top of the oral cavity is bounded by, from front to back, teeth, hard palate and soft palate. The pharyngeal cavity is bounded by the root of the tongue (radix) at the front, and the posterior pharyngeal wall at the back. The oral cavity–pharyngeal cavity boundary is considered to be located at the faucal pillars, two vertical strips of tissue that can be seen, using a mirror, at the back of the open mouth either side of the pendulant uvula. One part of the pharynx is the laryngo-pharynx, or larynx-tube, which can extend from the vocal cords to the epiglottis.

One of the major speech functions of the oral and pharyngeal (and nasal) cavities is to act as acoustic resonators. What this means will be explained in detail in Chapter 8 on the acoustics of speech production, but it can be noted here that during speech, especially in the production of vowels, the air in the cavities of the vocal tract vibrates in a complex way that is determined by the shape of the cavities, and that this vibratory response determines in large part the quality of the speech sound. So for a vowel like 'ee' [i] as in *he*, the oral cavity is relatively small, and the pharynx is relatively large. This difference in size determines in part the vibratory response which sounds like 'ee'. With a large oral cavity and small pharynx the vibratory response sounds like the vowel 'ah' [ɑ] in British English *father*.

Nasal cavity

Figure 6.5 shows the third large supralaryngeal cavity that is of great functional importance in speech – the nasal cavity. Unlike the oral and pharyngeal cavities, which are highly plastic and capable of considerable deformation, the nasal cavity is rigid. The only parts of it that can be deformed are its outlets. The external nostrils can be flared and narrowed a little, and the narrowing accompanies some (rare) speech sounds. The internal orifice is more important. Access to the nasal cavity is by means of the soft palate or velum, which is a soft tissue continuation of the roof of the mouth. The uvula dangles at its end.

The velum acts as a valve. When it is open, the nasal cavity is coupled to the oral and pharyngeal cavities, air can flow through it and out through the nostrils, and the cavity can then act as an extra resonator. The muscles that are partly responsible for lowering the soft palate and thus opening the nasal port (palato-glossus) are part of the structure notionally separating the oral and pharyngeal cavities – the anterior faucal pillars – mentioned above. When the soft palate is closed, pressed up hard against the pharyngeal wall, there is no air flow through the nasal cavities, and there is effectively no nasal resonance. In Figure 6.5, the typical sickle-shape of the velum can be seen to be tightly pressed against the posterior pharyngeal wall, thus isolating the nasal cavity. Figure 6.6 shows the author's soft palate in a lowered position, typical of quiet breathing.

The nasal cavity has a complex cross-sectional structure and a large surface area. It is divided by a vertical mid-line septum into two passages, which are usually not symmetrical. Much of its internal volume is taken up by three convoluted bony protrusions extending from the lateral walls of the nasal cavity and partially dividing it into three passages on either side. The mucous membrane covering the internal surface is acoustically absorbent, and subject to inflammation and infection, which can affect its volume and geometry.

The nasal cavity is also connected bilaterally to several auxiliary cavities, or sinuses. The sinuses are again usually not exactly symmetrical. The most important of these are the frontal sinuses, which lie above the nose behind the forehead, and the maxillary sinuses to either side in the upper jaw bones.

Like the oral and pharyngeal cavities, the nasal cavity acts as a resonator, but exclusively so. The way the nasal cavity functions in the production of speech sounds is easy to understand. When the soft palate is up, closing off the nasal cavities, there is no air-flow through them, and consequently effectively no resonance.

Forensic Speaker Identification

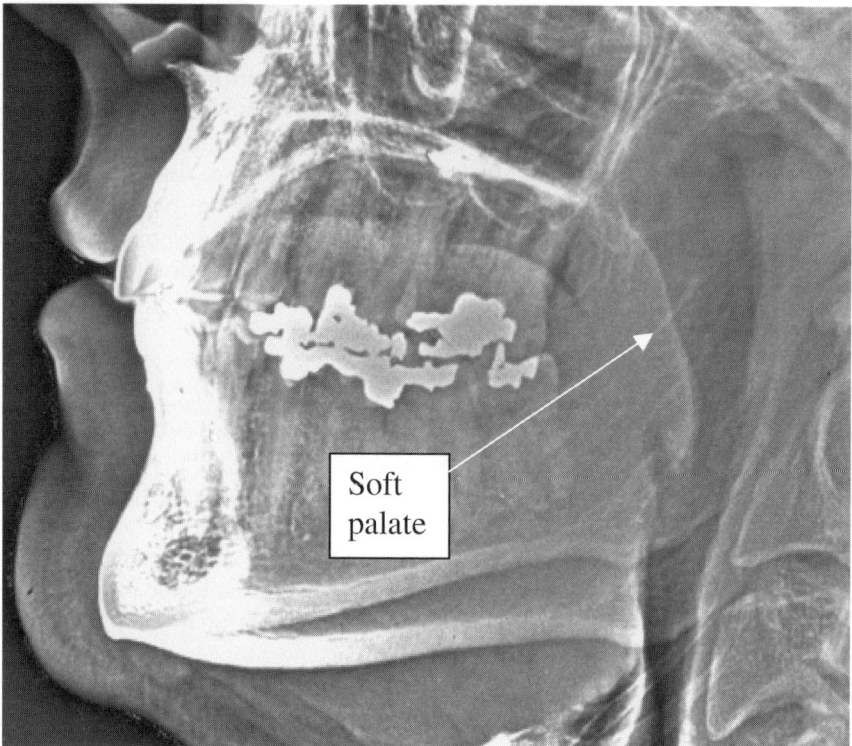

Figure 6.6 Soft palate in lowered position

When the soft palate is down, air can flow through the nasal cavities, and be caused to resonate.

Nasal and nasalised sounds

Two different circumstances are commonly found in the production of sounds with the soft palate down, depending on whether there is a closure in the mouth at the same time, i.e. whether there is air-flow through the oral cavity. If the mouth is closed – the closure can be at any position from the lips to the uvula – there is only flow through the nasal cavity, and such sounds are called nasals. Nasal sounds are usually produced with voicing (i.e. with concomitant vocal cord vibration, see above), and are very common in the world's languages.

English has three important nasals: 'm' [m] as in *simmer* (a bilabial nasal), where the closure is made at the lips; 'n' [n] as in *sinner* (an alveolar nasal, with tongue closure just behind the teeth at the alveolar ridge); and 'ng' [ŋ] as in *singer* (a velar nasal, with tongue closure at the velum).

Although, as mentioned above, these nasals are usually voiced, it is perfectly possible to produce voiceless nasals. Closing the lips and blowing air through the nose, for example will produce a voiceless bilabial nasal [m̥] (the voicelessness of the sound is indicated in the phonetic transcription by the little circle diacritic under the m).

Description of speech sounds

Such sounds are found, albeit much more rarely, in the world's languages. In Burmese, for example [na] with a voiced alveolar nasal means *pain*, whereas [n̥a] with a voiceless alveolar nasal means *nose*. Voiceless nasals are also sounds that may involve deliberate control of the external opening of the nasal cavity, by narrowing the nostrils to generate greater turbulence.

The second large class of speech sounds produced with the soft palate down is where there is no obstruction in the oral cavity. This means concomitant air-flow through the oral cavity, as well as through the nasal cavity, and such sounds are called nasal*ised*. The most common sounds thus produced are vowels. Nasalised vowels are usually phonetically transcribed with a tilde over the vowel, thus: [ã]).

From a linguistic point of view, of course, vowel nasalisation can be contrastive in some languages, signalling a different word depending on whether a vowel is nasalised or not. French is a well-known language with a set of contrastively nasalised vowels, with many examples like 'dos' [do] *back*, 'dent' [dɔ̃] *tooth*. But many other languages have contrastively nasalised vowels, for example Portuguese and many Chinese dialects.

Forensic significance: Nasals and nasalisation

Nasal consonants like [m] or [ŋ] are forensically relevant because their acoustic characteristics are assumed to be among the strongest automatic speaker identification parameters. This is because the relative rigidity of the nasal cavity ensures a low within-speaker variation in the acoustic features associated with the cavity's acoustic resonances, and its internal structure and dimensions are complicated enough to contribute to relatively high between-speaker variation.

One relevance of nasalisation to forensic phonetics lies in the fact that it is another parameter that easily shows tripartite functioning at linguistic, paralinguistic and extralinguistic levels (Laver 1991b: 184–5). The linguistic exploitation of contrastive vowel nasalisation was mentioned above. Extralinguistically, some speakers are known to characteristically speak with a 'nasal twang'. Paralinguistic function of nasality is quoted for the Bolivian language Cayuvava, where it is used by an individual of lower status towards one of higher status.

Squeezing the supralaryngeal vocal tract tube: Vowels and consonants

The sections above have described how, in the production of speech, air is pushed out by the lungs and flows through the glottis, where it can be modulated by the vibration of the vocal cords. The air-flow is directed backwards and upwards through the laryngo-pharynx, and into the oro-pharynx, where it travels vertically. If the soft palate is closed, it then can go forwards through the oral cavity to exit the lips. If the soft palate is open, the air stream can continue upwards through the naso-pharynx and forwards through the nasal cavity to exit through the nostrils. We have thus effectively introduced the role of some of the vocal tract anatomy in specifying speech sounds. An 'm' [m] is a voiced nasal for example. A lot more happens in the SLVT in speech than this, however.

Forensic Speaker Identification

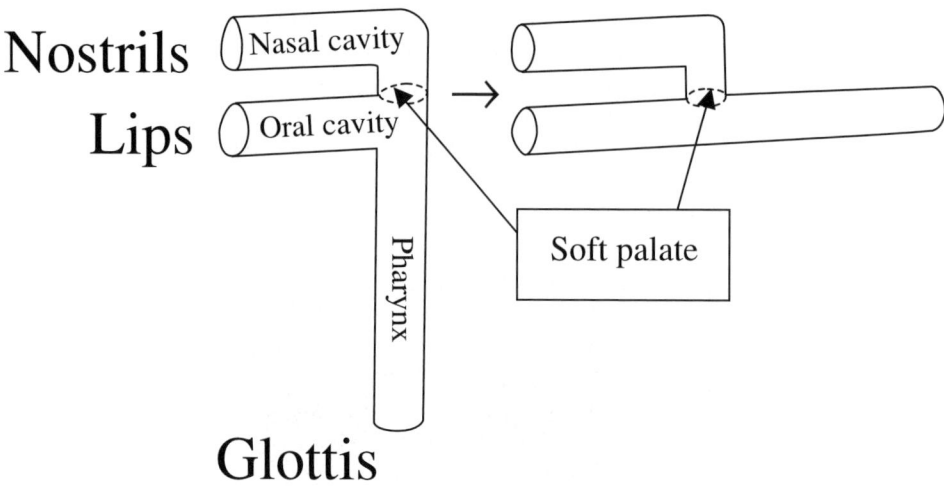

Figure 6.7 The vocal tract schematised as a tube

The SLVT can be conveniently thought of (and is indeed so conceptualised in speech acoustics) as a single straight tube (the L-bend can be straightened out in the imagination, it makes no difference), with vocal cords at one end and lips at the other. The nasal cavities constitute a side branch, which can be optionally coupled to this tube in the middle by action of the soft palate. This configuration is shown in Figure 6.7.

A more detailed picture of how speech is produced can now be described as involving the squeezing, or constriction, of this tube at various locations and to various degrees. This idea of supralaryngeal vocal-tract squeezing is used to draw, first of all, an important distinction between the two major classes of *vocalic* and *consonantal* sounds. Consonants are speech sounds produced by radically constricting the tube. Either it is totally constricted so that air-flow is (for a short time) totally blocked, or such a narrow constriction is produced so that air flowing through it usually becomes turbulent. Sounds like the 'p' and 't' in *pit*, or 'b' and 'd' in *bad* involve total closure. Sounds like 's' or 'z' in *sit* and *zit* involve narrowing and turbulence. All these sounds are by this definition consonants. More complex consonantal constrictions are possible, e.g. trills or laterals, and will be dealt with below.

If the tract is not radically constricted, a vowel sound is produced. Sounds like 'ee' [i] in *meet*, 'a' [ɑ] in British English *hard*, 'e' [ə] in *herd* qualify as vowels. In [i] the front half of the tube is squeezed, in [ɑ] the back part, and in [ə], no part is squeezed, with the tract retaining a uniform cross-sectional area along its length. Figure 6.8 shows a three-dimensional picture of the vocal tract tube of a Russian speaker in the shape for his [i] vowel. The figure is by courtesy of Drs Barlow and Clermont, of the Australian Defence Force Academy, and represents part of their current research (Barlow *et al.* 2001) reconstructing three-dimensional representations of the vocal tract from physiological and anatomical data, e.g. X-rays. It shows nicely how the front part of the tube (the oral cavity) is squeezed, and the back part (the pharyngeal cavity) is expanded.

Unfortunately, although they are produced in the same mouth, vowels and consonants are traditionally described differently, and therefore have to have separate exposition. Consonants will be described first.

Description of speech sounds

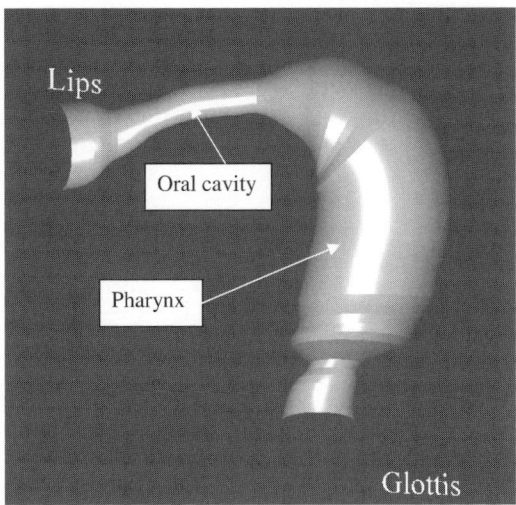

Figure 6.8 Vocal tract tube for an [i] vowel.

Place and manner description of consonants

As mentioned above, consonants are sounds that are produced with radical constriction of the vocal tract tube. In order to produce different consonants, the vocal tract can be constricted at different positions, and for (almost) each position it can be constricted in different ways. Thus consonants are traditionally described by referring to these two effectively independent parameters. The location of the constriction is termed Place (of articulation). The type of constriction is termed Manner (of articulation).

To illustrate this, we can take two places and see how the same manners are to be found at both. We will use predominantly English examples, but also examples from some other languages. The reader will find it useful to attempt to make the sounds silently, i.e. with no phonation, in order better to feel how they are articulated.

Let us focus on the bilabial and alveolar places. The former term means that the consonant is produced with the two lips; the latter that the consonant is produced on or approximating the alveolar ridge. This is just behind the front teeth: its location is shown in Figure 6.5.

Manners at the alveolar place

At the alveolar place, English has four different consonantal manners: plosive, fricative, lateral, and nasal. At the bilabial place, English has two: plosive and nasal.

Plosives

The sounds 't' and 'd' as in *too* and *do*, are both produced with the same manner. This involves the vocal tract being quickly and totally closed off (in this case, by the tip of the tongue tip sealing against the alveolar ridge, the sides of the tongue against the side teeth, and the soft palate closing off the nasal cavity), and air flow interrupted. After a short hold phase (of the order of several hundredths of a second), the closure is abruptly released, and the overpressure that has built up in the air behind the closure is quickly dissipated as the air flows out of the mouth. Sounds of this type are called plosives, or stops, and therefore the consonants 't' and 'd' as in *too* and *do* are described as alveolar plosives.

Fricatives

The sounds [s] and [z] as in *Sue* and *zoo* are both produced by moving the front of the tongue quickly to create a narrow constriction between it and the alveolar ridge. The constriction is so narrow that air becomes turbulent as it passes through, thus creating acoustic noise. After a brief hold, the tongue is then quickly removed. Sounds of this type are called fricatives, and therefore the consonants [s] and [z] are described as alveolar fricatives.

In the case of [s] and [z], there is an additional source of turbulence downstream, as the jet of air encounters the upper incisors. This second origin of turbulence is not necessary to the definition of fricatives, but does define an important sub-set of sounds called sibilants. Thus the sounds [s] and [z] are more specifically described as sibilant alveolar fricatives.

Affricates

Affricates are rather like a plosive followed by a fricative, both made at the same or similar place of articulation. The affricate is started in the same way as the corresponding plosive, with total closure and subsequent build-up of air pressure behind the closure. Upon release, however, the articulator is not removed quickly and totally, but is decelerated in such a way as to generate a short period of turbulence like a fricative. The sound at the beginning and end of the English word *church* is an affricate, although it is not strictly speaking made at the alveolar place. It is transcribed [tʃ]. It can be felt that its production involves something like the combination of a 't' followed by 'sh', although it can also be felt that the 't' is made slightly further back than the alveolar ridge. The place of articulation of this sound is thus called post-alveolar. The voiced affricate counterpart to this in English occurs at the beginning and end of the word *judge* [dʒ].

Alveolar affricates are found in German, for example in the word *Zeit* [tsait] 'time'. Alveolar affricates also occur in English, for example in the word *seats* [sits]. However, we can see from the fact that this word is made up of two bits – the root *seat* and the plural suffix *s* – that in English, unlike German, alveolar affricates occur as the result of a 't' and an 's' coming together. For this reason they are not considered structurally as a single sound in English.

Nasals

Nasals have already been discussed above, but the reader should check, by saying the [n], [t] and [d] in 'no', 'toe' and 'dough' that [n] does in fact have the same (alveolar) place as in [t] and [d].

Description of speech sounds

Approximants

The last major manner category to be mentioned involves a lesser degree of stricture between the articulators than fricatives – so much so that turbulence is only generated when the sound is voiceless and there is consequently a higher rate of air-flow. These sounds are called approximants, implying that the articulators are only approximated.

The category of approximant includes quite a few sounds, one of which – the 'l' in *left* or *love* (transcribed [l]) – occurs at the alveolar place in English. It should be possible to feel the same alveolar place for the [l] sound in *low* as in the other alveolars in *no*, *toe* and *dough*. The difference between [l] and the previous alveolars described is that in [l] there is a mid-line closure, at the alveolar ridge, but there are openings, starting at about the molars, at either side of the tongue through which the air stream passes. (These lateral exits can be seen in a mirror if the lips are pulled out of the way, especially if the jaw is lowered as well.) Because the air passes to the sides of the tongue, [l] is called a *lateral approximant*, or often just a *lateral*. In English, the lateral is also voiced.

The American and Australian English 'r' [ɹ] in *red*, *root*, *read* is also an approximant, although not alveolar. In [ɹ] the tongue is usually approximated to just behind the alveolar ridge. Unlike the lateral [l], it is a *central approximant*, because the air flows centrally. (It is also a rhotic and shares acoustic similarities with trills, taps and flaps.) Other central approximants are the 'y' [j] at the beginning of *yet* or *you*, and the 'w' [w] at the beginning of *we* or *want*.

It has thus been demonstrated how, at the same alveolar place, five different ways, or manners, of constricting the vocal tract can be used to produce five different types of speech sound: plosive, fricative, affricate, nasal, and lateral approximant.

Manners at the bilabial place

The independence of the manner parameter vis-à-vis the place parameter can now be demonstrated by moving forwards to the bilabial place, and running through (nearly) the same manners (two parameters that are independently related in this way are called orthogonal).

Bilabially, (voiceless and voiced) plosives [p] and [b] as in English *pit* and *bit*, can be produced, as also a nasal [m]. Bilabial fricatives [Φ] (voiceless) and [β] (voiced) do not occur in English, but can be found in a few other languages. For example, Ewe, a West African Language, has both [Φ] in the word 'efe' [əΦə] *five* and [β] in 'Eve' [əβə] *Ewe*. Some Japanese speakers, also, are said to have [Φ] in the word 'fune' [Φɯne] *boat*, for example.

A labial affricate, consisting of bilabial plosive followed by labio-dental friction [pf], occurs in German: 'Pfennig' [pfɛnɪk] *penny*. It is probably not possible to produce a bilabial lateral without external digital help to keep the lips together in the mid-line but apart at the sides, and not surprisingly this speech sound is not attested.

Other manners of articulation

This does not exhaust the list of manners. In addition to the categories exemplified above (plosives, fricatives, affricates, nasals, and (lateral) approximants), there are also trills, taps and flaps.

Table 6.1 Sounds illustrating some different Manners at the bilabial and alveolar Places

Place	Manner						
	Plosive	Nasal	Fricative	Affricate	Lateral	Tap	Trill
Bilabial	b	m	β	bβ	–	–	ʙ
Alveolar	d	n	z	dz	l	ɾ	r

Trills, taps and flaps

Many people can 'trill their r's' (and many cannot but still know what a trill is). Such alveolar trills occur as *bona fide* speech sounds in various languages, for example Spanish 'perro' [pero] *dog*, or Arabic [mʊdarɪs] *teacher*.

The tongue tip is not the only thing that will help produce a trill. Although used in English to signal 'I'm cold', trills with the lips (bilabial trill, transcribed [ʙ]) are also found as speech sounds, albeit rarely, as are trills with the uvula. (The vibration of the vocal cords is also strictly speaking a trill).

In taps, the tongue tip quickly moves upwards to tap the part of the mouth near it, usually the alveolar ridge. The Japanese 'r' [ɾ] in 'arigato' [aɾiŋatɔ] *thank you* is usually a tap. In a flap [ɽ] the tongue is first drawn back, and then flapped forwards, striking just behind the alveolar ridge on its way past. Such flaps are therefore usually classified as post-alveolar. Examples of flaps can be found in Yolngu (an aboriginal language of Australia), for example 'maadi' [maːɽi] *crayfish*, and Hindhi 'aṛi' [aɽi] *cooked*.

Most trills, taps and flaps belong within a larger class of rhotics, or 'r-like sounds', which can be classified together on the basis of shared acoustic, as opposed to articulatory features.

Since quite a lot of new sounds have been introduced above, it is useful to have an overview of some of them in a table. Table 6.1 accordingly gives the sounds for different manners mentioned above at the bilabial and alveolar place. To avoid complexity only voiced sounds are shown.

Place of articulation

It was noted above that consonants – sounds made with radical constriction of the vocal tract – are described and analysed in terms of two major parameters: the kind of constriction (manner), and the place where the constriction is made. This section describes the important places of articulation in the production of consonants. More detail can be found in Ladefoged and Maddieson (1996: 9–46).

Active and passive articulators

In the phonetic description of place of articulation it is necessary to refer not only to where the articulation is made (this is called the passive, or stationary articulator), but also what (part of) what speech organ it is being made with (the active articulator).

Watch yourself make the fricative sounds [f] and [v] at the beginning of the words *fat* and *vat*: your lower lip moves to position itself near or touching the upper incisors. The lower lip is the active, because moving, articulator, and the upper incisors are the passive articulator. When you make the plosive sounds [p] and [b] at the beginning of the words *pat* and *bat*, however, you can see that, although the lower lip is still the active articulator, a different passive articulator is involved: the upper lip. The same active articulator can thus make different sounds depending on what passive articulator it articulates with. This is why a full description of place needs to nominate both passive and active articulators.

In fully specifying the place of articulation of a consonant, the term for the active articulator, which is often a latinate adjective in -o, precedes the passive term. Thus the place of [f] and [v] is termed *labio-dental*. In this place descriptor, *labio-* means that the active articulator involves the lips, and *dental* means the passive articulator involves the teeth. The active articulator term is often omitted.

Active articulators

The most common active articulators involved in making consonants – the movable parts of the vocal tract – are the lower lip (*labio-*), the tongue tip (*apico-*), the tongue blade (*lamino-*), and the tongue body (*dorso-*). The latter is subdivided if necessary into front (*anterodorso-*) and back (*posterodorso-*). Less commonly encountered active articulators are the underside of the front of the tongue (*sub-lamino-*), the tongue root (*radico-*), and the epiglottis (*epiglotto-*). These active articulators are labelled, in capitals, in Figure 6.9. If one does not want to distinguish between tip and blade of the tongue, the term *coronal* is used, meaning both tip and blade.

Passive articulators

The list of commonly encountered passive articulatory locations is somewhat greater, but mostly self-explanatory, and best demonstrated with the aid of Figure 6.9, which illustrates nine different passive places of articulation. Starting from the front of the mouth, these are:

- the upper lip (*labial*)
- upper front teeth (*dental*)
- the short area behind the upper incisors called the alveolar ridge (*alveolar*)
- just behind the alveolar ridge (*post-alveolar*)
- the hard palate (*palatal*)
- the soft palate (*velar*)
- the uvula (*uvular*)
- the posterior pharyngeal wall (*pharyngeal*)

Note that the glottis is sometimes considered a place of articulation (thus one refers to *glottal* fricatives or stops), although the notion of passive and active articulators is not coherent here.

Thus describing a consonant as *apico-alveolar* means that it sounds as if it has been made with the tip of the tongue articulating with the ridge just behind the top

Forensic Speaker Identification

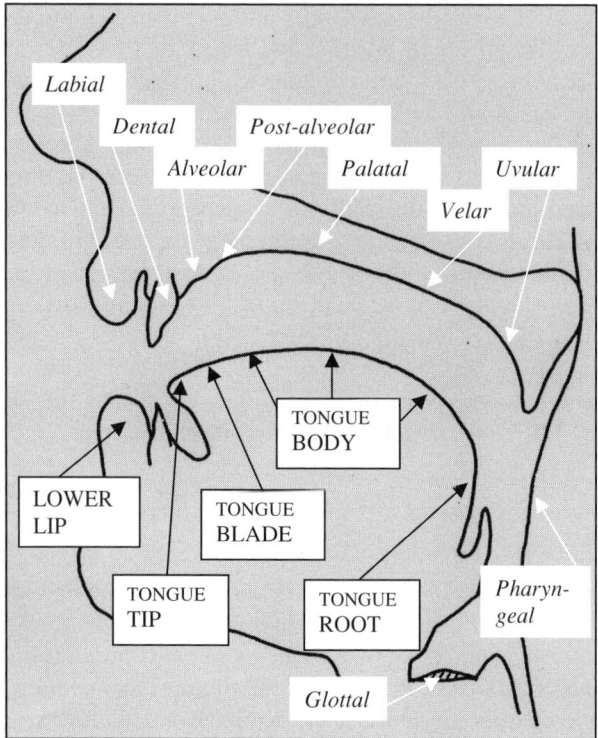

Figure 6.9 Mid-sagittal section of vocal tract showing some common active (capitals) and passive (italics) articulators

teeth (as in the sounds at the beginning of the words *toe dough no* and *low* [t] [d] [n] and [l]. A *dorso-velar* consonant is made with the top of the tongue body articulating with the soft palate as in the sounds at the end of the words *back* [k], *bag* [g] and *bang* [ŋ]. In practice, other names are often used for some common places. The most common are *bilabial* for (*labio-labial*), and *palato-alveolar* for post-alveolar affricates and fricatives.

English consonants

Almost enough of the architecture of basic articulatory description has now been covered to present, in Table 6.2, a table of the commonly occurring consonantal sounds in English. It is possible to do this because there is not very much variation in consonants across the main varieties of English (Ladefoged 2001: 47).

In Table 6.2, the sounds are arranged in a Place–Manner matrix (Place horizontally, Manner vertically). For each sound, a word is given that typically contains it. Some cells in the matrix contain symbols for more than one sound. These sounds differ along other phonetic parameters, the most common of which is voicing. Thus of

Description of speech sounds

Table 6.2 Major consonantal sounds of English

MANNER ↓	← PLACE →						
	Bilabial	Labio-dental	Dental	Alveolar	Post-alveolar	Velar	Glottal
Plosive	p (pat) b (bat)			t (tip) d (dip)		k (cat) g (goose)	
Fricative		f (fat) v (vat)	θ (think) ð (this)	s (sip) z (zip)	ʃ (ship) ʒ (Persian)		h (help)
Affricate					ʧ (chip) ʤ (judge)		
Nasal	m (simmer)		n̪ (tenth)	n (sinner)		ŋ (singer)	
Lateral approximant				l (leaf) ɫ (milk, feel)			
Central (rhotic) approximant				ɹ (red)			

the two sounds in the labio-dental fricative cell [f] is voiceless and [v] voiced, and of the two sounds in the bilabial plosive cell [p] is voiceless and [b] voiced. As far as the English plosives go, it is somewhat of an oversimplification to say there are only two phonetic variants, and that they differ in voicing. This will be taken up later.

Table 6.2 shows that phonetically English has plosives at three places: bilabial, alveolar and velar, plus affricates at the post-alveolar place (these are commonly referred to as palato-alveolar, or even palatal, affricates).

Phonetically, there are fricatives at five different places; nasals at four places; one lateral; one rhotic ('r' sound); and two semivowels. Except for the higher number of fricatives at the front of the mouth (labio-dental, dental and alveolar), this is a fairly typical phonetic consonantal inventory.

The lateral requires additional comment. The symbol [ɫ] stands for a velarised lateral (a non-velarised, or plain lateral would be transcribed [l]). This introduces an additional complexity called secondary articulation. It is possible with consonants produced with the tip of the tongue simultaneously to independently control the position of the tongue body to produce different sounds. In the case of the velarised lateral, in addition to its apico-alveolar place (and lateral manner), it is typically made with the body of the tongue raised towards the soft palate or velum.

The tongue body is not the only articulator that can be independently controlled in secondary articulation. Lip activity is also common: the 'sh' sound in English words like *sheet shut shark* is usually produced with slightly rounded lips, for example. The reader can also try producing a voiceless alveolar fricative ([s]) with and without rounding their lips to notice the difference in sound (rounding lowers the pitch). The ending -ised, as in velar*ised*, is often used to refer to secondarily articulated sounds; thus sounds secondarily articulated with lip rounding are called labialised.

143

Vowels

Vowels are important in forensic phonetics, for several reasons. As will be shown in Chapter 8, their acoustical properties are relatively robust and easy to quantify. This is not only because they tend to last longer – have greater duration – than consonants, but also because of their relatively well-defined acoustic structure. (Vowels belong to the class of speech sounds called sonorants that are defined by precisely this characteristic.) Vowels are the speech sounds from whose acoustics the imprint of the vocal tract that produced them can be most easily extracted. Some vowels have been shown to have considerable individual-identifying potential.

More importantly perhaps, vowel quality can be one of the more obvious ways in which a speaker's accent is realised. *Accent* refers to speech patterns that are typical of a given geographical location (a dialect), or typical of a particular social group (sociolect). Vowel quality can therefore be of potential forensic use when profiling an unknown speaker as a member of a particular social group, or as being associated with a particular geographical area. As exemplified at the beginning of this book, vowel quality can also be crucial when comparing samples with respect to accent.

Primary parameters of vowel description

It was stated above that the production of vowels involves less radical modifications of the vocal tract tube than consonants. These modifications to the shape of the vocal tract tube are accomplished by gestures of the tongue-body and lips: moving the body of the tongue from front to back and from high to low, and altering the size of the mouth opening by spreading or rounding the lips.

In order to illustrate this, recall the simplified picture of the vocal tract as a uniform tube with a right-angle bend (left hand side of Figure 6.7). Since the body of the tongue forms the lower boundary of the oral cavity and the front boundary of the pharyngeal cavity, moving the body of the tongue will alter the sizes of the oral and pharyngeal cavities. If there is no movement of the tongue – the so-called neutral position – and if the lips are neither spread nor rounded, the vocal tract approximates a tube with a uniform cross-sectional area. This will produce a vowel similar to that in British English *heard*. This vowel is written with an inverted and reversed 'e' ([ə]) and is called *schwa*. This is shown in the bottom left-hand part of Figure 6.10.

Now, if the body of the tongue is raised from the schwa position towards the hard palate, and fronted somewhat, it will constrict the front part of the tube. This is what is involved in the production of an 'ee' [i] vowel as in *heed*. It is schematised in the top left part of Figure 6.10. (The tongue body can be raised and lowered by moving the body itself, or by keeping the tongue still, and raising and lowering the lower jaw. There might also be a slight amount of lip spreading involved as well.) Figure 6.11 shows the real thing: an X-ray of the tongue position for an [i] vowel. Note the narrow oral cavity, the wide pharynx, and the raised soft palate. The small oral cavity and large pharynx were also clear in the model of the vocal tract tube for an [i] in Figure 6.8.

If the tongue body is backed toward the posterior pharyngeal wall, it will constrict the back part of the tube: this is what is involved in the production of the 'ah' vowel

Description of speech sounds

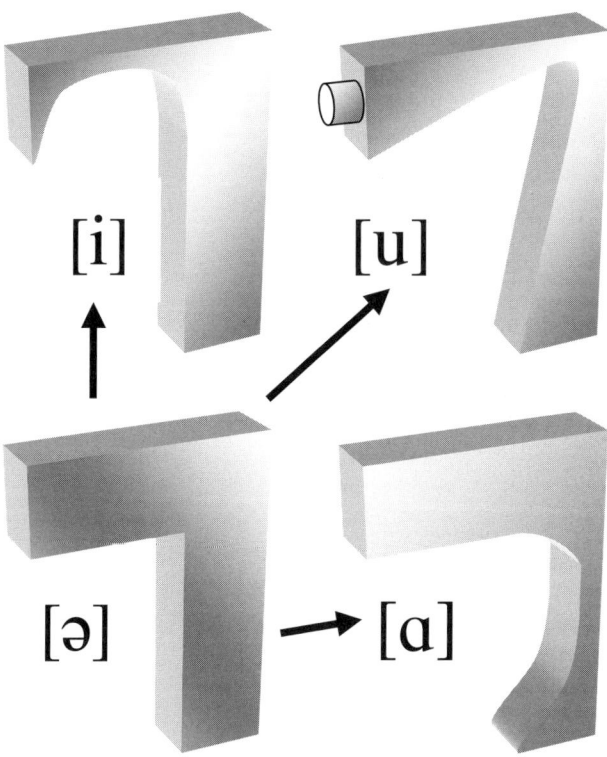

Figure 6.10 Schematic mid-sagittal section of the supra-laryngeal vocal tract showing movement of tongue body and lips from the neutral vowel [ə] to [i] (upper left), to [ɑ] (lower right), and to [u] (upper right)

([ɑ]) as in British English *father*, or *path*, or General American *father* or *hot* (schematised in the bottom right part of Figure 6.10). Figure 6.12 shows an X-ray of the tongue position for a vowel that was said with the tongue almost as far back as in an [ɑ] vowel). Note the narrow pharynx and large oral cavity.

If it is raised towards the soft palate, the tongue will constrict the tube in the middle: this, together with a considerable rounding of the lips, is what is involved in the production of an 'oo' [u] vowel, as in German *gut* (Figure 6.8, top right). An X-ray of a [u] vowel was shown in Figure 6.1.

This then is basically what is involved (with the addition of voicing!) in the production of the four vowels [i], [u] [ə] and [ɑ]. The tongue body does not have to be moved the whole way in any direction, of course, and different degrees of raising and backing are used to produce different vowels.

Using English to demonstrate this is not particularly easy. This is because the vowel system in English is complex, involving more than differences in height and backness, and is not particularly systematically distributed. Most languages have far simpler vowel systems. However, some idea of how different vowels can be produced by different positions of the tongue body (in particular exploiting the dimensions of height and backness) can be obtained in the following way. Saying the following

145

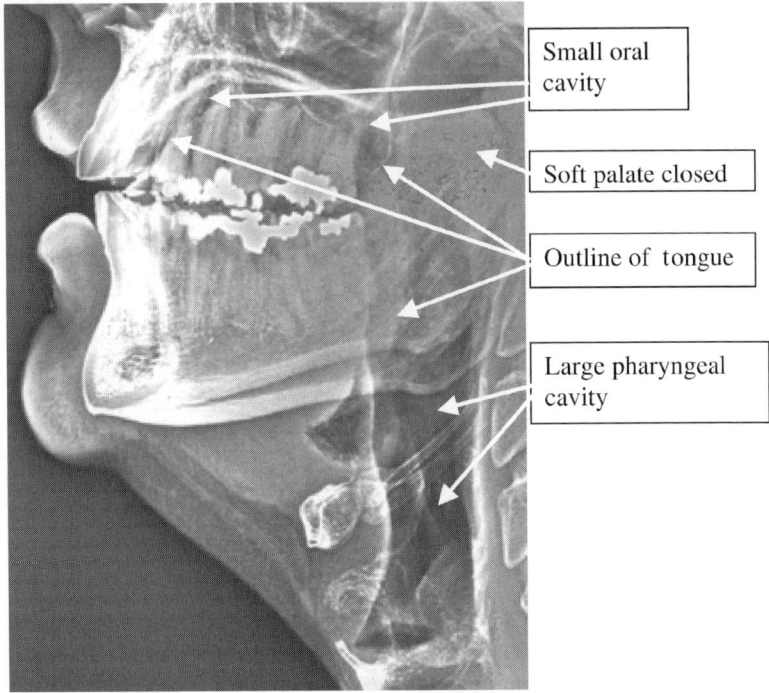

Figure 6.11 X-ray of [i] vowel

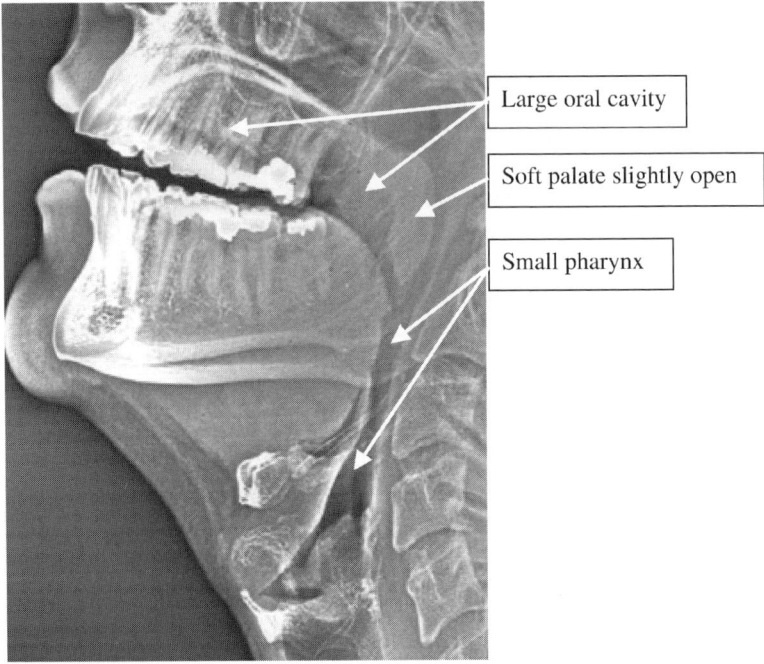

Figure 6.12 X-ray of vowel similar to [ɑ]

sequence of words should allow you to feel movement in the height dimension, from low to high: *had – head – hid – heed*. Tongue body height can also be controlled by positioning of the jaw, so, in order to feel the tongue itself moving, anchor the jaw by holding a pencil, or finger, in between the teeth. As suggested above, if the exercises are done silently, i.e. with no phonation, the articulatory movements can be felt more clearly.

Demonstrating different degrees of movement in the front to back dimension is a bit more tricky because of differences between different accents of Standard English. Speakers of British English and General American should be best off here, since for open vowels they will probably have all three positions: front, central and back. Speakers of British English can switch from the vowel in *part*, to the vowel in *putt*, to the vowel in *pat*. If possible, look in the mirror while doing it. It should be possible to feel and see which is the front vowel in this sequence, which is the central, and which the back.[2] General American speakers can get the same effect with *hot*, *hut* and *hat* – which is which?[3] Australian speakers can probably only experience front versus central: try *hut* and *hat* – which is the central vowel?[4]

In addition to tongue body placement, it was noted that the position of the lips is also important in vowel production. Lip position constitutes the third major parameter of vowel production/description, and is independent of tongue body position. (The independence is worth noting: it is possible to change the lip configuration independently of the position of the body of the tongue, and vowels differing solely in lip position are found in many languages, although Standard varieties of English are not one of them.) From the point of view of vocal tract squeezing, therefore, the mouth aperture constitutes an additional location where constriction can take place. In the production of a [u] vowel, for example, there is constriction of the tube at the mid point and also at the lips (this was shown in Figures 6.5 and 6.1).

Changing the shape of the lips alters the geometry of the aperture of the mouth opening. The most commonly found ways of doing this are by rounding and spreading the lips, and this parameter is called lip-rounding. Different lip configurations can be experienced by alternating between the 'ee' vowel in *heed*, and the vowel in *hoard*. (This alternation also involves tongue-body changes in backness and slight changes in height, but the changes to the lips should be easily felt.) Again, silent alternation will enhance the ability to feel the changes involved.

Height, backness and rounding

Tongue body placement in vowel production is described in traditional articulatory phonetics in terms of two dimensions: *Height* and *Backness.* Each dimension is divided up into smaller locations for convenience.

Four locations, or values, are commonly recognised on the height dimension: *high* (or *close*); *half-close* (or *close-mid*); *half-open* (or *open-mid*); and *low* (or *open*). Thus

---

2 The vowel in British English *pat* is front, the vowel in *putt* is central, and the vowel in *part* is back.
3 General American *hot* has the back vowel, *hut* is central and *hat* has the front vowel.
4 Australian *hut* is central, *hat* is front.

describing a vowel as half-close means that the body of the tongue is assumed to be fairly high in the mouth. Three self-explanatory locations serve on the Backness dimension: *front*, *central* and *back*. Thus a vowel described as *front half-close* would mean that the body of the tongue is assumed to be fairly high in the front of the mouth. (This would be very like the vowel in the word *bid*). A *low back vowel* would mean that the body of the tongue is assumed to be low in the back of the mouth. This would be very like the vowel in American *hot* or British English *heart*.

Although various configurations of the lips are possible, phonetic description is usually content with a binary specification of a vowel as either *rounded* or *unrounded*, so descriptions like *high back unrounded* or *low back rounded* will be found.

Secondary parameters of vowel description

In addition to the three major parameters of vowel production (Height, Backness, Rounding), there are several minor parameters. Of these, *nasalisation*, *length* and *dynamicity* are the most important.

Nasalisation

The idea behind vowel nasalisation is simple, and has already been mentioned in the section dealing with the nasal cavity. Normally, vowels are produced with the oral and pharyngeal cavities acting as resonators. We have seen that the soft palate governs access to the nasal cavity. Lowering the soft palate in vowel production means that the additional resonances of the nasal cavity will be combined with those from the pharynx and oral cavities, and a set of very different sounding *nasalised* vowels can be produced. Traditional phonetic description uses the label of nasalised (as opposed to *oral*) for a vowel that sounds as if it has been produced with the soft palate down. One nasalised vowel that is common in English can be heard at the end of some people's pronunciation of the word *restaurant*. Other examples of nasalised vowels were given above.

The implementation of nasalisation in reality is considerably more complex than this, and digressing a little to explain how offers the chance to say something informative about speech perception. Because the tongue body is attached, by muscles, to the soft palate, a low position of the tongue body, as in open vowels, tends mechanically to pull the soft palate open (this can be seen by comparing soft palate height in the [i] and [ɑ] vowel X-rays, Figures 6.11 and 6.12 above). This introduces the acoustic effects of nasalisation. However, because a certain amount of acoustic nasalisation is expected to occur automatically with open vowels as a function of the mechanical coupling between tongue body and soft palate, the perceptual mechanism ignores it when deciding whether to perceive the vowel as nasalised or not. Consequently, an open vowel requires a relatively much larger soft palate opening for it to sound nasalised, even though it may clearly be acoustically nasalised.

This demonstrates that speech perception, as opposed to the perception of non-speech sounds, is mediated by the brain's knowledge of how the vocal tract operates. This is known as Speech Specific Perception – perceiving sounds as if they had been produced by a vocal tract – and is crucial to the correct perceptual decoding of speech acoustics. This is addressed further in Chapter 9.

Description of speech sounds

Monophthongs, diphthongs and triphthongs

Consider the vowel in the British English[5] word *kit*, transcribed [ɪ]. Only a single vocalic target, i.e. the ensemble of tongue and lip position, is involved. To produce this word, the tongue moves from the position for the initial velar plosive consonant [k] to the (front half-close) position for the vowel, and then to the position for the final alveolar plosive consonant [t]. The lips remain unrounded. (There is also the appropriate synchronised vocal cord activity at the larynx, of course: the cords have to be apart for the two voiceless consonants and together and vibrating for the vowel.) Single vocalic targets like the one in *kit* are called monophthongs.

Now contrast the word *kite*. Here there is a complex vocalic target, transcribed [aɪ], involving tongue (or jaw), and possibly also lip movement. The tongue body moves away from an open central position as in [a] towards, but not reaching, a close front position as in [i]. (The vowel in *kit* is used to transcribe the position it actually reaches.) The lips may also become slightly more spread. A more complex example, because it involves clear changes in lip-rounding, would be the vocalic target in the word *boy*, transcribed [ɔɪ]. In *boy*, the vowel again involves tongue body movement: from a low back towards a high front position. But there is also a clear change in lip-rounding, the lips changing from rounded in the first vocalic target to unrounded for the second.

The involvement of two vocalic targets *in a single syllable*, as in *boy* and *kite*, is called a diphthong. It is possible to get even more complex, with three targets involved in a single syllable, and such cases are called triphthongs. One example would be the word *why*. In this word the tongue body and lip movement is from high back rounded for the [w] towards low central unrounded for the [a] and thence towards high front [i]. A similar trajectory occurs in the Thai word 'duây' [duɛɪ] *together*, although the time spent on the targets is different. The word *fire* in British and General Australian English is considered by some to have a triphthong, the targets of which the reader might try to specify in terms of tongue height and backness, and lip rounding.[6]

There is not an accepted term to describe the dimension within which such vowels (i.e. monophthongs, diphthongs and triphthongs) contrast: a possible one would be dynamicity.

Vowel length

A speaker obviously has control over how long they wish to hold a particular vowel target – in terms of the above description, how long they wish to maintain their tongue body and lips in a particular position before moving to the next articulatory target. It is no surprise therefore that length is a parameter in vowel production, with some languages having both long and short vowels. Japanese is a good example, with systematic contrasts like 'toki' [tɔki] *time* and 'tooki' [tɔːki] *pottery*; 'teki' [tɛki] *enemy*, 'teeki' [tɛːki] *transport pass*. As can be seen the symbol [ː] is used to transcribe a long

---

5 By British English I mean 'Standard Southern British'. This is characterised as 'an accent of the South East of England which operates as a prestige norm there and (to varying degrees) in other parts of the British Isles and beyond' (IPA 1999: 4).
6 The tongue body and lip movement is from open central unrounded [a] towards high front [i] to central mid schwa [ə]: [faɪə].

Forensic Speaker Identification

vowel. If it is necessary to transcribe a short vowel, [ˇ] is used, thus: [ă]. The terms long and short are also used in phonetic description.

Length is also a consonantal parameter, albeit a rarer one, and some languages like Japanese, Finnish and Tamil contrast long and short consonants.

English vowels

It would be nice to be able to describe the vowels of English in the same way as the consonants. However, there is too much phonetic variation between different varieties to do this, and it is therefore better for me to use the descriptive detail presented in the previous sections to describe one of the varieties I am most familiar with – Australian English. Australian accents, like most varieties of English, have rather complex vowel systems. Different varieties of Australian English are recognised, defined predominantly on differences in vowel quality. Here only one variety, so-called General Australian, will be described.

Australian vowels come in two groups: long and short. Moreover, the phonetic length distinction is reflected by a difference in the way the two sets of vowels are distributed in words. Whereas long vowels can occur at the end of words, short vowels cannot, and must be followed by at least one consonant. Thus there is a word *he*, with a long [iː] vowel, but a word ending in the short vowel in *hid* is not possible. From the point of view of this distribution, Australian diphthongs pattern with the long vowels in being able to occur word-finally, as for example in *hay, high, boy, here, there*, etc. This bipartite grouping of vowels according to a phonological basis is characteristic of many varieties of English (Wells 1982: 118–19; 168–78), and seems to be a typological characteristic of Germanic languages (i.e. languages descended from proto-Germanic). We look at the system of Australian short vowels first, then the long vowels and diphthongs.

Short vowels

There are seven short vowels in Australian English. They occur for example in the words *pit pet pat putt pot put* and in the first syllable of *potato*. The General Australian short vowels are shown in Figure 6.13. The phonetic symbols for the vowels have been placed on a conventional rhomboidal vowel chart with axes of height (vertically) and backness (horizontally). A word containing the vowel is shown in italics near to the symbol. Thus it can be seen that the vowel in the word *hid* has a relatively high and front location. (As explained above, this means that the body of the tongue for this vowel is assumed to lie relatively high and front in the mouth.) This vowel is unrounded, and the absence of rounding is encoded in the symbol [ɪ]: if this were a rounded vowel, it would have a different symbol (there is strictly speaking no part of the symbol which indicates the rounding status of the vowel: you have to learn which symbols indicate rounded vowels and which do not). Another transcriptional point is that I have used the symbol [a] to represent a central vowel. I mention this because in the system commonly used for describing vowels devised by the English phonetician Daniel Jones (called the Cardinal vowels) [a] is used to indicate a vowel which sounds front, not central (IPA 1999: 12).

Description of speech sounds

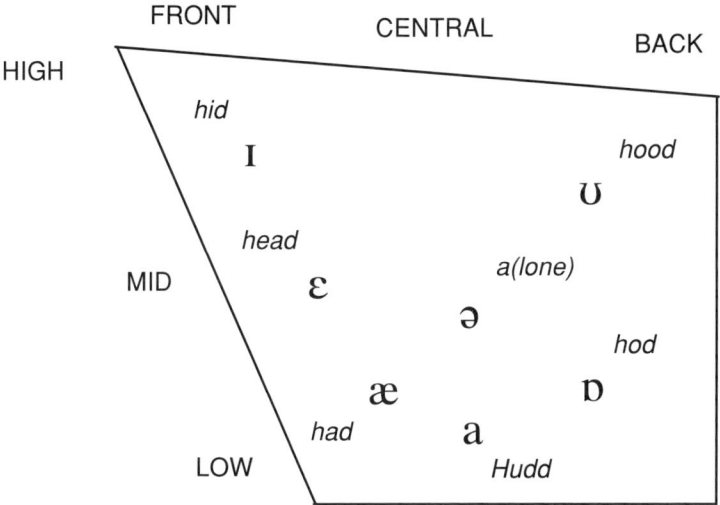

Figure 6.13 The seven short vowels of General Australian

The Australian short vowel configuration is fairly typical, with several degrees of height at front and back; back vowels being rounded; a low central vowel; and a schwa in the middle. (This last is of a different status to the other vowels, since it does not occur in stressed syllables).

Long vowels

There are five long vowels in General Australian, exemplified in the words *beat Bart bought Bert boot*. Their symbols, position and typical words are shown in Figure 6.14. The high vowels are shown with a superscript schwa onset, which means they typically have a slight diphthongal pronunciation. In contrast to the short vowels, this long vowel configuration is rather atypical, lacking both a high back vowel, and non-high vowels in the front, and only having one back vowel.

Diphthongs

It is rare for languages to have lots of diphthongs, but General Australian English has no less than seven. Many other varieties of English also have more than the usual number of diphthongs. Five of the General Australian diphthongs involve tongue body movement from a relatively open to a close position and are therefore called *closing diphthongs*. Two, with tongue body movement towards a central, schwa, position, are called *centring diphthongs*. The closing diphthongs are exemplified in the words *high, hay, boy, how* and *hoe*, the centering diphthongs in the words *here* and *hair*. Figure 6.15 shows the approximate trajectories of these diphthongs on the height–backness vowel chart. It can be seen, for example, that the tongue body in the diphthong in *high* moves from a low central position towards a high front position.

151

Forensic Speaker Identification

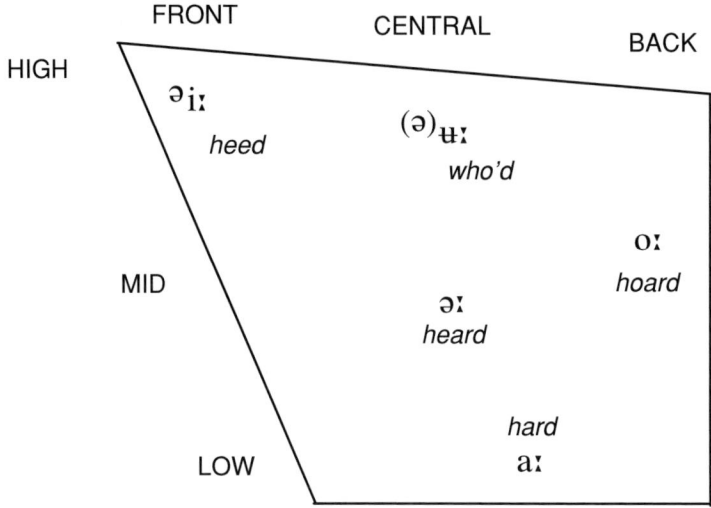

Figure 6.14 Long vowels in General Australian

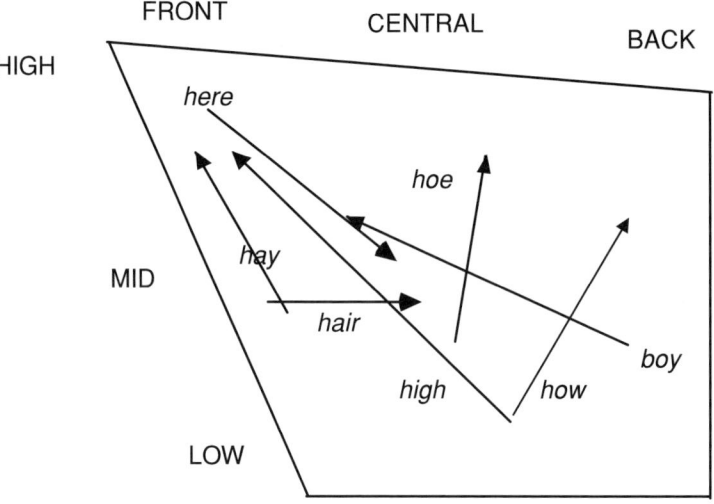

Figure 6.15 Diphthongs in General Australian

The componentiality of speech sounds

The examples given above illustrate an important point about speech sounds: *they have structure*. Speech sounds are not single, unanalysable wholes, but have the important property of componentiality (Lass 1984: 42). They can be both analysed and described as if made up of, or factorisable into, smaller parts. Thus the alveolar sounds [z] and [s] are not just different, but different by virtue of the fact that the former has the component of voice which is not present in the latter. Moreover, the same difference characterises many other pairs of sounds like the bilabial pair [v] and [f], or the nasals [m] and [m̥]. Or take the difference between the stops [p] and [t]. The former is bilabial, the latter alveolar, and this recurrent difference can be extracted from the voiced stops [b] and [d], the trills [ʙ] and [r], the fricatives [Φ] and [s] etc. As far as vowels are concerned, [i] and [æ] share the property of frontness, for example, and [o] and [u] the property of rounding.

Speech sounds can thus be analysed and described in terms of their components: the [d] in *do* is appropriately described as a *voiced apico-alveolar stop* after the components [voice], [apico-] [alveolar] and [stop], which also respectively occur in, *inter alia*, [b], [l], [n] and [k]. These speech sound components are termed features, and one part of linguistics, called feature theory, has been concerned with determining what the features are that serve to signal the difference between the words of individual languages (Ladefoged 1997: 589–92).

Note, finally, that the structure of speech sounds is the result of concerted activity from many different parts of the vocal tract. In [d], the voicing is contributed by vocal cord activity, and the apicality comes from the active articulator – the tip of the tongue. The alveolar part of [d] is the passive articulator, and the stop part is a manner specification, involving many other structures, including for example the soft palate, which has to be up for stops.

Summary: Consonants and vowels

The sections above have described the main parameters in the production of vowels and consonants. Consonants are produced by appreciably obstructing the flow of air through the vocal tract. They are described by reference to where in the vocal tract the obstruction is located (*place of articulation*), and what kind of a barrier is presented to the air-flow (*manner of articulation*).

Vowels are produced by modifying the shape of the supralaryngeal vocal tract in non-radical ways. By primarily moving the tongue body and changing the lip aperture, the vocal tract is effectively squeezed at differing locations along its length, creating vocalic resonators of different dimensions. By lowering the soft palate an additional set of nasalised vowels can be produced, and vowels can also be produced both long and short. More than one vocalic target can be produced in a syllable, giving rise to diphthongs and triphthongs. These modifications can be described phonetically in terms of the vowel's *height*, *backness*, *rounding*, *nasalisation*, *length* and *dynamicity*. Speech sounds are analysable and describable in terms of smaller components called features.

Suprasegmentals: Stress, intonation, tone and pitch accent

In the previous sections, the production of vowels and consonants was described. These sounds are referred to as segmental, the metaphor being that they are strung together like segments in a word. A second major class of speech sounds – that

Forensic Speaker Identification

containing *intonation, stress, and tone* – is called *suprasegmental*, the inference being that they in some sense exist on top of the segmental sounds. This inference is somewhat misleading because intonation, stress, and tone are mostly associated with laryngeal activity, which, if you are speaking standing up, even in Australia, takes place below the other speech-producing organs.

Stress has to do with the relative prominence of syllables within a word. Tone has to do with the use of different pitch shapes to signal word identity, and intonation has to do with the use of pitch to signal features of utterances longer than the word. These, together with another suprasegmental category called pitch accent, will be discussed and exemplified below.

The main characteristic of suprasegmentals – of tone, stress and intonation – is that their realisation is variable with respect to segments (Nolan 1983: 32). Thus the *same* rising intonational pitch of a question is distributed over one (vowel) segment in one syllable in the word *two?* but over two syllables and several segments in the word *ninety?* and over three syllables and even more segments in the word *seventy?* This state of affairs can be represented as follows:

```
  two?      ninety?    seventy?
   tʉ       naɪnti      sɛvənti
   /\        / |         |  |
   LH        L H         L  H
```

The rising intonational pitch of the question is represented by a sequence of the two symbols L and H. These stand for a low pitch target and a high pitch target respectively. The lines show how the pitch targets are associated with the segmental material. Note how the same suprasegmental intonational LH sequence, or rise in pitch, is present in all three words, and is independent of the number of segments, or syllables, in the word. Another property of suprasegmentals that will become evident is that they are traditionally described in more clearly perceptual terms, like stress and prominence. This is in contrast to the inferred articulatory reality of the description of segmental production.

Stress

In English words of more than one syllable, one of the syllables usually sounds more prominent for native speakers than the others. For example, in *Boston, Michigan, Canterbury* it is the first syllable that sounds more prominent: BOSton, MICHigan, CANterbury. In *New York, Manhattan, Connecticut*, the second syllable sounds prominent; in *Tennessee* and *Chattanooga* the third; and in *Kalamazoo* the fourth. Table 6.3 shows these place names together with some other words displaying the same stress patterns. Looking along the rows, it can be seen that in English, the stress can fall on any syllable in a word.

This prominence is put to many different linguistic uses. It can be contrastive, i.e. serve to distinguish two words that are otherwise (i.e. segmentally) the same, as for example in the (British English) pair be**low** and **bil**low, or in the pairs per**mit** (verb)

Table 6.3 Examples of English words with stress on different syllables

	Final	Penultimate	Antipenultimate	Preantipenult
2-syllable word	*New York* *extreme*	*Cambridge* *puddle*	–	
3-syllable word	*Tennessee* *chimpanzee*	*Manhattan* *forensics*	*Iowa* *syllable*	–
4-syllable word	*Kalamazoo* *provocateur*	*Massachusetts* *intonation*	*Saskatchewan* *asparagus*	*Canterbury* *algorithm*

and **per**mit (noun). In the latter pair, an additional function of the prominence is to signal differences in word class: cf. **in**sult, **im**plant (noun) vs. in**sult**, im**plant** (verb). When words are said together, moreover, the location of prominence can signal other grammatical differences, for example the difference between a phrase (black **bird**, white **house**), and a (compound) noun (**black**bird, **White** House). When many words are said together, the recurrent prominence peaks contribute a rhythm. In all these examples, the prominence is functioning linguistically, and such linguistically significant prominence is termed *stress* (Lehiste 1970: 119–20). Stress is considered to be a property of syllables (in that syllables can be stressed or unstressed), and is often notated by a superscript tick before the stressed syllable, thus: ['ɪmplant] *implant* (noun), [ɪm'plant] *implant* (verb).

One of the features of stress is that it is culminative – only one syllable is singled out per word to carry the main stress (Hyman 1977: 38–9). Possibly because of this, stress is often conceived as a kind of accent (in the musical sense, as in 'the accent is on the third beat'), and the term *stress-accent* is often encountered. In any case, English is often characterised as a *stress language* or *stress-accent language*. Other examples of stress languages are Modern Standard Arabic, Russian and German.

Although most naive native listeners feel that stress has something to do with loudness (as indeed it can, but not necessarily so), the main perceptual dimension in which stress is signalled is pitch (Lehiste 1970: 153). Compare the noun **in**sult with the verb in**sult**, both said as a statement. In both words the stressed syllable has a higher pitch. In English, however, it is not high pitch *per se* that signals stress. It is a greater pitch deviation, either up or down, from the pitch on surrounding unstressed syllables. To see this, compare the noun **in**sult said as a statement, with the same word said as a question: '**in**sult?' In the latter, the pitch of the stressed syllable is no longer higher than the adjacent unstressed syllable, but lower.

Another correlate of stress is length: stressed syllables are often, though not invariably, longer than unstressed syllables. Segmental quality is another: vocalic targets, especially, tend to be most closely approximated on stressed syllables. The opposite of this also occurs, namely that vowels tend to be less differentiated in unstressed syllables. This is also reflected in the large number of schwa ([ə]) vowels in English unstressed syllables. Compare, for example, the vowel quality in the second, unstressed syllable of the word in**ton**ation [ɪntə'neɪʃən] with that in the stressed second syllable in in**tone** [ɪn'toʊn].

Forensic significance: Stress

This last characteristic is one of the reasons why stress is forensically important. Vowels in unstressed syllables tend to be subject to greater deformation by surrounding sounds, whereas sounds in stressed syllables tend to be more stable: they in fact cause that deformation. Thus there tends to be an increase in within-speaker variation in parameters associated with unstressed vowels, and some types of speaker recognition give better results if they restrict themselves to the acoustic qualities of vowels in stressed syllables. Forensically, one would certainly not assume that parameters are comparable across stressed and unstressed syllables, and stress is therefore one category in terms of which comparability across samples is defined.

Intonation

A familiar use of pitch for the English speaking reader will be to signal the difference between a statement and (some types of) question. For example, the sentence *Are you going?* in its most basic version will have a rising pitch on *going* (probably realised by a low pitch on *go-* and a low rising pitch on *-ing*). In the statement *He's going*, on the other hand, *going* will have a falling pitch.

Pitch can also signal boundaries between syntactic units. For example, the sentence *My neighbour said the prime minister was an idiot* has two readings, depending on whether my neighbour or the prime minister is doing the talking (Lyons 1981: 61). If the prime minister is doing the talking he will be in a clause surrounded by a syntactic boundary: *My neighbour, said the prime minister, was an idiot*. If it is my neighbour who is doing the talking, there will be no boundary signalled: *My neighbour said the prime minister was an idiot*. Although the difference is marked in writing with punctuation, it is signalled in speaking by intonational pitch. This difference, then, reflects syntactic structure – the way words go together – and will be signalled by pitch.

The use of pitch to signal syntactic information of this kind (question vs. statement; sequences of coherent linguistic structure) is called *intonation*. Intonation is also usually taken to include not just the signalling of syntactic information by pitch, however, but also the signalling of affectual content, for example the emotional attitude of the speaker (bored, excited, angry, depressed). All languages have intonation in these senses.

Intonation has been the focus of a lot of phonetic and linguistic research in the past 20 years, and a considerable amount is now known about it, both as a general phenomenon and how it is implemented in individual languages (Ladd 1996). Good descriptions exist for several languages; see, e.g., Hirst and Di Cristo (1998).

Examples of intonation

Some examples of common Australian intonational pitch patterns are given in Table 6.4 using the disyllabic word *hello*. Many will also be applicable to other varieties of English. Since the use of current methods of intonational description for

Description of speech sounds

Table 6.4 Examples of intonational variation on Australian *hello*

	Name	Transcription	Example
1	Rise	[L.LH]	Answering the phone
2	Fall	[L.HL]	Reading the word off the page
3	Rise–fall	[L.LHL]	Smug, conceited, sleazy
4	Fall–rise	[L.HLH]	Questioning if someone is there (insistent)
5	Call	[L.HH!]	Questioning if someone is there (neutral)
6	Spread fall–rise	[HL.LH]	Greeting (enthusiastic)
7	Spread call	[H.H!]	Greeting (neutral)

these data would introduce unnecessary complications, the pitch patterns are simply described in terms of contours (rise, fall, etc.), and transcribed with Hs and Ls indicating high and low pitch. Pitch contours then result from a sequence of these; for example, a fall in pitch is transcribed as [HL]. A full-stop shows a syllable boundary, and an exclamation mark shows a slight drop, or *downstep* in pitch.

Table 6.4 shows that the simplest forms are a fall ([HL]) and rise ([LH]) on the stressed syllable, with a low pitch on the unstressed. The latter *hello*, with [L.LH], is a very common way of answering the telephone; the former might be used if one were reading the isolated word off the page. Using a rise–fall pitch on the second syllable ([.LHL]) sounds smug or conceited: it is the *hello* of a sleazy male spying a good-looking female. Two forms can be found for questioning whether someone is there. One has a fall-rise ([.HLH]), and one a downstepped high ([.HH!]) on the stressed syllable. Both have low pitch on the unstressed syllable. Interestingly, if these two pitch patterns on the stressed syllable are spread across both syllables, i.e. [HL.LH] and [H.H!], a greeting intonation is signalled.

Intonation and stress

In English and many other languages, intonation and stress interact in the sense that stress provides the framework for the realisation of intonation (Cruttendon 1986). A simplified account of what this means can be given by taking words with stress on different syllables, as in Table 6.3, and saying them with two contrasting intonational tunes such as a rising [LH] to signal a question, and a falling [HL] to signal a statement. For example: *Kalamazoo?* (= did you say Kalamazoo?); *Kalamazoo* (= Yes, that's right, Kalamazoo). Note how the intonational pitch is distributed over the words with different stress locations. This is shown in Table 6.5, with four words of four syllables with different stress patterns. These patterns are shown, using the conventional notations of sigma (σ) as a syllable, and ' to indicate stress, both above and below the individual words. The falling intonation of the statement [HL] is shown on top, and the rising intonation of the question [LH] is shown beneath.

Table 6.5 shows that the intonational pitch is aligned with the word quite regularly, with reference to the stressed syllable. The first tone of the intonational sequence is aligned with the stressed syllable of the word, and the second is aligned with its final

157

Forensic Speaker Identification

Table 6.5 Alignment of intonational pitch with respect to the stressed syllable of a word

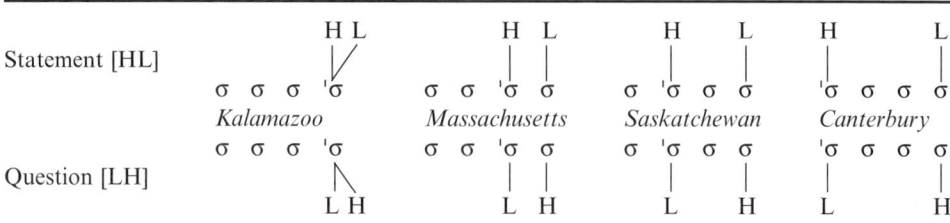

syllable. The pitch transition between the two tones – the fall for the statement and the rise for the question – is realised, or spread, over the intervening segmental material. Thus in *Kalamazoo*, with final syllable stress, the pitch fall for the statement or the pitch rise for the question have to take place on the final syllable, whereas in *Canterbury*, with initial stress, the pitch fall/rise can take place over the whole word. It is in this sense that the stress serves as a framework upon which to hang the intonational pitch.

Tone

A rather large subset of the world's languages – about a third – use pitch to signal the identity of words (Hombert 1977: 21). As already illustrated with Cantonese data in Chapter 3, in these languages each word (or smaller meaningful unit called a morpheme) has one of a small number of characteristic pitch shapes, called *tones*. This use of pitch is called *tone*, and languages with tone are called *Tone Languages*. Many of the world's tone languages, for example Chinese, Burmese, Thai, and Vietnamese, are found in East and South East Asia. Sub-Saharan Africa is the other tonal hot-bed. There are also a few so-called restricted tone languages in Europe and India, for example Serbo-Croatian, Swedish and Panjabi.

Table 6.6 gives some examples of tone in three different languages: Modern Standard Chinese (also called Putonghua, or Mandarin), the Southern Thai dialect of Pakphanang (Rose 1997: 191–2), and Zulu (Rycroft 1963). The languages are shown in the leftmost column. The two main parts of a word are its meaning and its sound. The meaning of the words is given in the second left column; the words' sounds are in the two rightmost columns, and these are divided into segmental sounds (i.e. consonants and vowels) and pitch. In the Chinese and Thai examples, each word consists of one syllable only. In the Zulu examples, each word consists of three syllables, the first of which is a prefix *i* indicating that the word belongs to a particular gender (like masculine and feminine in French, but more complicated). In order to show the alignment of syllables and pitch, syllable boundaries are shown with a period. Thus in the word *feather* the first syllable *i* has a high pitch, the second syllable *pa* has a falling pitch and the last syllable *pu* has a mid pitch.

Table 6.6 shows that Modern Standard Chinese has four contrasting tones. One tone is realised by a high level pitch, one by a rising pitch, one by a low dipping

Description of speech sounds

Table 6.6 Examples of tones in different languages

Language	Word		
	Meaning	Sounds	
		Segmental sounds	Pitch
Modern	*mother*	ma	high level
Standard	*numb*	ma	rising
Chinese	*horse*	ma	low dipping
	to swear at	ma	high falling
Southern	*leg*	ka	very high convex
Thai	*to kill*	ka	upper mid level
Dialect	*crow*	ka	mid convex
	mark	ka	lower mid level
	thatch grass	ka	mid falling
	value	ka	low dipping
	trade	ka	low falling
Zulu	*tea*	i.ti.ye	high . high . low
	bone	i.ta.mbo	high . fall . low
	feather	i.pa.pu	high . fall . mid
	jersey	i.je.zi	high . low . mid
	duck	i.da.da	high . low . low

pitch and one by a falling pitch. Thus a word consisting segmentally of [ma] said with a high level pitch means *mother*, but if the pitch on [ma] is dipping, a completely different word is signalled: *horse*. Modern Standard Chinese is a fairly typical Asian tone language in both the number of tones and their distinctive realisation – the pitches are all fairly different, so the tones are easy to recognise. Tone languages can get more complicated, both in number and realisation of tones, however. The Southern Thai dialect of Pakphanang in Table 6.6 is a good example. It has seven contrasting tones, the pitches of two of which (upper and lower mid level) are quite close and more difficult (for a non-native listener!) to distinguish.

Whereas the Thai and Chinese examples featured segmentally minimal or near-minimal forms, the Zulu words in Table 6.6 are not minimally distinguished by the tonal pitch: due to the greater segmental complexity of Zulu words and syllables, it is far more difficult to find words that differ solely in pitch. However, each word still has its own appropriate pitch. Zulu has fewer contrastive tonal pitch shapes (high, low, falling) than Thai and Chinese, and there are restrictions on their occurrence, but tone interacts with consonants to produce other pitch shapes.

It can be seen in all the examples from Table 6.6 that in a tone language *pitch features are just as much part of the sound of the word or morpheme as the segmental features*. For example, in the Modern Standard Chinese word *to swear at* the falling pitch of the tone is just as much part of the word's sound as the bilabial nasal [m] or the vowel [a].

159

Forensic Speaker Identification

This is different from the intonational use of pitch in English, where a falling pitch on, say, the word *yes* is not part of the inherent sound of the word *yes*. A tone language is thus more properly characterised as a language where pitch is assumed to be part of the sound of words (or smaller meaningful parts of words called morphemes).

Pitch accent

One further linguistic use of pitch is usually recognised: pitch accent. Modern Standard Japanese is a prototypical *pitch-accent language*. Pitch accent, like tone, has to do with pitch features that are in some sense part of the word. Some examples from Modern Standard Japanese will show what this means.

Japanese words can have a single accent, or no accent. If they have an accent, it can be located on any syllable and is *signalled by a fall in pitch onto the following syllable*. So in the disyllabic word ha.shi H L *chopsticks*, with a high pitch on the *a* and low on the *i*, the fall in pitch signals an accent on the syllable *ha*. This can be notated iconically with a bracket thus: ha˧shi. In the word ko.ko.ro L H L *heart*, the location of the pitch drop shows the accent is on the second *ko*: koko˧ro. These words also illustrate the rule that, unless the first syllable carries a pitch accent, its pitch will be low. If that is the case, then the pitch on the following syllable will be high, irrespective of whether it carries an accent or not.

Since location of accent is shown by a drop in pitch onto a following syllable, if the final syllable carries a high pitch – for example ha.shi L H *bridge*, or sa.ka.na L H H *fish* – there is no way to signal whether an accent is present or not. The presence of an accent only becomes clear if more material follows.

To illustrate this, the grammatical postposition *ga* can be added (marking the word as subject). Thus in the phrase ha.shi ga L H L *bridge* subj., the pitch falls onto the *ga* particle and therefore this word does have an accent on the final syllable: hashi˧. In the phrase ha.shi ga L H H *edge* subj. there is no fall in pitch onto the *ga*, and so there is no accent on the word: hashi. If the words *edge* and *bridge* are said without the following syllable, they both have L H pitch and are homophonous. This 'hidden' word-final accent is a nice demonstration of the abstract complexity of sound in language: two words can sound the same, under some circumstances, but be clearly differentiated under others.

The examples above show that pitch accent is both like stress and like tone. It is like stress in that, if there is an accent, it is culminative: there is only one per word. It is like tone in that its pitch realisation is an invariant property of words. This also means that there is not such a drastic influence from intonation on the pitch realisation of words. For example, in Japanese there is no change in the realisation of the pitch-accent pitch pattern under intonational difference between question and statement (Abe 1972). In the question: kore wa ha˧shi desu ka *Are these chopsticks?* the word ha˧shi *chopsticks* still retains its H L pitch realisation (seen, e.g. in the statement: kore wa ha˧shi desu. *These are chopsticks*). In English the pitch on *chopsticks* would normally change from H L (chop H sticks L) to L H (chop L sticks H) in a question, as demonstrated in the intonational section above.

Unlike tone, however, the pitch of the word in a pitch-accent language is predictable from a diacritic, like the '˧' above. So it is not the pitch itself, but the location of

Table 6.7 Suprasegmental categories in different languages

Suprasegmental category	Language			
	Mandarin, Thai, Zulu	Japanese	English, German, Russian, Arabic	Korean
Intonation	×	×	×	×
Tone	×			
Stress (-accent)	×		×	
Pitch accent		×		

the accent that is considered as part of the inherent sound of the word. If we know the location of the accent, the pitch of the whole word can be specified. This is considered to be the main difference between a tone language and a pitch-accent language (McCawlcy 1978).

Typology of suprasegmentals

The way the three suprasegmental categories of tone, intonation and accent (both pitch accent and stress accent) might relate in languages is schematised in Table 6.7. This table shows that some languages are tonal, some have pitch accent, and some are neither, but all have intonation. Of the non-pitch-accent languages, some have just stress accent, like English, and some have both stress accent and tone. Examples of these are Modern Standard Thai, and Zulu. Korean is an example of a non-pitch-accent language lacking both tone and stress accent.

Since intonation, tone and pitch accent are all predominantly realised in the dimension of pitch, it is clear that pitch in many languages is a highly complex medium, simultaneously realising several suprasegmental linguistic systems. In Modern Standard Chinese, for example, or Modern Standard Thai, pitch features will simultaneously realise tone, intonation and stress (Kratochvil 1968: 35–47, 1998; Luksaneeyanawin 1998).

Forensic significance: Suprasegmentals

Tone, intonation, pitch accent and stress accent are forensically important because they are primarily signalled by pitch. As explained above, pitch is ultimately related, via its acoustical correlate of fundamental frequency, to the mass and length of the speaker's vocal cords. Females, for example, because they have shorter and less massive cords, will have overall higher fundamental frequencies than males. Thus part of the speaker's anatomy will be encoded in the suprasegmentals, which consequently carry the imprint of the speaker, and have individual-identifying potential.

Moreover, the acoustical correlate of pitch, called fundamental frequency (F0) is one of the more robust acoustical parameters: it is transmitted undistorted under most common adverse conditions, e.g. telephone conversations, high ambient noise. This has the interesting consequence that in tone languages it is often possible to identify the tone on a particular syllable when the segmental content is obliterated and to form a hypothesis as to the possible content of the recording. The measurement of fundamental frequency is also relatively unproblematic, although sometimes laborious.

However, it is clear from the account of suprasegmentals given above that it is necessary to understand their complexity and interaction in order to be able to interpret fundamental frequency data forensically. It will be seen that *the understanding of the linguistic structure encoded in speech is in fact a general forensic-phonetic principle*, and therefore it is worth illustrating here.

Consider comparing two forensic speech samples with respect to fundamental frequency. It would be necessary to ensure that the samples were comparable with respect to factors that are known to influence F0. In a tone language this would include tonal composition. It would not be correct, for example, to compare two sample sentences with respect to their F0 if one sentence consisted of words with predominantly high tones, and the other consisted of words with predominantly low tones. The two sentences would clearly differ in F0, with the first being overall higher than the second, but this could not be attributed automatically to provenance from different speakers. In the same way, intonation with respect to sentence types would also need to be considered: a sample containing predominantly questions and commands would have a higher F0 than a sample containing declarative utterances.

It can be seen from these simple examples that *it is important to have knowledge of factors affecting variability in the acoustic parameters used to compare forensic samples*. The list will not be restricted to linguistic factors either: it is possible, for example, that speakers will use higher pitch to make themselves heard against a noisy background.

Timing of supralaryngeal and vocal cord activity

The sections above have shown how the vocal cords and the SLVT independently contribute to the production of speech sounds. For example, different tones are produced by making the cords vibrate at different rates, whereas different consonants and vowels are made by obstructing the SLVT in different ways and differentially changing its shape.

Although they contribute independently, the production of speech requires precise alignment of the activity in these two domains. If you have to produce a voiceless alveolar fricative like [s], it is obviously important that the laryngeal activity – pulling the cords apart to make sure the sound is voiceless – is timed to coincide with the supralaryngeal activity – using the tongue to create a narrow channel at the alveolar ridge to cause turbulence. Since these gestures typically only last a few hundredths of a second, the degree of precision needed in their timing is clearly very high: a slight misalignment and a different sound might be produced.

Vocal cord and SLVT activity thus have to be timed precisely with respect to one another in order for the intended sound to be produced. It can also be noted that it is precise *differential* timing that is often important. Different speech sounds are made by differentially timing the activity in the two domains. Two aspects of this differential timing that are commonly exploited across languages can be used to illustrate

Description of speech sounds

this. One is called *voice onset time*, or VOT (Lisker and Abramson 1964); the other is production of tone and intonation.

Voice onset time

Three different types of consonant can be made by exploiting the differential timing of vocal cord and supralaryngeal activity. These three types are traditionally called 'aspirated', 'voiceless unaspirated' and 'voiced'. Thus we find consonants described as 'aspirated bilabial stop', or 'voiceless unaspirated alveolar affricate' or 'voiced velar fricative'.

To explain what is meant by voice onset time, we take a bilabial stop. As explained above, this involves three temporally defined events: the lips coming together to completely block off the air-flow, a short hold phase during which the oral air-pressure builds up behind the bilabial closure, and then the abrupt release of the closure venting the overpressure. It is the release of the closure that is the important supralaryngeal event for VOT. The important event for the vocal cord activity in VOT is the instant when the cords come together to start vibrating. This is called voice onset, and the difference in time between the release of supralaryngeal stricture and onset of voice is called the voice onset time.

The idea, then, is that these two events, the onset of vocal cord vibration and the release of supralaryngeal stricture, can be differentially timed to produce three different types of consonant. The production of the three types – *aspirated, voiceless unaspirated, and voiced* – is described below.

Aspirated consonants

If vocal cord vibration starts a long time after the release of stricture (a long time in articulatory phonetics is about ten hundredths of a second), an aspirated consonant is produced. Aspirated stop consonants occur in English at the beginning of the words *pool, tool* and *cool*. They are normally transcribed with a superscript *h*, thus: [pʰ tʰ kʰ], and [kʰ] would therefore be described as an aspirated velar stop. If we chose to transcribe VOT details, *pool, tool* and *cool* could be transcribed phonetically as [pʰʉɫ tʰʉɫ kʰʉɫ]. Aspirated consonants are said to have *VOT lag*, because the onset of voice is timed to lag the release of stricture.

Voiceless unaspirated consonants

If vocal cord vibration starts at about the same time as the release of supralaryngeal stricture a voiceless unaspirated consonant is produced. Voiceless unaspirated stop consonants occur in English at the beginning of words after *s*: *spool, stool* and *school*. They are normally transcribed thus: [p t k], and [t] would therefore be described as a voiceless unaspirated alveolar stop. *Spool, stool* and *school* could be transcribed as [spʉɫ stʉɫ skʉɫ]. Voiceless unaspirated consonants are said to have *coincident VOT*, because the onset of voice is timed to occur at about the same time as the release of stricture. The reader should compare how they pronounce the bilabial stops in *pool* and *spool* to verify that they are indeed not the same sound. The *p* in *pool* will be accompanied by what feels like a puff of air, which is lacking in the *p* in *spool*. The same difference obtains for the alveolar stops in *tool* (puff) and *stool* (no puff), and the velar stops in *cool* (puff) and *school* (no puff).

Voiced consonants

If vocal cord vibration starts before the release of stricture, a voiced consonant is produced. Voiced stop consonants can occur in English at the beginning of words like *boo*, *do* and *goo*, especially if a vowel or nasal precedes, thus: *a boo, I do, some goo*. They are normally transcribed thus: [b d g], and [g] would therefore be described as a voiced velar stop. Thus *a boo* [ə **bʉ**], *I do* [aɪ **dʉ**], *some goo*[səm **gʉ**]. Voiced consonants are said to have *VOT lead*, because the onset of voice is timed to occur before the release of stricture. The reader should compare how they pronounce the bilabial stops in *a boo* and *pool* and try to feel the vocal cord vibration in the *b*.

For reasons to be described in the following chapter, the difference between voiced and voiceless unaspirated stops is not very easy to hear for phonetically naive speakers of English. Thus most will not at first be able to hear the difference between nonsense syllables like [bi] with a voiced stop and [pi] with a voiceless unaspirated stop. That they are indeed phonetically different sounds is shown by languages that use all three VOT types to distinguish words. Probably the best-known example of this is Thai. Table 6.8 gives some Thai word triplets minimally distinguished by the VOT in word-initial stops. It can be seen that the difference between the words for *cloth*, *aunt* and *crazy* lies solely in the stop VOT. For Thai speakers there is all the difference in the world between a [b] and a [p], a [d] and a [t].

Although voice onset time is most commonly thought of in conjunction with stops and affricates as illustrated above, it can also apply to other consonants. For example the difference between a voiceless and voiced alveolar fricative like [s] and [z] can be understood as one of coincident VOT versus VOT lead. The third VOT value, VOT lag, can also be found in fricatives, although it is much rarer: voiceless aspirated fricatives, e.g. [s^h], occur in Burmese for example (Ladefoged 1971: 12).

Contrastive pitch

Another example of differential timing between laryngeal and supralaryngeal activity occurs with contrastive use of pitch in suprasegmentals. Here is an example from English intonation. Try saying *hello* in a matter-of-fact way with a low pitch on the

Table 6.8 Examples of three-way voice onset time (VOT) contrasts in Standard Thai stops

	Aspirated/ VOT lag		Voiceless unaspirated/ coincident VOT		Voiced/VOT lead	
Bilabial stop	p^{h}aː	*cloth*	paː	*aunt*	baː	*crazy*
Alveolar stop	t^{h}am	*to do*	tam	*to pound*	dam	*black*

Description of speech sounds

first syllable and a high falling pitch on the second (this was the kind of intonation transcribed as L.HL in Table 6.4), thus:

Now say *hello* in a smug self-satisfied way with a rise–fall pitch on the second syllable: (keep the pitch low until you start the 'o', then rise and fall). This kind of intonation was transcribed as [L.LHL] in Table 6.4.

Both these hellos involve a change from a low pitch [L] to a high pitch [H] to a low pitch [L] thus:

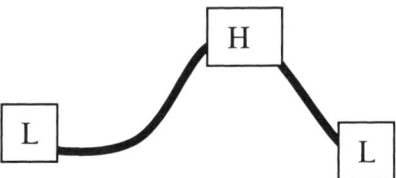

Pitch is the result of rate of vocal cord vibration, therefore the [LHL] pitch sequence in both is being produced by controlling this rate – first low, then increasing to high, then decreasing to low.

The main difference between these two intonations on *hello* is that the change in vocal cord vibration rate from low to high is being differentially timed with respect to supralaryngeal activity, in particular the articulation of the [l]. To signal the *hello* with the low–high fall intonational pitch, the change from a low to high vocal cord vibration rate starts quite early and takes place mainly during the [l], so that by the time the second vowel starts, the rate is already quite high. To signal the *hello* with the low–rise–fall pitch, the low vibration rate is prolonged and the change from low to high vibration rate starts quite late, often not until the end of the [l]. The same phenomenon can occur in tone languages with a contrast between falling and convex pitch. For example in the Chinese dialect of Zhenhai [weɪ le] with low pitch on the [weɪ] and falling pitch on the [le] means *to return*, whereas [weɪ le] with low pitch on the [weɪ] and convex pitch on the [le] means *able to come* (Rose 1989: 61, 62–5). Both these words involve the same changes in rate of vocal cord vibration, but differentially timed with respect to the segmental structure.

Forensic Speaker Identification

Non-linguistic temporal structure

The sections above have mentioned some linguistic aspects of articulatory timing. How long a segment lasts can be linguistically relevant, as for example in contrasts between long and short vowels, or long and short consonants. Differential timing between laryngeal and supralaryngeal activity was exemplified with voice onset time, and tonal and intonational pitch contrasts. There are also non-linguistic ways in which speech samples can differ with respect to articulatory timing, however. Perhaps the one that most readily comes to mind is how fast (or slow) we talk, the other is how fluently. These two aspects are termed rate and continuity (Laver 1994: 534–46). They are both internally complex, and some of their components have potential as forensic-phonetic parameters. Rate is discussed first.

Rate

Phonetics textbooks are fond of using the professional pianist as a yardstick for the virtuosity of the natural speaker. According to Laver (1994: 1) speech is 'both faster and more complicated' than the concert pianist's 'rapid arpeggio'. Ladefoged (2001: 169) says a speaker's rapid, precise movements are 'as skilled as the finer movements of a concert pianist'.

Since it is not really possible to compare pianist and speaker with respect to skill, or complexity or number of muscle movements, a more useful comparison might be in terms of notes and speech sounds per second. In keeping with the realistic orientation of forensic phonetics to unelicited phenomena, Figure 6.16 shows the cadenzal part of

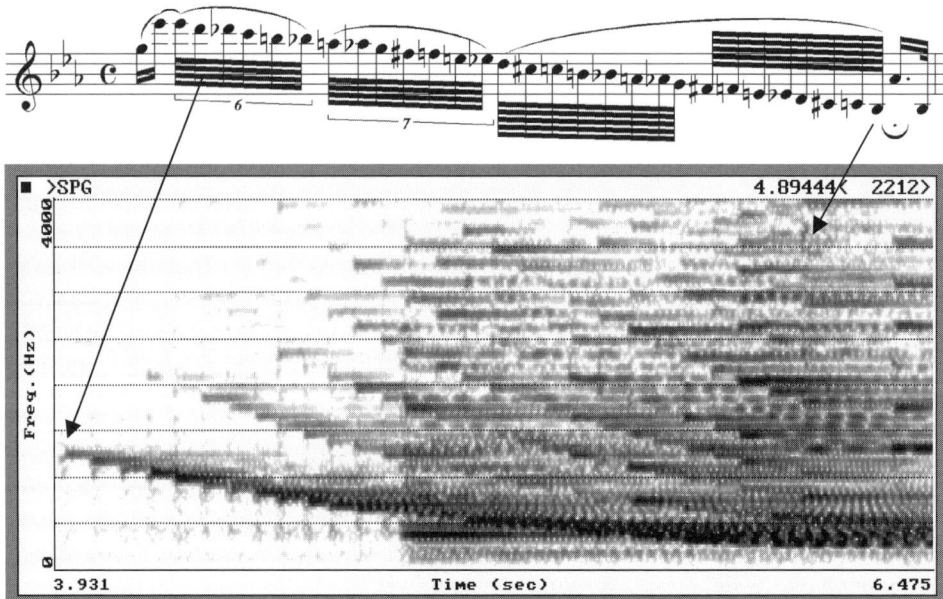

Figure 6.16 Musical structure and acoustical realisation of professionally played piano cadenza

measure 9 in the first movement of Beethoven's C-minor piano sonata opus 13, the *Pathétique* (Schenker 1975: 143). It consists of a run of over two octaves from a D to a B-flat, with 28 notes in all. Below the music is a spectrogram of an actual performance by Emil Gilels (a famous Beethoven interpreter). The spectrogram has been made so that it is easy to see the onset of the notes, and they can be picked out fairly easily. The elapsed time from the D at the start to the B-flat at the finish can be measured easily with the help of the software package, and is about 2.05 seconds. This indicates that Gilels accomplished this cadenza at a rate of about 13.5 notes per second. Measurements made on the same music played by Wilhelm Kempff (another famous Beethoven interpreter) gave effectively the same result of 13.3 notes per second. (It can be noted in passing that a rather important forensic-phonetic truth is illustrated in the reference to both music and acoustics in this figure. This is that it is not possible to make sense of the acoustic patterns unless the structure they realise is known.)

For comparison with Figure 6.16, Figure 6.17 shows spectrograms of stretches of uninterrupted actual speech taken from two intercepted telephone conversations in case-work. One is in Australian English, saying 'Fuck this for a joke I mean he fuckin' says to me don't do anything'; the other is some Cantonese instructions: 'léih gwojó kìuh daihyatgo gāaiháu jynjó joi jynjó' ('*The first intersection after you cross the bridge turn left then left again*'). The Cantonese utterance sounds to be slower and more deliberate than the English, which sounds to be neither particularly fast or slow.

The speech sounds of both utterances have been transcribed phonetically. The speech sounds in these examples show changes in their realisation that are typical of real connected speech. For example, when said carefully, *fuck this for a* contains the sounds: [fak] (*fuck*) [ðɪs] (*this*) [fɔɹə] (*for a*). But in the actual speech ([fakɪʂɹə]), some sounds, e.g. [ð] and [f] are lost; and others are changed, e.g. [ɔ] and [s] change to [ə] and [ʂ]. The English utterance contained 37 speech sounds and took about 2.64 seconds, a rate of 14 speech sounds per second. The more deliberate-sounding Cantonese utterance had 34 speech sounds and took about 2.76 seconds: about 12 sounds per second. These figures are just a little bit faster than the values of over ten vowels and consonants per second demonstrated by Ladefoged (2001: 169–70), and the average rate of 10 sounds per second in conversation cited by Calvert (1986: 178).

This can only be a very crude comparison for several reasons. For example, there is not an unambiguous answer as to what counts as a speech sound. Perhaps I should also have included the tones in the Cantonese speech, for they are clearly speech sounds? And if so, then it would be difficult to justify excluding the intonational tones in the English utterance. Nevertheless the examples still give a useful idea of what rates everyday normal speakers are capable of: both these utterances show speech sounds being encoded at rates comparable to those of two of the world's greatest pianists. Considering the fact that such finely coordinated cadenzas probably will have taken a longer time, and more assiduous practice, to acquire than the few years it takes a human to painlessly and automatically acquire speech, perhaps the yardstick for pianistic virtuosity should be the humble speaker.

Although the paragraphs above quoted fixed rates, speed of articulation is not of course constant, and it is a remarkable fact that we can both produce and understand speech at very different tempos (Künzel 1997: 48–9). Not only do we speak at different

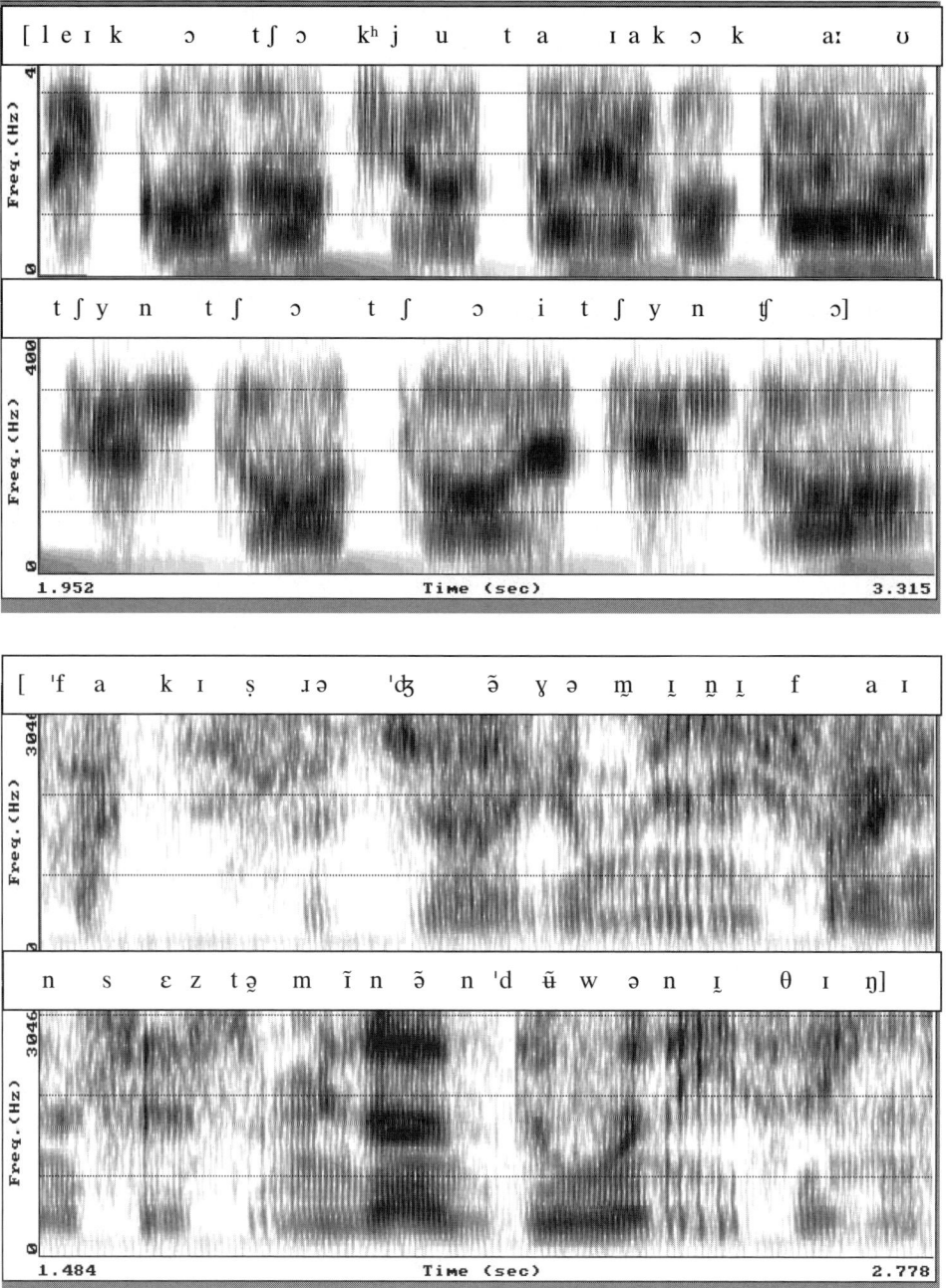

Figure 6.17 Annotated spectrograms of connected speech in Cantonese (above) and English (below), showing rate at which segments are produced

speeds on different occasions, but people characteristically speak at different speeds. Thus there are paralinguistic as well as extralinguistic aspects to rate (Laver 1994: 534). It is paralinguistic in the sense that different rates can have conventional cultural interpretations. For example, a fast rate can be interpreted as signalling impatience. Rate is extralinguistic to the extent that different speakers can habitually have different rates. Both these aspects are forensically important, of course, since in order to evaluate differences between samples in rate it is essential that the samples are first comparable with respect to the paralinguistic features.

Articulation rate and syllable/speaking rate

It is necessary to distinguish two different measures of rate, although their precise definition varies a little. Although the examples just given quantified rate in terms of segments per second, a more common, and tractable, unit is the syllable. The first measure, called articulation rate (AR), quantifies 'the speed at which an individual utterance is produced' (Laver 1994: 539). This is basically the number of syllables in an utterance divided by the utterance duration *excluding pauses*. As will be explained below, two common types of pause are filled and silent, and sources, e.g. Laver (1994: 539), Künzel (1997: 56), differ as to whether filled pauses are to be included in the articulation rate. In the English example in Figure 6.17 there are 15 syllables, giving an AR of 5.68 syllables per second, which is at the upper end of the typical range for medium AR in English (Laver 1994: 541).

The second measure of rate quantifies the overall tempo of performance, and includes all pauses. It is called speaking rate (Laver 1994: 541), or syllable rate (Künzel 1997: 56) – both terms helpfully abbreviated as SR – and is basically the number of syllables in a set of utterances divided by the utterances' duration *including pauses*. Since people tend to pause during speech, syllable/speaking rates are usually lower than articulation rates.

Forensic significance: Rate of utterance

Künzel (1997) investigated SR and AR for a group of ten speakers in speech of three realistically different degrees of spontaneity (spontaneous, semi-spontaneous and read out), and both direct and over the phone. He found that different speakers did indeed have different articulation rates, and claimed (p. 79) that articulation rate showed some promise as a forensic-phonetic parameter. This was because, although it had a relatively low discrimination power (the equal error rate was 38%), this was offset by its inertia with respect to different realistic experimental conditions. (Another study (Butcher 1981: 148), however, has shown different results, with AR differing significantly between spoken and read out speech, and so its invariance should not be overstated.) Syllable rate, on the other hand, was found to be of less value.

Despite its potential, the forensic use of AR obviously still needs to be evaluated with greater numbers of speakers and speech material from real case-work. In particular, it needs to be evaluated from a Bayesian point of view. We need to know what the probabilities are of observing a particular difference in AR between speech samples assuming same-speaker and different-speaker provenance.

Continuity

Speakers can differ not only with respect to how fast they speak, but also in fluency. The main thing that contributes to fluency is the number and types of pauses a speaker makes. It is clear that people pause when they speak, and that they do this for several reasons. Although there is no consensus on the function of speech pauses (Künzel 1997: 51), several types of pauses can be generally recognised on the basis of their form and function, and it is also possible for a single pause to be simultaneously fulfilling several functions.

Filled and silent pauses

Pauses do not have to consist of silence but can be 'filled' with phonological material. Common British English filled pauses, often represented orthographically as *um* . . . and *er* . . . are [əmː] and [əː]. Other languages can use other fillers (Kunzel 1997: 51). Pause-filling can also involve prolongation of segments. It would be possible to pause on the word *account*, for example, by prolonging the [a]: [əkʰaːːːʊnt], or the [n]: [əkʰaʊnːːːt], or even the [k]: [əkːːːʰaʊnt]. It is also possible for a pause to be partially filled and partially silent.

Juncture pauses

Juncture pauses occur primarily at boundaries between intonational phrases. Since intonational phrases also tend to reflect syntactic constituents, juncture pauses also tend to reflect syntactic boundaries, or junctures. A possible reading of the last but one sentence with juncture pauses would be:

> Juncture pauses <pause> occur primarily at boundaries between intonational phrases <pause>.

Syntactically, this sentence consists of a subject noun phrase (NP) *juncture pauses*, and a verb phrase (VP) *occur primarily . . . phrases*. The subject NP might be signalled as a separate intonation phrase by a fall–rise pitch, with the fall on the word *juncture* thus: [dʒaŋ H tʃʰə L], and the rise on *pauses* thus [pʰɔ L zəz H]. The verb phrase might be signalled as a separate intonation phrase by a gradual falling of pitch from *occur* to *intonation*, with a final pitch fall on *-nation*, thus: [neɪ HL ʃn̩ L]. The first pause occurs at the boundary of these two intonational phrases, which is also at the boundary between the two major syntactic constituents of the sentence. The second pause occurs at the end of the sentence – a major syntactic boundary par excellence. Juncture pauses are said to enhance the syntactic and semantic structure of the speech flow. A useful simplifying assumption is that juncture pauses are not filled (Künzel 1997: 68).

Hesitation pauses

Pausing can sometimes be indicative of some cognitive (verbal) planning in process: the speaker's brain needs to make decisions in the episodes of speech and so holds up speech production while this occurs (Laver 1994: 536). Such *hesitation pauses* can be silent, filled, use prolongation, or manifest combinations of these. Künzel (1997: 68) acknowledges the problem of identifying hesitation pauses, and suggests (p. 70) that, since juncture pauses are assumed to be silent, a filled pause may be taken to indicate a hesitation pause. Another indication of a hesitation pause might be the fact that it has not occurred at an intonation phrase or syntactic phrase boundary.

Respiration pauses

It is a good idea to continue breathing while you speak, and speakers also make use of pauses to take a breath. Sometimes the breath will be too low in amplitude to be picked up by a microphone, in which case a respiration pause cannot be identified as such.

Turn pauses

As part of the convention of normal conversation regulation, speakers can use pauses as one of the cues that they have finished saying what they want to say, and that their partner can now respond. These can be called turn pauses. Conversation analysis (Jefferson 1989) has suggested that they appear to be culture-specific in their duration: some cultures (e.g. Japanese) allow longer pauses than others (e.g. Anglo), and in some (e.g. 'Romance' cultures, like French and Italian) there are even negative pauses, where one speaker starts to speak before the other has finished (there are cues in speech before the end of an utterance that the speaker is approaching the end of their turn that allow this.) This is reminiscent of the cultural relativity of personal space (Morris 1978: 131–2).

Such turn pauses are not part of a speaker's fluency: for one thing, it is clear that a turn pause belongs in a sense to both conversational participants. It can be imagined, however, that turn pauses might have both positive and negative forensic relevance. A speech sample might be characterisable in terms of pause behaviour that is inappropriate for the culture implied by the language used. This could happen, for example, when one speaker, from a different culture to their partner, interrupts them when they should be waiting for them to finish. Also, of course, if speakers are regularly interrupting each other at the end of conversational turns, that reduces the amount of speech material available for analysis, since it is difficult to analyse overlapping speech.

Fluent and hesitant speech

Here is an example, taken from case-work involving telephone fraud, illustrating some of these fluency features. Intonational phrases are enclosed by square brackets, and pauses occurring at these boundaries are assumed to be juncture pauses. Hesitation pauses, assumed to be such because of their occurrence within intonation phrases/syntactic constituents, are marked with italics. Two of them are silent, but two – <er *pause*> and <is *pause*> – are examples of filled pauses. The <er *pause*> consists of a prolonged schwa: [əːː]. The case of the <is *pause*> is more complicated, and consists of the prolongation of the inherently short vowel in the word *is*: [ɪːːz].

[yeah] <pause> [<er *pause*>they've already paid that] <pause>[but I was wondering if I could <*pause*> pay the money towards <*pause*> my first job when it arrives] <pause>[the amount of the bill] <pause> [<is *pause*> two hundred and fifty-nine dollars and ninety-one cents]

Laver (1994: 535–9), who calls the fluency aspect of speech behaviour *continuity*, makes use of the distinction between juncture and hesitation pauses to classify speech as either fluent or hesitant. He describes speech as *fluent* if the linguistic material of its intonational groups is uninterrupted. If pauses are present in fluent speech they are juncture pauses and do not interrupt the linguistic structure. Speech that contains pauses that interrupt the linguistic structure – hesitation pauses – is described as *hesitant*.

This scheme thus allows for speakers, and speech samples, to be classified into two continuity categories: *fluent* and *hesitant*. (Laver also draws a less useful distinction between *continuous* and *non-continuous* speech on the basis of whether it contains pauses or not. This is not very useful for forensic purposes, where the sheer amount of data will almost certainly guarantee that speech is non-continuous.) The actual example given above contains several pauses, and so it is an example of non-continuous speech. Since pauses occur not only between but in the middle of intonational groups, and thus interrupt the linguistic structure, it is an example of hesitant speech.

Forensic significance: Continuity

There has not been very much research into how well same-speaker samples can be discriminated from different-speaker samples on the basis of continuity categories, although certain properties of pausing appear to be promising. In particular, the forensic importance of filled pauses and hesitation has been repeatedly emphasised:

> In der Tat ist die phonetische Realisierung von Häsitationsphenomänen in hohem Maße sprecherspezifish, und zwar nicht nur hinsichtlich der phonetichen Qualität, sondern unter bestimmten Bedingungen auch hinsichtlich ihrer Verteilung, d.h. der Häufigkeit des Auftretens, und zwar auch im Verhältnis zur Anzahl der sog. stillen Pausen . . .
>
> Künzel (1987: 37; cf. also 1997: 51, 70–71)

> [In fact the phonetic realisation of hesitation phenomena is highly speaker-specific, not only in its phonetic quality, but also under certain circumstances with respect to distribution, i.e. its frequency of occurrence particularly in relationship to the number of so-called silent pauses . . .]

Künzel (1997: 68–70) quantified the degree of hesitancy a speaker shows as the proportion of filled pauses to all pauses, under the assumption that the proportion of filled to silent hesitancy pauses remains constant. The resulting proportions (table 6, p. 69) tended to show rather large differences between speakers and relatively small within-speaker differences and are therefore possible forensic-phonetic parameters.

Description of speech sounds

Speakers can also differ in the percentage of overall speech duration taken up by pausing, as well as the average duration of the pauses (Künzel 1997: 65–71). The discrimination power of these parameters remains to be determined, however.

Summary: Non-linguistic temporal structure

Speakers can differ in the non-linguistic aspects of temporal structure, and therefore derived rate and continuity parameters are of potential forensic-phonetic value. Currently the most promising parameters are articulation rate, realisation of filled pauses, and hesitancy.

Chapter summary

This chapter has focused on the production and description of segmental speech sounds like vowels and consonants, and suprasegmental speech sounds like tone, intonation and stress. It has shown that all the time during fluent speech the vocal cords are being stretched and relaxed to control pitch, which signals tone, stress and intonation. At the same time they are being pulled apart and brought together to signal voicing, and all this vocal cord activity is being precisely timed with all the highly complex and coordinated movements of the different speech organs of the supralaryngeal vocal tract, like the tongue tip, the tongue body, the soft palate, and the lips. This activity is being coordinated and executed at speeds resulting in rates upwards of 10 speech sounds per second.

An important insight from this chapter is, however, not the sheer complexity of normal speech production, but that speech sounds are componentially structured entities, their structure resulting from complicated gestures in different parts of the vocal tract. More important still is the fact that speech sounds can be described and transcribed using these components: a 'voiceless unaspirated bilabial stop', for example, or a 'high front nasalised creaky-voiced vowel'. This is what makes auditory forensic comparison possible.

7

Phonemics

Speech sounds in individual languages

The preceding chapter introduced the general speech-sound producing capabilities of humans. It was intended to give a brief answer to the question: what is the articulatory nature of the speech sounds that humans use?

However, forensic phonetics does not look at speech sounds context-free, as in the previous chapter, but as they occur in the speech of someone speaking a particular language. There is a profound difference between these two ways of looking at speech sounds. It is a fact of the structure of Language that speech sounds operate on two different levels, and because of this dual structure it is not enough, indeed it is not possible, to describe the similarities and differences between two or more samples of speech in terms of the articulatory (or perceptual) nature of the sounds alone: to say, for example, that sample A has such and such an inventory of sounds compared with sample B.

The two different levels at which speech sounds exist are called the *phonetic* level and the *phonemic* level, and the difference between them is often called the etic-emic difference. The former, phonetic level, has been covered in Chapter 6: it describes speech sounds *tout court*, without reference to a particular language. The latter, *phonemics*, can be seen as a *conceptual framework within which to describe the sounds of a given language, or dialect*. It is thus language-specific, although its analytical principles are supposed to be applicable to all languages. This chapter introduces those principles. It describes how forensic phoneticians, and indeed most descriptive linguists (linguists who have to do with the description of languages), conceptualise the structure of the sound component of Language. It also describes how speakers can differ in phonemic structure.

Most introductory textbooks on linguistics or phonetics, for example Fromkin *et al.* (1999: 247–69), Crowley *et al.* (1995: 48–114), Finegan *et al.* (1992: 33–74) and Clark and Yallop (1990: 116–51) contain chapters on phonemic analysis that can be consulted for further examples. Specialist, theoretical accounts of phonemics can be found in some of the very many textbooks on the linguistic sub-discipline of phonology, of which Lass (1984: 11–38) is particularly recommended.

Phonemic contrast

It is best to start with a concrete example. Consider the English word *sat*. Most English speakers will feel, possibly on the basis of the spelling, that it consists of three bits: two consonants *s* and *t*, and a vowel *a*. Let us call these three bits speech sounds (another term is *phones*). From the previous chapter it is known that they can be described, *qua* speech sounds, as a voiceless alveolar fricative [s], a voiceless alveolar stop [t], and a low front unrounded oral vowel [æ].

The [s] in the English word *sat* can be substituted by other speech sounds to produce other, different English words. Thus changing [s] to [f] gives *fat*. The phones [s] and [f] are said to *contrast*, because substitution of one for the other, in the same environment, i.e. before *at*, produces a new word (here, *fat* from *sat*).

Contrast is the central notion of phonemic analysis. Speech sounds that contrast in the sense just described are said to *realise different phonemes*. Representing phonemes, as is customary, by symbols in oblique slashes, we can write that the speech sound [s] in *sat* realises the phoneme /s/. Since it can be shown that, in English, [t] realises /t/ and [æ] realises /æ/, we write *sat* phonemically as /sæt/, and understand and convey thereby that the English word *sat* consists of three phonemes: /s/, /æ/, and /t/. The reader should now try to show with a pair of minimally contrasting words that the phonetic consonants [m] and [n] contrast in English and therefore realise different phonemes.[1]

Most languages have a relatively small number of phonemes: 20 consonant phonemes, say, and five vowels, which function contrastively, to distinguish words. What the phonemic inventory is for a given language has to be determined by phonemic analysis. How to do phonemic analysis is beyond the scope of this book, but instructions can be found in the textbooks referred to above.

Phonemes and allophones

The nature of language makes it necessary to distinguish the phoneme, which is an abstract contrastive unit of sound, from its realisation – the sound itself – which is called an *allophone*. An allophone is written, like a phone, in square brackets, and the relationship between phoneme and allophone is symbolised as follows: /x/ → [y]. This is to be read as *the phoneme x is realised by the allophone y*. For example, we write, using now the correct symbols, /æ/ → [æ]: 'The (English!) phoneme æ is realised as the allophone æ.'

This may look redundant ('æ is realised as æ?!') but it is not, because the sounds exist on two different levels of structure: one of the æs is a phoneme, and one is its realisation. This becomes clearer when it is seen that a phoneme can be realised by more than one sound (i.e. it can have more than one allophone). In this case it is necessary to stipulate the environment in which a particular allophone occurs, thus:

---

1 Possible word pairs demonstrating this would be *mat* and *gnat*, where [m] and [n] contrast word-initially, or *simmer* and *sinner*, where they contrast word-internally, or *comb* and *cone* (word-final contrast).

/x/ → [y] / a

→ [z] / b

This is read as *the phoneme x is realised by the allophone y in the environment a, and by the allophone z in the environment b*. A very simple example of a single phoneme being realised by two allophones can be found with the alveolar and velar nasal consonants in Italian, the phonemic analysis of which is given in the next section.

An Italian example

Like English, Italian has both an alveolar nasal [n] and a velar nasal [ŋ] (like the *ng* in English *sing*). Now, these two sounds never occur in the same environment in Italian. [ŋ] always occurs before a [k], or a [g] (in terms introduced in the previous chapter, before velar stops), as in [liŋgwa] *language*, or [staŋko] *tired*. [n] never does.

This means that the two sounds [n] and [ŋ] cannot contrast in Italian, since by definition two contrastive sounds have to be permutable *in the same environment*. Since the two sounds are phonetically similar (they are both nasals), and because they cannot contrast, they can be considered as different allophones, or positional realisations, of a single contrastive phoneme.

$$/n/ \rightarrow \begin{cases} [\eta] \ /\_ \text{ velar stop} \\ [n] \ / \text{ elsewhere} \end{cases} \qquad (7.1)$$

These observations can be formalised as at (7.1), which says that the Italian alveolar nasal phoneme has two allophones, or realisations. Before velar stops (i.e. [k] and [g]) it is realised as a velar nasal [ŋ]; in other environments the /n/ is realised as an alveolar nasal [n].

The two Italian words *tired* and *language* given above would be written *phonemically* thus: /stanko/, /lingwa/ (this is assuming that all the other phones in these words represented different phonemes). The allophonic statement at (7.1) relates the phonemic representations (/stanko/, /lingwa/) to their phonetic representations ([staŋko], [liŋgwa]). Going from the phonemic to the phonetic representation, it states how a phoneme /n/ will be realised as [ŋ] before a velar stop and as [n] elsewhere. Going from the phonetic to the phonemic representation, it states how the phone [ŋ] occurring before a velar stop is the realisation of the phoneme /n/.

Quite often the particular realisations of a phoneme can be seen to occur because of the influence of surrounding sounds. The velar realisation of the /n/ phoneme in Italian is the result of the velar place of articulation of the following stop. It is as if the Italian speaker is aiming for an alveolar place for the nasal, but the alveolar target is perturbed backwards to a velar place by the following sound. This is known, self-explanatorily, as anticipatory assimilation in place of articulation: the alveolar nasal phoneme assimilates in place of articulation to the velar stop.

Assimilation, especially anticipatory assimilation, is quite common in languages. Persevatory assimilation, when a sound following another sound changes to be more

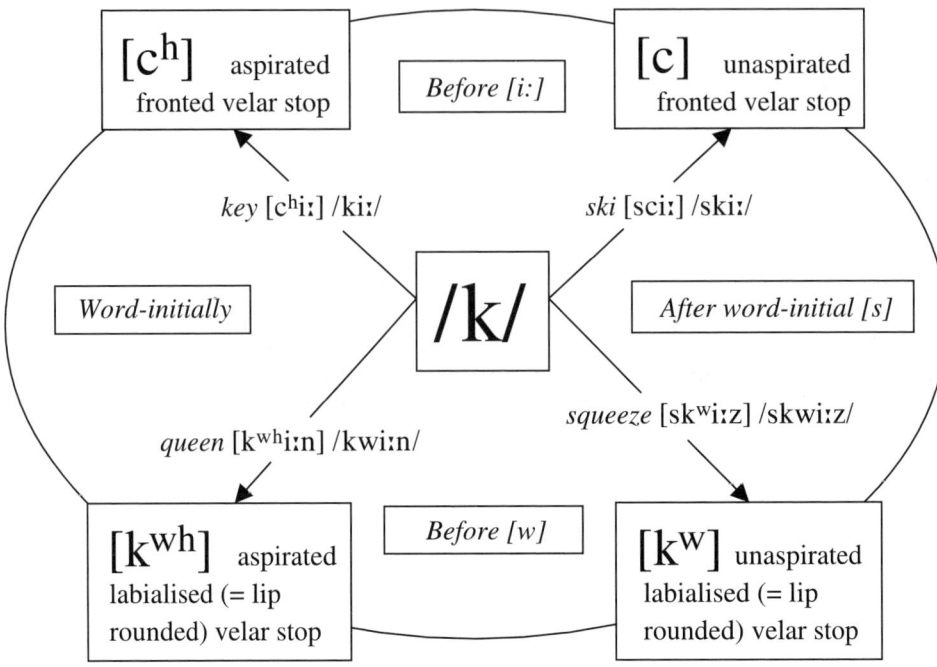

Figure 7.1 Some allophones of the English voiceless velar stop /k/

like the first sound, is also found. Similar sounds can also change to be unlike each other. This is called dissimilation, and also has its anticipatory and perseverative types. It is partially because of processes like these that phonemes can have more than one allophone. When there is a plausible phonetic reason for a particular allophone, it is said to be an intrinsic allophone.

A phoneme may have more than two allophones. For example, the voiceless velar stop /k/ phoneme in most varieties of English actually has many more than two. Four of them are shown in Figure 7.1, which shows the phoneme in the centre, with four allophones, and their descriptions, at the periphery. The conditioning environments are shown in boxed italics, and an example word, together with its phonetic and phonemic representation, is shown on each realisation arrow. Thus it can be seen that before an [iː] vowel and after a word initial [s], the phoneme /k/ is realised by an unaspirated fronted velar stop [c], as in the word *ski*, which is phonemically /skiː/, but phonetically [sciː]. (An alternative transcription for the fronted velar is [k̟].)

The figure makes it possible to see the effect of the four individual environments. Thus occurrence before an [iː] evinces a fronter allophone, word-initial occurrence evinces an aspirated allophone, occurrence before a [w] evinces a lip-rounded allophone, and occurrence after a word-initial [s] evinces an unaspirated allophone. This is typical of the way different allophones arise, and it can be appreciated that because of this it is possible for phonemes to have many allophones.

These different allophones of /k/ are intrinsic, that is, they are plausibly phonetically conditioned by its particular environment. It can be appreciated for example that it is the lip-rounding on a following [w] that causes the anticipatory lip rounding on the stop, the front position of the tongue body in [iː] that pulls the body of the tongue forward in the stop, and so on.

Comparison with English: The phonemic insight

We have seen that in Italian the velar nasal [ŋ] is an allophone, or realisation, of the alveolar nasal phoneme /n/. In English it is not. This is evident from English word pairs like *sin* [sɪn] and *sing* [sɪŋ], or *sinner* [sɪnə] and *singer* [sɪŋə]. In these examples, the alveolar and velar nasals are *contrastively* distributed: substitution of one by the other has the potential to produce a new word. Therefore they realise different phonemes: /n/ and /ŋ/. We have to assume for English therefore that:

/n/ → [n]

/ŋ/ → [ŋ]

The examples with [n] and [ŋ] in Italian and English show that phonetically the *same sound* – in this case the velar nasal [ŋ] – is phonemically a *different sound* in English from in Italian. Put another way, the phonetic difference between a velar and an alveolar nasal is contrastive in English, but not in Italian. We can generalise from this to linguistic structure, and state the phonemic insight, or principle, that *some phonetic differences are contrastive in some languages but not in others* (Anderson 1985: 65). It is the linguistic fact that *speech sounds do not all behave in the same way in all languages* that requires the additional phonemic level in linguistic structure.

A Canadian English example

A slightly more complicated, but instructive, example of one phoneme having two allophones occurs in Canadian English. One very common American stereotype of a Canadian English accent is the pronunciation of words like *out* and *about*. The diphthong in these words in Canadian English typically onsets at a mid position, before moving to a high back rounded second target. For simplicity, we will transcribe it as [əʊ], so *out* is phonetically [əʊt]. Interestingly, however, the diphthong in the word *loud* in Canadian English is not the same phonetically. Its onset sounds to have a much lower tongue body height: [aʊ], so *loud* is [laʊd]. If we look at the environments in which these two diphthongs occur in lots of words, we note that the one with the mid onset [əʊ] occurs before voiceless stops, and the one with the low onset [aʊ] occurs in other positions (Wells 1982: 494). This enables us to analyse the two diphthongs differing phonetically in onset vowel height as allophones of the same /aʊ/ phoneme as in (7.2):

/ aʊ / →   [əʊ]   /_ voiceless consonants   (e.g. *out, mouth, couch, house*)   (7.2)

[aʊ]   / elsewhere   (e.g. *loud, houses, sound, how*)

Forensic Speaker Identification

More instructive is the parallel example of diphthongs in Canadian English words like *right* and *ride*. Again there is the same phonetic difference in onset vowel height, with one diphthong having a mid onset: [əɪ], and one a low onset: [aɪ]. Not only is there the same phonetic difference as in the previous example, however, but there is also the same conditioning environment. The mid onset occurs in diphthongs before voiceless consonants as in *right*, *price*, *life*, and the low onset occurs elsewhere, as in *ride*, *knives*, *high*, so *right* is [ɹəɪt] and *ride* is [ɹaɪd]. The same phonemic analysis applies as at (7.2), and is shown in (7.3).

$$/ \text{aɪ} / \rightarrow \begin{cases} [\text{əɪ}] \ /\_ \text{ voiceless consonants} & (\text{e.g. } right, price, life) \\ [\text{aɪ}] \ / \text{ elsewhere} & (\text{e.g. } ride, knives, high) \end{cases} \quad (7.3)$$

It can be noted that the [əʊ] and [əɪ] allophones in this Canadian example are extrinsic, since it is not easy to give a plausible phonetic explanation for the mid realisation of their initial target before a following voiceless consonant. It may be the case that these mid forms represent a relic, i.e. historically correct form, and that the more open allophone is the result of a sound change.

The Canadian diphthong example is important, because it shows a common phenomenon in the phonemic structuring of language. We can see in fact that (7.2) and (7.3) are two examples of the same thing, namely that diphthongal phonemes are realised with mid onset targets before voiceless consonants, and have low onset target allophones elsewhere. We expect languages to behave symmetrically like this. So that if a particular allophonic realisation is found to obtain in a given language for, say, a voiceless bilabial stop phoneme /p/, we expect to find it *mutatis mutandis* in other voiceless stops /t/ and /k/. (For example, perhaps /p/ is realised as a voiceless aspirated stop [pʰ] word-initially. We therefore expect to find aspirated allophones of /t/ and /k/ ([tʰ] and [kʰ]) word-initially as well.)

This assumption of symmetry is an analytical postulate based on frequency of occurrence across the world's languages. It does not have to occur in every case. However, it can be appreciated that it constitutes a very useful base-line for evaluating similarities and differences in the allophonics of speech samples. For example, suppose a questioned sample contained examples of a phoneme /p/ being realised as aspirated [pʰ] word-initially, but no words with initial /t/ or /k/, and the suspect sample lacked examples of word-initial /p/, and only had words with initial /t/ and /k/. If the /t/ and /k/ in the suspect sample had *un*aspirated allophones word-initially, this could legitimately be considered an area of disagreement between the two samples.

Types of distribution

Two types of distribution of phonetically similar phones have been illustrated above. One type of distribution is called *contrastive*. This is where two phonetically similar phones can occur in the same environment and the substitution of one for the other can signal a different word, like [s] and [f] in *sat* and *fat*. The other type is called *complementary* distribution. This is where two phonetically similar sounds occur in mutually exclusive environments, and therefore cannot contrast, like [n] and [ŋ] in Italian or [aʊ] and [əʊ] in Canadian English. A third type of distribution of phonetically similar phones is found in language: *free variation*. This is where two phonetically similar sounds can occur in the same environment, but can freely interchange without changing the meaning of a word.

Consider for example the English lateral phoneme /l/, when it occurs before a consonant in the same word, e.g. *help, milk*. In this environment, the realisation of the /l/ is usually complicated. It is normally realised with a velarised lateral allophone [ɫ]. This means that, in addition to contact being made with the tip of the tongue at the alveolar ridge, and the air flowing round the sides of the tongue as for a plain lateral, the tongue *body* is also raised towards the soft palate, similar to the position for a high back vowel [ʊ] (the vowel in *hood*). Thus *help* and *milk* are phonetically [hɛɫp] and [mɪɫk].

Now, in this environment, the phonemics of English do not *require* there to be an alveolar contact for the /l/. The absence of contact results therefore in a short vowel like that in *hood* [ʊ]. Thus both [hɛɫp] and [hɛʊp] are heard for *help*, and [mɪɫk] and [mɪʊk] for *milk*. The point here is that, although they can occur in the same environment, the phonetic difference between the similar sounds [ɫ] and [ʊ] is not contrastive, since [hɛɫp] and [hɛʊp] both mean *help*. Thus [ɫ] and [ʊ] are said to be in free variation in this environment. This can be stated formally as: /l/ → [ɫ] ~ [ʊ] /_C# (*the phoneme /l/ can be realised as either [ɫ] or [ʊ] before a word-final consonant*). The rule is actually somewhat more complicated, but this will suffice as an illustration.

Thus it can be seen that there are two conditions of non-contrast for phonetically similar sounds: complementary distribution and free variation.

The reality of phonemes

There is always debate as to the ontological status of linguistic units: whether they have some reality or whether they are just analytic conveniences, and phonemes are no exception. Whereas it is clear the phonemic model fits some languages better than others, there is some evidence for the psychological reality of at least phoneme-like units.

Cross-linguistic perceptual experiments, at least as far as consonants are concerned, have shown that 'listeners reflect the phonological [i.e. phonemic] categories of their native language by their classification of phonetic segments' (Studdert-Kennedy 1974: 2352). For example, *the same phonetic continuum* is categorised differently by listeners according to the phonemic contrasts in their language. Listeners whose language has a single contrast, e.g. /p/ vs. /b/, reveal in perceptual tests a single categorical boundary dividing the phonetic continuum into two categories. Listeners whose language has two contrasts, e.g. /p^h/ vs. /p/ vs. /b/, impose two boundaries, with three categories (Clark and Yallop 1990: 272–4).

It is a language's phonemic structure – the number and kind of sound contrasts that are made, and how they are realised – that determines to a large extent the foreign accent of its speaker when they try to acquire another language later in life. In trying to understand the foreign accent of a speaker of Australian English, for example, it is necessary to know the phonemic system of the speaker's native language, and that of the target language. This kind of knowledge can have obvious forensic uses – in identifying, for example, whether someone is speaking Australian English with a Malaysian or Vietnamese accent. It hardly needs to be said that in order to make such an identification the phonemic structure of both languages needs to be taken into account.

Native speakers are usually well aware of phonetic differences that are contrastive in their own language, but are generally not aware of phonetic differences that are not. For example, if the reader is a native speaker of British or American English, they will probably not be aware that the plosive sounds at the beginning of the words *cup* and *keep* are phonetically different. The stop in *cup* is velar: [k] (I will leave out the aspiration [ʰ] to simplify things). If the stop in *keep* is isolated – make as if to say *keep*, but just say the initial stop – it will easily be appreciated that it is made with a place of articulation in front of velar. This sound can be transcribed [c] or [k̟]. The two words are phonetically then: [kap] and [ciːp].

As explained above, these two plosive sounds, [c] and [k], are allophones of the same phoneme /k/ in English, the fronter allophone being conditioned by the height and frontness of the following [i] vowel. This is formalised at (7.4).

$$/k/ \quad \begin{array}{l} \rightarrow \quad [c] \quad /\_ \text{ high front vowel} \\ \\ \rightarrow \quad [k] \quad / \text{ elsewhere} \end{array} \quad (7.4)$$

Unlike speakers of English, speakers of Irish Gaelic are well aware of the phonetic difference between the sounds [c] and [k]. Why might this be? It is because they realise different phonemes in Irish: they can distinguish words. For example, [c] and [k] contrast before -a in the words capall [kapəɫ] *horse* and ceacht [caχt] *lesson*. This is formalised at (7.5).

$$\begin{array}{l} /c/ \rightarrow [c] \\ /k/ \rightarrow [k] \end{array} \quad (7.5)$$

Phonotactics

Languages differ not only in phonemic inventory and allophonics, but also in the way that words are composed of phonemes. Each language may have its bag of phonemes with their associated allophones, but not every phoneme can occur anywhere. An English word cannot consist of five consonant phonemes in a row, for example, nor is /pslɪt/ a possible English word, even though /splɪt/ (*split*) is.

The structure of the word in terms of phonemes – how the phonemes can be strung together to form words – is called *phonotactic* structure, and phonotactic analysis is the other part of phonemic analysis alongside determining the phonemes and their allophones. Some well-known aspects of English phonotactic structure are that its velar nasal phoneme /ŋ/ cannot occur at the beginning of a word, and that its glottal fricative phoneme /h/ cannot occur at the end of a word. The reader might like to determine for themselves another well-known phonotactic rule of English: what must the first phoneme be in words beginning with three consonant phonemes?[2] The syllable is an important phonotactic unit: strictly speaking in phonotactic analysis one works out what the structure of the syllable is in terms of phonemes and then works out what the structure of the word is in terms of syllables.

Phonotactics, just like phonemes and allophones, are language-specific. The two English phonotactic rules concerning the restricted occurrence of /ŋ/ and /h/ just mentioned are part of English phonemics. Other languages might be different. In Indonesian, for example, there is a phoneme /ŋ/, but unlike English it *can* occur word-initially, and there is a phoneme /h/, but unlike English it *can* occur word-finally.

English phonemes

It was said above that languages generally have a small number of contrasting sounds or phonemes, and, since it is useful to have an idea what a phonemic inventory of a language looks like, that is what this section will do for English. Table 7.1 presents the phonemes for one particular variety of English, Australian English, together with some words containing them. There is very little variation between the consonantal phonemes of different varieties of English, so those in Table 7.1 can be taken as representative for most varieties. The vowel phonemes, their allophones, and their incidence in words, do differ across different varieties of English, however, so care must be taken in extrapolating to other varieties. A good reference work describing the phonological structure of many different varieties of English is Wells' (1982) book *Accents of English*. American and British English vowel allophones are compared in Jassem and Nolan (1984: 39–44), and Ladefoged (2001: 43–46) compares their acoustics.

In Table 7.1 the consonantal phonemes are arranged in a place–manner chart; the vowel phonemes are simply listed. For the latters' position on a height-backness chart, see Figures 6.11–6.13. Künzel (1994: 140) reports a case where forensic-phonetic evidence was given by someone who was apparently not aware of the difference between sounds and letters (let alone the difference between (allo)phones and phonemes!). Therefore words illustrating vowel phonemes in Table 7.1 have been chosen to display the many different ways that English phonemes can be spelled. It can be seen for example that the vowel phoneme /ʉː/ is spelled with *o*, *ou*, *oo*, *ue* and *ew*, some of which, e.g. *oo* and *u*, are also used for other phonemes like /ʊ/ and /a/. English orthography is clearly a very long way from a phonemic orthography. The one-to-many relationship between vowel phonemes and English orthography (and also consonant phonemes and orthography) should help to point up the fallacy of confusing letters with speech sounds: *the relationship between spelling and sounds is never ever one-to-one*.

---

2 It must be /s/: cf. *splash, strain, screw*. . . .

Table 7.1 Phonemes of Australian English

	Consonants (24)					
	Bilabial	Labio-dental	Dental	Alveolar	Post-alveolar/Palatal	Velar/Glottal
Plosive/ affricate	p (pat) b (bat)			t (tip) d (dip)	tʃ (chip) dʒ (judge)	k (cut) g (gut)
Fricative		f (fat) v (vat)	θ (think) ð (this)	s (sip) z (zip)	ʃ (ship) ʒ (Persian)	h (help)
Nasal	m (simmer)			n (sinner)		ŋ (singer)
Lateral approximant				l (leaf)		
Central approximant					r (red) (post-al.) j (yet) (pal.)	w (wed)

Vowels (19)	
Short Vowels (7)	
/ɪ/	hid sit injury kick finish
/ʊ/	hood good should put
/a/	hut tough among loves blood
/ə/	alone tucker melody horses
/ɛ/	head bed said says
/æ/	had adder pack sang bladder
/ɒ/	hod crock oral auction
Long Vowels (5)	
/iː/	heed believe people seize meat key
/ʉː/	who'd through move food crude sued chewed
/aː/	hard path serjeant car balm galah
/ɜː/	heard birth pert word fur burr
(/æː/	sad gladder)
/oː/	hoard fought taught saw whore awe chalk
Diphthongs (7)	Examples
/aɪ/	hide guy lie sight
/aʊ/	how out lout doubt
/ɛɪ/	hate may paid
/ɔɪ/	Hoytes groin coy
/aʉ/	hoe crow boat
/ɪə/	here ear beer
/ɛə/	hair bear there share

The reader must be prepared to encounter some variation in the symbols used to represent vowel phonemes (and to a much lesser extent consonant phonemes), both for different varieties of English like British and American, or within a particular variety, like Australian English. This variation is partly due to slightly different analyses, partly to different transcription traditions, and partly due to the use intended for the phonemic transcription. Thus the vowel in words like *heed, sleep, me* may be found phonemically transcribed as /iy/, or /i/, or /iː/ reflecting different analyses; the choice of the symbols /tʃ/ or /č/ for phonemic affricates in words like *church* reflects American vs. British English traditions in transcription; and the choice of /u/ vs. /ʉː/ in Australian English words like *who* reflects how much one wants to make explicit comparisons with the British English system, or how much one wants to use phonemes as a base to teach phonetics. The vowel symbols in Table 7.1 are actually a compromise between the three sets listed in Durie and Hajek (1994: 104).

Forensic significance: Phonemic structure

It has been explained above how the conceptual framework of phonemics, working as it does in terms of basic abstract contrastive speech sounds (phonemes), their organisation (phonotactics) and their realisations (allophones), is used both to describe the sound structure of a language and to compare languages with respect to their sound structure.

Phonemics, however, is also crucial for the description and comparison of forensic speech samples. This is for three reasons. Firstly, and most obviously, it is simply the way Language works, and since forensic speech samples are samples of Language, it makes sense to use the appropriate model for comparing speech sounds in two or more samples on the linguistic level alluded to in earlier chapters. Secondly, phonemic information is simply logically necessary to establish comparability between sounds in different samples. Thirdly, there is both within- and between-speaker variation in phonemic structure, and the phonemic conceptual framework provides a principled and precise way of talking about it, and evaluating it, when it occurs in forensic speech samples. These last two points are important and need explaining. This is done in the next two sections.

Establishing between-sample comparability

Assume that a questioned forensic sample contained the words *white tide life die* pronounced in the following way: [wɑɪʔ] [tɑɪd] [lɑɪf] [dɑɪ], and the suspect sample contained the words *lie fight cry time* pronounced: [laɪ] [faɪt] [kɹaɪ] [tãĩm]. It would be possible to neatly characterise the difference between the two samples by saying that they differ in the backness of the onset to the /aɪ/ phoneme, with a backer allophone ([ɑ]) in the questioned sample, compared to the central allophone ([a]) in the suspect's sample. It would also be possible to say that the questioned sample has a glottal-stop allophone to /t/ word-finally ([wɑɪʔ] for /waɪt/). The details of the comparison are summarised in Table 7.2.

Since it is a predictable allophone conditioned by the following nasal, the nasalised realisation of the suspect's /aɪ/ in *time* ([ãĩ]) would only warrant comment if the

Forensic Speaker Identification

Table 7.2 Comparison between speech samples in the realisation of phonemes

	Questioned sample	Suspect sample
Words in sample	*white side life die*	*lie fight cry time*
Phonetic transcription	[waɪʔ] [taɪd] [laɪf] [daɪ]	[laɪ] [faɪt] [kɹaɪ] [tãĩm]
Realisation of /aɪ/ phoneme	/aɪ/ → [aɪ]	/aɪ/ → [ãĩ] / _ m → [aɪ] / elsewhere
Realisation of /t/ phoneme	/t/ → [ʔ] / _ #[a] /t/ → [t] elsewhere	/t/ → [t]

[a] # Means word-finally

questioned sample did not have a nasalised allophone in a comparable environment. This would be, for example, if the questioned sample contained words with nasals after the /aɪ/ like *fine* or *crime* pronounced with an non-nasalised allophone, e.g. [faɪn] [kɹaɪm].

It is important to understand that *none of this comparison could have been made without prior phonemic knowledge*. Without a phonemic analysis to inform it, there is nothing to say that the phonetic diphthong in the questioned sample's *life* [laɪf] is comparable to that in the suspect sample's *fight* [faɪt]. Both diphthongs are phonetically different, both occur in different words! What justifies the comparison of the two samples with respect to these two diphthongs, then? Because it is known that they are both allophones of the same diphthongal phoneme.

The role of the postulate of symmetry in allophonic realisation as indicating areas of between-sample comparability has already been mentioned above.

The phonemic model thus allows the between-sample differences (and similarities) to be characterised neatly. The evaluation of such differences as between- or within-speaker would, of course, depend on further information.

Between-speaker and within-speaker differences in phonemic structure

Now that the basic idea of phonemic structure has been explained, it is possible to describe here the several ways in which speakers are understood to differ with respect to their phonemic structure. These are actually the same as the ways in which languages are said to differ from each other in phonemic structure and are known, after Wells (1982: 72–80) as *systemic*, *phonotactic*, *incidental* and *realisational* differences (Nolan 1983: 40–2). These categories can also be taken to include both between-speaker and within-speaker differences.

Systemic differences

Speakers may differ *systemically*, i.e. have phonemic systems with a different number of (a sub-set of) phonemes. Thus speakers of British English have one more short vowel phoneme (/ɒ/ as in *hot*) than speakers of General American (Wells 1982: 182). Northern Irish English speakers have in addition to the consonant phonemes listed in

Table 7.1 an extra consonant: /ʍ/. This is a voiceless labial-velar fricative – a kind of voiceless [w] – and serves to distinguish words in minimal pairs like /waɪn/ *wine* and /ʍaɪn/ *whine*; /wɪtʃ/ *witch* and /ʍɪtʃ/ *which*. A more drastic example of a systemic between-speaker difference occurs in some Northern Wu dialects of China (dialects spoken around Shanghai). In these dialects, younger-generation speakers lack whole subsets of nasalised high and mid vowel phonemes, for example /ĩ ũ ẽ and õ/, though they are present in older-generation speakers. Thus for younger speakers words like *heaven* and *for* are both /tʰi/ and homophonous, while for their parents they are different: *heaven* is /tʰĩ/ with the nasalised vowel, and *for* is /tʰi/ (Rose 1981: 82–3).

Systemic *within-speaker* variation is found in the speech of bidialectals and bilinguals. For example, in Thailand and Laos most educated speakers command at least two varieties of Thai that often differ vastly in their tonemic structure: their local dialect and the standard language. A speaker of a Southern Thai dialect, for example, will switch between their seven-toneme Southern Thai system and the five-toneme Standard Thai system with ease depending on the social circumstances (Diller 1987; Rose 1994). There are even tridialectal speakers, who switch between three different tone systems (Rose 1998b).

Phonotactic differences

Speakers can differ *phonotactically*, that is, in what phoneme sequences their language permits. The example of post-vocalic *r* is a case in point. Many varieties of English, but not Australian or British English, permit the phoneme /r/ to occur after the vowel in the same syllable. Thus /kar/ is a possible phoneme sequence for the word *car* in American English, but not Australian, where it is /kaː/.

Phonotactic within-speaker variation may also arise from sociolinguistic factors. In the early, heady days of (Black) American influence on British pop music, British pop singers like Cliff Richards, or Ray Davies of the Kinks, could be heard using post-vocalic *r* in their songs. As is often the case with imitated accents, which are complicated, systematic things, mistakes were made, and *r*s were inserted where they had no place. Thus we were treated (indeed still are treated) to the erratic rhoticism of 'Son, you be a**R** batchelo**R** boy' (first *r* wrong, second correct), or 'gone back to her ma**R** and pa**R**' (incorrect omission of *r* in *her*; correct, but vacuous *r* in *ma* (this would occur as a so-called intrusive *r* in British English anyway); incorrect *r* in *pa* (Trudgill 1983: 146–9).

There is an ironic post-script: Black American English, which was presumably the target of imitation, is actually not rhotic, so *all* of the above *r*s except that in *her*, are incorrect.

Incidential differences

Speakers can differ *incidentially* in their phoneme structure. This refers to variation in the incidence of phonemes in particular words, and is common. Well-known examples of incidential differences are /ɛ/ vs. /ɒ/ in the word *envelope*; /aɪ/ vs. /iː/ in the word *either*; or /iː/ vs. /ɛ/ in the word *economics*. Wells (1999) includes many examples of between-speaker incidential differences from current British English, like /ʒ/ vs. /ʃ/ in *Asia*, /θ/ vs. /ð/ in *booth*, /s/ vs. /z/ in *newspaper*, and /æ/ vs. /ɔː/ vs. /ɒ/ in *falcon*. An

187

Australian example is whether one says the word *dance* with the phoneme /aː/, as is typical of speakers from Melbourne and Adelaide, or with the phoneme /æ/. Australian speakers also vary in use of the phonemes /ɪ/ and /ə/ in the unstressed syllables of certain Australian English words like *paddock* and *stomach*: both /padɪk/ and /padək/, /stamɪk/ and /stamək/ are heard.

The most important within-speaker incidental differences are probably among those that reflect the way someone changes their speech in response to their perception of aspects of the situation – who they are speaking to, how formal the situation is, etc. For example, an Australian speaker may say the word *city* as /sɪdiː/, with a /d/ in informal conversation, but /sɪtiː/ with a /t/ in more formal speech. Another common example in many forms of English is verbal forms in -ing, like *going* and *eating*, which may end in /n/ under informal speech, but /ŋ/ formally.

In linguistics, variation of this sort is described in terms of a so-called socio-linguistic variable (marked by different brackets), which has different realisations under different sociological circumstances. The two cases just mentioned are examples of the English socio-linguistic variable {t} being realised as /d/ and /t/, and the socio-linguistic variable {n}, realised as /n/ or /ŋ/.

Realisational differences

From the description of phonemic structure above, it can be appreciated that an important part of the sound structure of a language has to do with the rules governing how phonemes are realised: the rules that specify what allophones occur in what environments. Of probably clearest application to forensic phonetics, then, are differences in *realisation* of phonemes, where speakers differ in the phonetic quality of the allophones of a phoneme.

The most important aspects of allophonic realisation for forensic phonetics have to do with three things: the phonetic nature of the allophonic target itself; how many different allophonic targets are involved for a given phoneme in a given position; and to what extent the realisation is affected by surrounding sounds. Since the first category involves more discussion, the last two will be dealt with first.

Choice in allophonic targets

The rules for an accent usually only allow one target allophone per environment for a phoneme: for example, in most varieties of English the voiceless bilabial stop phoneme /p/ has to be realised as a voiceless aspirated stop [pʰ] word-initially in a stressed syllable as in *Peter*. However, a language may also allow speakers a choice for some realisations. For example, the rhotic phoneme /r/ in British English can now be realised by a post-alveolar approximant [ɹ], or a labio-dental approximant [ʋ], so that some speakers may say [ɹaɪt] and others [ʋaɪt] for *right*. (A labio-dental approximant [ʋ] can be made by trying to say a [v] without generating friction between the lower lip and the upper incisors.) There is thus between-speaker free-variation in the allophones of this /r/ phoneme.

The possibility of a linguistic choice in allophonic realisation has obvious relevance for forensic phonetics. It will be recalled, for example, that the linguistic difference between [ʋ] and [ɹ] was one feature distinguishing the speech of a pair of English speaking identical twins who were otherwise very difficult to distinguish (Nolan and Oh 1996).

Speech defects and deviant speech

The conventional, rule-governed nature of language means that speakers of a language usually have a good idea of what constitutes deviant allophonic realisations. Thus, whereas both [ɹaɪt] and [ʋaɪt] are now acceptable for *right* in British English, another realisation, however, with the labial-velar approximant [w], is not (and neither [ʋaɪt] nor [waɪt] are acceptable for *right* in Australian English).

Coarticulation

In speech, the articulatory mechanism is always having to get from the set of targets for one segment to those for the next. For example, in the word *saw*, one has to get from a voiceless alveolar fricative [s] to a voiced mid back rounded vowel [oː] or [ɔː]. Some aspects of the transition are prescribed by the language. Thus the vocal cords have to start coming together for the voicing of the vowel at the time when the front of the tongue begins to move away from the alveolar constriction for the [s]. Some aspects of the transition, however, may not be prescribed. Thus in *saw* the lips may begin to round in anticipation of the roundedness of the vowel at any time during the [s]. This phenomenon is called coarticulation: the /s/ is said to show anticipatory coarticulation with the lip rounding of the /oː/. Looked at from an allophonic point of view, the /s/ will have different allophones depending on the degree of lip rounding implemented. (It is also worth emphasising here that coarticulation is generally a language-specific phenomenon. For example some Chinese dialects *do* specify full anticipatory lip rounding from a rounded vowel onto a preceding [s], so the choice is not available to speakers.)

Now, if the language does not specify how you get from articulatory target A to target B, individuals may be free to find their own articulatory solution, and this can therefore give rise to between-speaker differences. One speaker might, for example, show a high degree of coarticulation and another minimal. Speakers have in fact been shown to differ in the extent to which they allow the same allophonic target to be affected or perturbed by surrounding speech sounds. Thus Nolan (1983: 74–120) demonstrated that British English speakers differed in the degree to which their word-initial /l/ showed anticipatory coarticulation with its following vowel, and was able to achieve moderate success in discriminating speakers on the basis of their degree of coarticulation (p. 113). To a large extent, however, such between-speaker differences are small and only demonstrable acoustically, and Nolan doubts (p. 115) that the method will be of great use in practical identification.

Allophonic differences and accent

When differences in allophonic realisation between groups of speakers are systematic, they contribute to differences in accent. For example, the realisation of the vowel phoneme in words like *path*, *far* and *calm* with a low central [aː] is typical of an Australian accent, as opposed to the Standard British English accent realisation with a low back [ɑː]. Accent differences of this kind are usually thought of as correlating with geographical location: they are part of what are recognised as different dialects.

However, accents can also correlate with non-geographical features. For example, as already mentioned, very little regional variation in accent exists within Australian English and the clearest examples of Australian English speakers differing in accent by virtue of differential realisation of phonemes are probably sociolectal or stylistic, with differences correlating with sex and age frequently cross-cutting these. Thus various types of so-called ethnic Australian accents, for example Italian or Vietnamese, can be distinguished on the basis of allophonic realisation (as well as phonotactics).

Below I shall illustrate how the differential realisation of the same phonemic system gives rise to different accents using an example of the three-way accent difference in Australian English known as 'Broad', 'General' and 'Cultivated' (Mitchell and Delbridge 1965). These terms have long been used to categorise variation within Australian English accents, and phoneticians can agree with a high level of consistency which accent type many speakers belong to (Harrington *et al.* 1997: 158).

Broad, General and Cultivated Australian accents

Cultivated Australian is nearest to a formerly prestigious British English accent known as Received Pronunciation, or RP. It enjoys a certain prestige in Australia, and is a minority type, estimated as being spoken by about 10% of the population. Two well-known speakers with a Cultivated Australian accent are Malcolm Fraser (a former prime minister) and Joan Sutherland. Broad Australian is recognised as showing the most extreme, or okker, Australian accent and is the most stigmatised. Paul Hogan, of *Crocodile Dundee* fame, is a speaker of fairly Broad Australian who will perhaps be known to non-Australians. General Australian is assumed to be spoken by the majority of the population, and can be heard in many accents on the radio, for example.

Although it has now been the subject of some fairly detailed acoustic analysis (Bernard 1967; Clermont 1996; Harrington *et al.* 1997; Cox 1998), the basis for the distinction between the three accent types is primarily auditory, and has to do with the realisation of certain vowel phonemes. The most important of these are the long high front vowel phoneme /iː/, and the closing diphthongal phonemes /aʊ/, /aɪ/, /ɛɪ/ and /oʊ/. Table 7.3 gives typical allophones for the first four of these phonemes in Broad, General and Cultivated Australian (/oʊ/ is omitted because its realisation is more complex).

Table 7.3 Typical realisations of vowel phonemes signalling differences in Broad, General and Cultivated Australian English. [˗] = non prominent target; [̟] = tongue body more forward than indicated by symbol, [̠] = tongue body backer than indicated by symbol; [a] indicates a central vowel

Phoneme	Word	Broad	General	Cultivated
/iː/	he tea be	[əĭ]	[ɜ̆i]	[ĭi] [i]
/aɪ/	high tie buy	[ɑĭ] [ɔĭ]	[aĭ]	[a̟ĭ]
/aʊ/	how tout bough	[æŭ] [ɛŭ]	[aŭ]	[a̠ŭ]
/ɛɪ/	hay Tate bay	[aĭ]	[ɛĭ]	[eĭ]

The table shows that the Broad allophone of /iː/ is a diphthong with a prominent schwa as its initial target (the prominence of the schwa, and by inference the non-prominence of the high front unrounded ɪ offglide is conventionally indicated by the breve mark on the latter: ĭ). The General allophone for /iː/ has a schwa as the first target, but the prominence is switched: the schwa is now a non-prominent onglide, and the second target is a prominent high front unrounded [i]. The Cultivated allophone may only have a slight non-prominent [ɪ] onglide to the second [i] target, or just a single target monophthongal allophone: [iː].

These differences in the realisation of /iː/ are one of the most frequently cited in reference to the accent distinction, and are often referred to as 'ongliding in /iː/'. The implication is that Cultivated has no, or only a small, onglide; General has a more noticeable onglide than Cultivated (the tongue moves from a more central, schwa-like position); and in Broad the onglide has become so large that it actually takes over the prominence from the second target. Prominence is largely signalled by length, so the differences could also be transcribed phonetically as [əːɪ] Broad, [əiː] General, and [ɪiː] or [iː] Cultivated.

Table 7.3 also shows a second often-noted accent difference in the initial target in the diphthongs /aɪ/ and /aʊ/. As the accent becomes broader, the initial target in /aɪ/ shifts backer and higher, from [a̘] to [a] to [ɑ] to [ɔ]; so that out of context a Broad *tie* might be mistaken for a General or Cultivated *toy*, a Broad *buy* for a General/Cultivated *boy*. Likewise, the initial target in /aʊ/ shifts fronter and higher, from [a̘] to [a] to [æ] to [ɛ].

Out of context, a Broad *hay* may sound like a General *high*, a Broad *make* like a General *Mike*. This reflects, with increasing broadness, a downwards and backwards shift in the first target of /ɛɪ/ from [e] to [ɛ] to [a]. Accent changes in the realisation of the /oʊ/ diphthong are more complicated, owing to individual and possibly regional variation in the frontness of its offglide. In some Broad varieties, the distinction between /ɛɪ/ and /oʊ/ (e.g. *bait* vs. *boat*) can be very small.

It is important to understand that the three categories of Broad General and Cultivated are not discrete, and accent variation is assumed to be continuous. Thus it is possible to find, for example, a speaker described as General to Broad, with the appropriate allophones underlying this classification specified. It is also possible to find mixed speakers with, say, General or even Cultivated allophones for /ɛɪ/ but Broad allophones for /aʊ/. Under these circumstances, a description of an Australian accent in the Broad/General/Cultivated continuum may be useful in highlighting similarities or differences between forensic speech samples, but its use is still limited by the fact that the details of the distribution of such types in the population are not known.

The classification is non-discrete in another way, which is probably more important for forensic phonetics. As described above, the differences are understood as underlying between-speaker differences. However, since the Broad/General/Cultivated variation is assumed to correlate with socioeconomic and stylistic factors, a single speaker may vary along its continuum:

> It has been observed that most Australians have the ability to vary their accent (along the Broad–Cultivated spectrum) and frequently do so when the social circumstances make such changes desirable. For example, speakers of General

Forensic Speaker Identification

> Australian may well find it appropriate to 'downgrade' their accent in the direction of Broad Australian at a football match, where associations of low social status, lack of education and in-group solidarity may be found desirable. By contrast, at a formal social function where they wished to present an image of educated prosperity, our speakers may find it appropriate to 'upgrade' their accent in the direction of Cultivated Australian.
>
> Fromkin *et al.* (1996: 292)

Thus a speaker may select General values in relatively formal circumstances, but switch to Broader allophones in less formal circumstances. Such variation, for example in /iː/ ongliding, has been heard in the speech of Bob Hawke (another former prime minister), for example. Knowing about the details of how individuals vary along the Broad/General/Cultivated continuum is of obvious use in the proper evaluation of forensic samples.

Morphophonemics

Strictly speaking, the phonetic and phonemic levels are not the only two levels on which speech sounds are organised linguistically. In addition to talking about how sounds are produced (articulatory phonetics), or how they signal the differences between words (phonemics), it is also part of linguistic structure to take into account how the smallest units of meaning in a language, called morphemes, are constituted in terms of sounds. One thus talks about the sound shape of morphemes. When sounds are considered from this aspect they are called morphophonemes.

It is an interesting fact of linguistic structure that the same meaningful unit, or morpheme, is not signalled by an invariant set of sounds. In English, for example, the word *cups* consists of two morphemes. Morphemes are often notated in braces, so one morpheme in *cups* is {cup} and the other is {plural}. The meaning of the morpheme {plural} in English is something like 'more than one' (in other, especially Australian and Polynesian languages, where there is a {dual} morpheme meaning 'two of something', {plural} will mean 'more than two'). The meaning of the English morpheme {cup} – for example, how it differs from the morpheme {mug} – is not so easy to specify and is a question for lexical semantics.

In addition to their names, e.g. {plural}, and meanings, e.g. 'more than one', morphemes also have their sounds. Consider {plural} in English. It is regularly realised by three different sets of sounds: by /s/ in words like *cups* /kʌps/; by /z/ in words like *mugs* /mʌgz/ and by /əz/ in words like *dishes* /dɪʃəz/. The choice of variant, or allomorph, is determined by the final sound of the preceding morpheme, as follows.

- The allomorph /əz/ is selected when the preceding sound is /s, z, ʃ, ʒ, tʃ, or dʒ/. This can be seen in the words *horse horses* /hoːs hoːsəz/, *maze mazes* /mɛɪz mɛɪzəz/, *dish dishes* /dɪʃ dɪʃəz/, *garage garages* /gæraʒ gæraʒəz/, *church churches* /tʃɜːtʃ tʃɜːtʃəz/, and *judge judges* /dʒʌdʒ dʒʌdʒəz/. The phonemes /s, z, ʃ, ʒ, tʃ, or dʒ/ that select the /əz/ allomorph share the phonetic property of sibilance and are known as sibilants.

- The allomorph /s/ is conditioned by a preceding voiceless non-sibilant, like /t/ (*cut cuts* /kʌt kʌts/) or /θ/ (*death deaths* /dɛθ dɛθs/).
- The allomorph /z/ occurs after other (i.e voiced, non-sibilant) sounds, like vowels (*door doors* /dɔː dɔːz/, *car cars* /kːa kːaz/), nasals (*song songs* /sɒŋ sɒŋz/), and voiced stops and non-sibilant fricatives (*mug mugs* /mʌg mʌgz/, *sieve sieves* /sɪv sɪvz/).

This is what is meant by saying that morphemes are not necessarily signalled by an invariant set of sounds. Many linguists like to analyse the variation between allomorphs by recognising a basic, or underlying sound shape for a morpheme, and then deriving the sounds of the allomorphs by rule. In the case of the {plural} morpheme just illustrated, its basic or underlying sound shape would be z, because this is the form that occurs in the most general, least conditioned, environment, and the three allomorphs /z/, /s/ and /əz/ would be derived by phonological rules. When considered as the sounds of morphemes, sounds are called morphophonemes, or systematic phonemes. Thus the sound shape of the morpheme {plural} is the morphophoneme z.

Allomorphic variation is not confined to grammatical morphemes like the {plural} inflectional suffix, but can occur with lexical morphemes too. For example, most speakers of English will say /livz/ for the plural of *leaf*, and thus the morpheme {leaf} has two allomorphs: /lif/ with a voiceless labiodental fricative /f/, which occurs when the noun is used in the singular, and /liv/, with a voiced labiodental /v/, which occurs in the plural. In many varieties of English only a small subset of nouns, for example *roof*, *wife*, have this allomorphic variation between voiceless and voiced fricatives; others, for example *cliff* /klɪf klɪfs/ only have one allomorph.

Between-speaker and within-speaker allomorphic differences

In the section above on morphophonemics it was pointed out that a morpheme can have different phonological realisations, or allomorphs. Not much is known of between-speaker and within-speaker variation in allomorphy. Whereas there is probably little phonological variation in the realisation of grammatical affixes, like {plural} in Australian English, there certainly is between-speaker variation in which words show the f ~ v alternation in the singular and plural described above. Informal counts (in class) show that some speakers have a plural for *dwarf* of /dwɔːfs/, for example, while others have /dwɔːvz/.

Wells (1999: 41) notes the existence of the same sort of between-speaker differences in morphophonemic variation in British English in the case of the word *youths*. About 80% of his respondents say they prefer the form /juːðz/, that is, with a separate allomorph for {youth} in the plural. That leaves 20% of speakers (more for younger speakers) who said they prefer the form /juːθs/, that is, with just the one allomorph for {youth} in singular and plural. Interestingly, Wells notes that this represents a very different state of affairs from American English, where overall 61% prefer the single allomorph. The caveat, once again, is clear: evidence from phonemic and morphophonemic phenomena needs to be interpreted with respect to the language in question.

The extent of within-speaker variation in cases like these is not known.

Accent imitation

The foregoing discussion has shown that an accent is a complex thing. Offenders often try to assume an accent as part of a disguise. A discussion of the sound structure of language is a good place to point out how difficult it is to imitate an accent correctly. It means at the very least getting the phonemes, their sequences, and their realisations right. Since accents can also differ in intonation, this too must be got right.

Given the systematic complexity of accents, and the way they can vary depending on context, it is relatively easy for a linguist who knows the phonological structure of a given language variety to detect a bogus accent when it is used for disguise. This means in turn that a forensic phonetician will be often able to identify those aspects of the accent-disguised speech that reflect the speaker's true accent. It is important to understand that this refers only to disguised segmental *accents*. Other forms of voice disguise, for example using overall articulatory settings like nasalised voice, might be more difficult to spot.

Chapter summary

This chapter has shown how the sounds of language must be analysed on two different levels: the phonemic level of contrastive sounds, and the allophonic level, on which the contrastive sounds are realised. It has also shown how the conceptual framework of phonemic analysis is indispensable for the auditory comparison of forensic speech samples, providing as it does the basis of comparison for speech sounds in the different samples. Because of this, a knowledge of the phonemic system of the language(s) of the speech samples is crucial.

8

Speech acoustics

As explained before, it is generally accepted that forensic-phonetic investigation includes both auditory and acoustic analysis, and the results of acoustic analyses will therefore normally constitute part of the evidence in forensic speaker identification (cf. the cases of Prinzivalli and others cited at the beginning of this book).

In order to understand and evaluate the acoustic part of a forensic-phonetician's report, it is important to have at least a basic understanding of speech acoustics: how the acoustics of speech are produced, how they are described, and how differences, both between-speaker and within-speaker, arise in acoustic output. One of the aims of this chapter is to facilitate this understanding. Just as importantly, it also provides the information necessary to properly understand the demonstrations of the method using the Bayesian likelihood ratio in Chapter 11.

This chapter gives a simplified account of the acoustic theory of speech production, using the overall approach, and specific examples, of introductory textbooks like Ladefoged (1962), Lieberman and Blumstein (1988) or Johnson (1997). The most up-to-date and advanced account of acoustic phonetics is Stevens (2000). For obvious reasons, however, this chapter also concentrates more than these works do on between-speaker and within-speaker variation in acoustics. It is divided into four parts. The first part presents some basic information on speech waves, their acoustic dimensions and spectral representations. The second part explains how these acoustics are produced. The third and fourth parts apply this knowledge to two forensically very important areas of speech acoustics: vowel acoustics and fundamental frequency.

Speech sounds

Speech sounds, just like any other sounds, are rapid fluctuations in air pressure. Speech sounds are generated when air is made to move by the vocal organs. When someone speaks, acoustic energy is radiated from their vocal tract. This acoustic disturbance, consisting of pressure fluctuations, causes the listener's ear drum to move rapidly in and out – in when the pressure is positive, out when negative. Thus acoustic energy is transformed into mechanical energy at the ear-drum. This mechanical energy, and the information it contains, go through several more transformations before arrival as patterns of neural energy at the listener's brain. The processing of the information in the listener's brain results in the percept of sound.

The acoustic properties of the radiated speech wave are of great importance from the point of view of forensics, since they constitute the basis for both the phonetician's acoustic analysis and the phonetician's auditory transcription. This section will describe some of the main properties of the radiated speech wave, and, more importantly, will show how they are related to the vocal tract that produced them. We will start with some important acoustic properties, and we will do this by looking at the least complex examples – oral vowels.

Speech waves

The speech wave is distributed, at any given instant, as a sound pressure wave in the air around the speaker, and can be looked at as pressure varying as a function of distance from the speaker. Think of the distribution of the height of waves in the sea. At any particular instant one could describe the height of the wave as a function of the distance from, say, someone standing in the water. In the same way it is possible to specify the pressure of the speech wave *at a given instant* as a function of distance, having such-and-such a pressure at such-and-such a point in space (including within the speaker's vocal tract). The mathematical equation that describes pressure fluctuations in this way is called a *distance–pressure function*.

In speech acoustics, however, it is more common to consider the air pressure in a speech wave as varying not as a function of distance but as a function of time. This is equivalent to saying that at such-and-such a point in space the air pressure varies in such-and-such a way over time. Go back to the surf and imagine plotting the height of the water at one point in space – say at your leg – as it goes up and down over time. In speech, one such point in space is commonly the microphone that transduces the acoustic signal. The mathematical equation that describes pressure fluctuations in this way is often referred to as a *time–pressure function*, and variations in air pressure shown as a function of time are called *time–pressure waves*. The top panel of Figure 8.1 shows what a typical time–pressure speech wave looks like. Time is shown horizontally, pressure vertically. The utterance is the two words *speaker verification*, said very carefully and slowly by the author.

Of course, what is depicted in Figure 8.1 is not the pressure fluctuations themselves. It is a general scientific principle that we can never observe reality directly, and this is a good example. The information in Figure 8.1 is already at some remove from the actual event. The pressure variations have been picked up by a microphone in front of the speaker and transformed into variations in the electrical analogue of pressure, namely voltage. These have been then recorded on a tape recorder, and the resulting patterns of magnetic energy on tape have been played into a computer, digitised, and displayed on a computer screen. It is important to remember that in such transformations, information is always lost and distorted. This will be even more acute for every

Speech acoustics

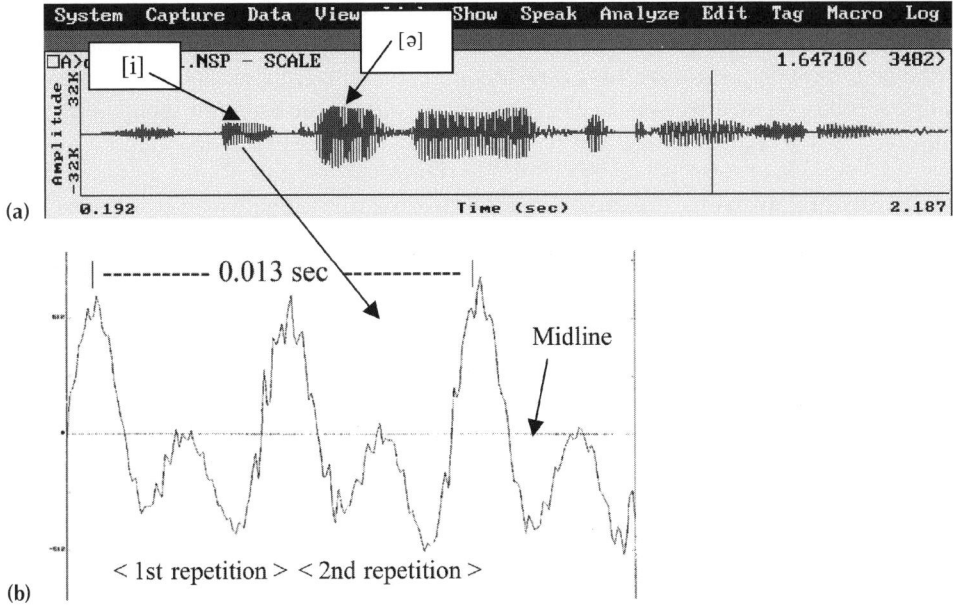

Figure 8.1 Wave-form of *Speaker verification* (above), and magnified portion of 'ee' vowel in *speaker* (below)

extra step in the transformation, as for example in the forensically typical case of transmission over the telephone.

The wave-form at the top of Figure 8.1 actually illustrates all four types of wave common in speech, but, because it is the least complex, we will concentrate on the type which corresponds to the oral 'ee' vowel [i] in the first syllable of *speaker*. Part of the speech wave of this vowel is therefore shown magnified in the bottom panel of Figure 8.1.

Frequency

It can be seen that the magnified speech wave in Figure 8.1b consists of rapid variations in air pressure as a function of time. Variations towards the top of the picture (away from the mid-line) represent positive increases in pressure relative to atmospheric pressure; variations towards the bottom are negative. The variations are periodic – they repeat – and are obviously complex, in that the air pressure can be seen to be varying simultaneously at several different frequencies. These frequencies can be roughly estimated visually as follows.

Fundamental frequency

Firstly, it can be seen that the magnified portion of the wave apparently consists of a little less than three repetitions of a complex pattern. (One repetition is taken to occur between recurring events in the wave, for example the peak values: the first and

197

second repetitions between the peak values are marked in the figure.) The first two repetitions can be seen to occur within 0.013 seconds. The elapsed time between recurring events is called the wave's *period*, and so the average period of this wave is 0.013 seconds divided by two, or 0.0065 seconds. The most common unit of acoustic duration is *milliseconds* (thousandths of a second, abbreviated msec or ms). So the average period of this wave is 6.5 msec. (Duration is also found quantified in *centiseconds* (hundredths of a second, abbreviated csec or cs), in which case the wave's period is 0.65 csec)

Frequency is expressed as number of times per second, or hertz. So if the wave has a period of 0.0065 seconds, in one second it will repeat (1/0.0065 =) 153.8 times, or 154 Hz (rounded off to the nearest Hz.). The frequency in Hz is thus the reciprocal of the period (1 divided by the period) in seconds (1/0.0065 sec = 154 Hz).

The rate of repetition of the complex wave is called its *fundamental frequency* (abbreviated F0, which is pronounced 'eff-oh' or 'eff sub-zero') and this wave therefore has an F0 of 154 hertz (Hz). Fundamental frequency is an extremely important measure in acoustic phonetics in general and forensic phonetics in particular, and will be discussed at much greater length later in this chapter. For the present, however, it can be noted that, at a first approximation, fundamental frequency correlates positively to a speaker's pitch: other things being equal, *the higher/lower the F0, the higher/lower the pitch*.

From the point of view of speech production, the F0 corresponds to the rate at which the vocal folds vibrate. Part of the production of the [i] in *speaker* in Figure 8.1 therefore involved my vocal cords vibrating for a short time at 154 Hz. From the point of view of speech perception, had you been in the vicinity of this wave-form, your ear-drum would have been going in and out 154 times per second as a response to these pressure fluctuations, and you would have experienced this as a particular pitch: about in the middle of my normal pitch range.

Higher-frequency components

Returning now to the magnified wave-form in Figure 8.1, within each single repetition of the basic pattern more fluctuations of pressure can be seen. There is one pattern of fluctuation that recurs twice in each period. Between two peak values, for example, the pressure first goes down, then up, then down, then up. These fluctuations in pressure are therefore occurring at about twice the F0, i.e. (2 × 154 Hz =) ca. 300 Hz. Finally there is a second pattern of pressure fluctuations that recurs about 15 to 16 times per period. These can most easily been seen in the second repetition. Starting from the beginning of the second main peak of the wave-form, for example, the pressure goes down a little, then up a little, then down a little, then up a little . . . about 15 to 16 times before arriving at the third main peak. Since there are between 15 and 16 of these little fluctuations per period, this means a frequency of between (16 × 154 =) ca. 2460 Hz and (15 × 154 =) ca. 2300 Hz.

These two higher frequency components above the fundamental, one at about 300 Hz, and one at about 2300 to 2460 Hz, *contribute to the phonetic quality of the vowel*. In this case a frequency at about 300 Hz and one at about 2400 Hz make the hearer's ear-drum go in and out at those frequencies, and make the wave-form sound like an 'ee' [i], as opposed say to a 'oo' [u] or 'ah' [a].

Thus the speech wave for a vowel contains two types of frequencies. One, the rate of repetition of the complex wave, or F0, corresponds to pitch; the other, the higher frequency components, corresponds to phonetic vowel quality. Since it is possible to say the same vowel on different pitches, and different vowels on the same pitch, these two sets of frequencies must be independently controllable. The acoustic theory of speech production, which is discussed later in this chapter, makes it clear how this happens.

Amplitude

In addition to frequency, there is a second obvious dimension in which variation in the speech wave occurs. It can be seen in the unmagnified speech wave in the top panel of Figure 8.1 that the vertical deflections in the part of the wave marked as [i] are not as great as those in the part marked [ə] (the 'er' vowel in *speaker*). This dimension is called *amplitude*: the wave-form in [ə] has a greater amplitude than in [i]. The main perceptual correlate of amplitude is loudness – other things being equal, wave-forms with greater amplitude will be heard as louder.

Spectral representation

By inspection of the magnified wave-form in Figure 8.1 above, the presence of acoustic energy at three different frequencies was demonstrated. There was energy present at the fundamental frequency; at a frequency twice the fundamental, and at a frequency 15 or 16 times the fundamental. (Energy is the capacity to do work: the work being done here is the generation of air pressure fluctuations at different frequencies.) However, it is difficult to see exactly how much energy is present, and, in reality there is also energy present at many more frequencies than one can actually see from the wave-form. For this, and many other reasons, speech acoustics often prefers not to quantify in terms of pressure variations over time, a so-called *time-domain representation*, but works instead in terms of the speech *spectrum*, or *frequency-domain representation*.

The spectral representation shows in a much clearer and mathematically tractable way *what amounts of energy are present at what frequencies*. It also makes the way the acoustics of speech are produced easier to understand. Below is a simplified account of how the energy present in the [i] wave-form in Figure 8.1 can be represented spectrally.

Fourier analysis

We have seen that one way of looking at the speech wave is as complex fluctuations in air pressure as a function of time. Another is as a spectrum, which shows exactly what frequencies are present with what amplitudes. But how does the spectrum relate to the time-domain speech wave? Fourier's theorem shows that any complex wave (like the speech wave in Figure 8.1) can be decomposed into, or represented as, a set of sine waves, also called sinusoids, each with its own frequency and amplitude. This is explained below.

Forensic Speaker Identification

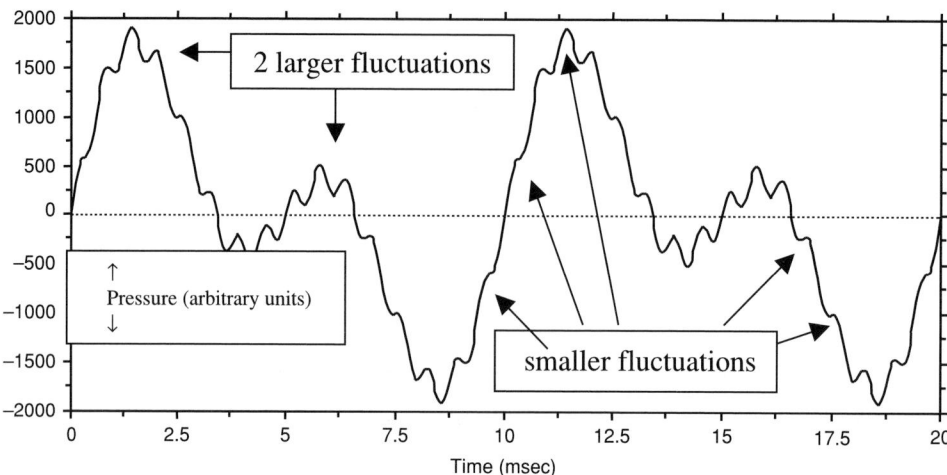

Figure 8.2 Two periods of a complex time-pressure wave

Figure 8.2 shows two periods, or two repetitions, of a complex wave that is very similar in shape to the one for [i] in Figure 8.1 (the time scale is in milliseconds; the pressure scale is arbitrary). The complex wave can be seen to repeat twice within 20 milliseconds, so its period is 10 milliseconds and its F0 is (1/0.01 sec =) 100 Hz. Within each 10 msec period can also be seen the two large fluctuations and 16 smaller ones already pointed out for Figure 8.1.

The complex wave in Figure 8.2 can be shown by Fourier analysis to be the result of adding together just the three sine waves shown in Figures 8.3a, 8.3b and 8.3c. The sine wave in Figure 8.3a has a frequency the same as the complex wave, i.e. 100 Hz, and an amplitude of 1000. The second sine wave, in Figure 8.3b, has a frequency double that of the complex wave at 200 Hz, and an amplitude also of 1000. The last sine wave, in Figure 8.3c, has a frequency 16 times that of the complex wave at 1.6 kHz, and a much smaller amplitude of 100.

Let us first show the summation graphically. To do this, all four waves – the complex wave and its three sinusoidal components – are shown together in Figure 8.4. The idea is that the complex wave is made up of the three sinusoids, such that the amplitude of the complex wave at any point in time, for example at point A at 1.25 msec, is the sum of the amplitudes of the three sinusoidal components at that point in time.

Since quantification is the name of the game in acoustic phonetics, we will introduce some here, and show how to calculate the pressure in the complex wave at a given point in time, namely when $t = 1.25$ msec (this is point A in Figure 8.4).

The calculation of the pressure in the complex wave at $t = 1.25$ msec can be illustrated as follows. In each of the sinusoids in Figures 8.3a, b and c, the value for the pressure at any given time is given by the formula at (8.1). In this formula, P is the pressure to be determined, A is the maximum positive amplitude of the sinusoid wave, t is the time, in seconds, at which the pressure is to be determined. ω, called the angular frequency, is 360 times the frequency of the wave, and is expressed as so many

200

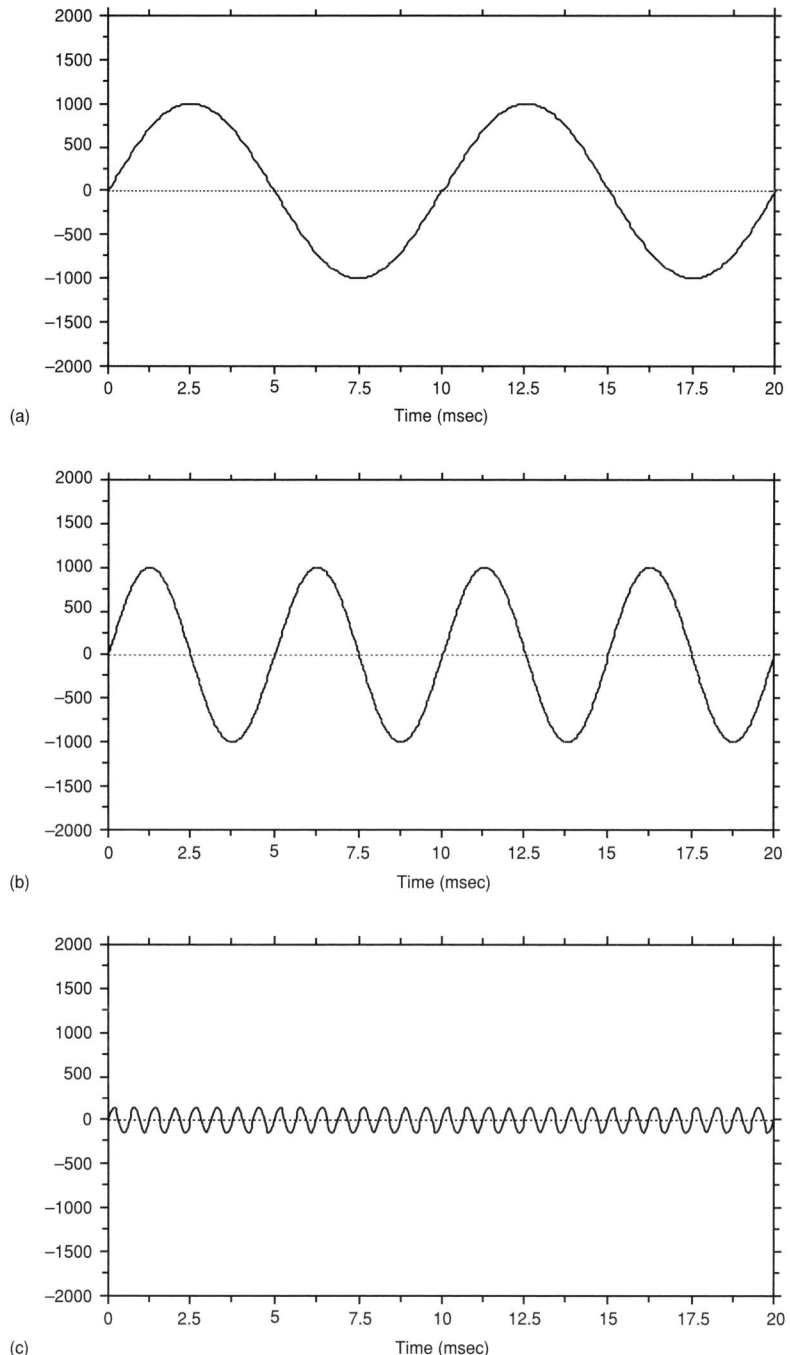

Figure 8.3 (a) Time-pressure sine wave with amplitude 1000 and frequency 100 Hz. (b) Time-pressure sine wave with amplitude 1000 and frequency 200 Hz. (c) Time-pressure sine wave with amplitude 100 and frequency 1.6 KHz

Forensic Speaker Identification

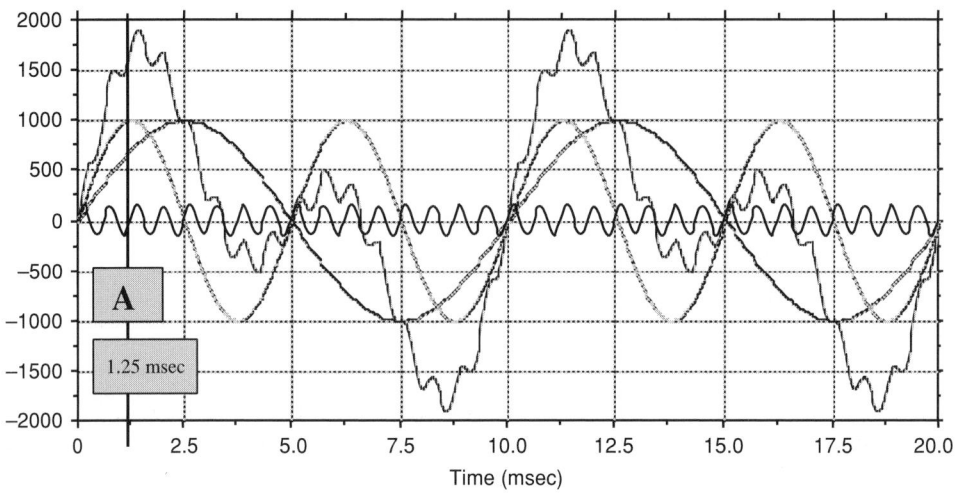

Figure 8.4 A complex wave and its sinusoidal components

degrees per second. Thus if the frequency of the wave is 2 Hz, the angular frequency ω is (360 × 2 =) 720 degrees per second. The number 360 and the use of degrees as a unit comes from the fact that the shape of a sinusoidal wave is related to circular motion, and 360 degrees represents one revolution in a circle and one full cycle of a wave. Sine (written 'sin' in formulae) is the name for one of the basic trigonometric functions relating the size of one angle in a right-angled triangle to the lengths of two of its sides, and the sine of an angle can be found on a calculator or looked up from a table: for example sin(90 degrees) is 1.

$$P = A \times \sin(\omega t) \tag{8.1}$$

Looking at the wave in Figure 8.3a, it can be seen that its pressure is zero when $t = 0$ msec, when $t = 5$ msec, and so on; it will have maximum positive pressure when $t = 2.5$ msec, or $t = 12.5$ msec, etc., and maximum negative pressure (−1000) at $t = 7.5$, $t = 17.5$ msec, and so on. The formula *Pressure = Amplitude × sin(ωt)* can be used to work out the pressure in this wave when time is 1.25 msec, as follows. The frequency of this wave is 100 Hz, that is, it repeats 100 times per second, and therefore ω will be (100 × 360 degrees/sec) = 36 000 degrees/sec. The maximum amplitude of the wave is 1000, so when $t = 1.25$ msec, the value of the pressure will be the maximum amplitude (1000) times the sine of the product of the angular frequency ω and the time (36 000 × 0.00125). This product is 45, and the sine of 45 (degrees) is 0.707, so the pressure at time = 1.25 msec is (0.707 × 1000 =) 707.

We do the same thing for the other two sinusoids in Figures 8.3b and c, and add up the results. Thus the pressure *P* in the complex wave in Figure 8.4 at any point in time is the sum of the three sinusoids each consisting of an amplitude term and a frequency term, as in the formula at (8.2).

$$P = [A_1 \times \sin(\omega_1 t)] + [A_2 \times \sin(\omega_2 t)] + [A_3 \times \sin(\omega_3 t)] \tag{8.2}$$

Speech acoustics

Table 8.1 Calculation of pressure of complex wave in Figure 8.4 when $t = 1.25$ msec

	Frequency (Hz)	Amplitude (arbitrary units)	ω (frequency × 360)	ωt	$\sin(\omega t)$	Pressure at 1.25 msec (amplitude × $\sin(\omega t)$)
1st sinusoid	100	1000	36 000	45	0.707	707
2nd sinusoid	200	1000	72 000	90	1	1000
3rd sinusoid	1600	100	576 000	720	0	0
Pressure of complex wave at 1.25 msec =						1707

The amplitude terms A_1 and A_2 of the first two sinusoids are both 1000, A_3 is 100; their frequencies are 100 Hz, 200 Hz and 1.6 kHz. Therefore ω_1 is 36 000, ω_2 is 72 000, and ω_3 is 576 000. Thus, the pressure of the complex wave at point A, where $t = 1.25$ msec, is

$$[1000 \times \sin(36\ 000 \times 0.00125)] + [1000 \times \sin(72\ 000 \times 0.00125)] +$$
$$[100 \times \sin(576\ 000 \times 0.00125)]$$

The calculations are worked out for the individual sinusoids in Table 8.1. The pressure of the complex wave at $t = 1.25$ msec is the sum of the pressures of the individual sinusoids, i.e. 1707. A glance at Figure 8.4 confirms that at the 1.25 msec time point the value of the pressure of the second sinusoid is 1000 units, that of the third is 0 units, that of the first is about 700 units, and the pressure of the complex wave is indeed the sum of all three, at about 1700 units.

The spectrum

The section above showed how a complex wave can be represented as the sum of a set of sinusoids, each with its own amplitude and frequency. The concept of spectrum can now be introduced. The spectrum corresponding to the complex wave in Figure 8.2 is shown in Figure 8.5. As can be seen, the spectrum is plotted on a frequency axis and an amplitude axis. *Each sinusoid is represented as a vertical line at its frequency with the length of the line corresponding to its amplitude.* This spectrum thus shows the wave to consist of energy at three frequencies: 100 Hz, 200 Hz and 1.6 kHz. It also shows the amount of energy present at each frequency: 1000 amplitude units at 100 Hz and 200 Hz, and 100 amplitude units at 1.6 kHz. Each of the lines – the spectral components – thus represents a sinusoid of a given frequency and a given amplitude. In this way, the spectrum shows *exactly how much energy is present at what frequencies* in the wave.

The Fourier process works both ways: it can be understood in terms of both analysis and synthesis. The section above has given an example of a Fourier analysis – the decomposition of a complex wave into a set of sinusoids – and it has also been shown how a complex wave can be synthesised to any desired approximation by summing the set of its component sinusoids.

Forensic Speaker Identification

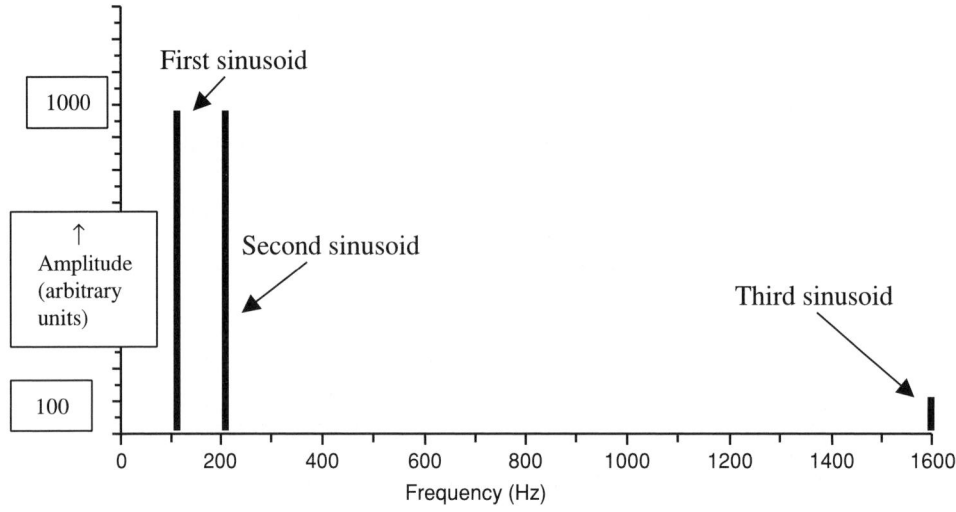

Figure 8.5 Spectral representation of wave-form in Figure 8.2

Unlike the time-domain wave-form, the spectrum no longer contains information about the variation of the wave over time. If the original wave is to be resynthesised, it is necessary to know one more thing: a specification of the alignment in time of the sinusoids. This is because, if the sinusoids are aligned differently in time, a different looking wave-form will result. The time alignment is given by a *phase* term, which shifts the time term t in each sine wave formula. Thus *if the frequency components in the spectrum in Figure 8.5 were summed with the appropriate phase relationships, the wave-form in Figure 8.2 would result*. Since it would introduce additional complexity, correct phase relationships will simply be assumed below.

The section above has shown the idea behind the spectral representation of a complex wave similar to the real one for [i] in Figure 8.1. It has shown in particular how the wave can be represented spectrally as a set of sinusoids each with its own amplitude and frequency. In reality, the situation is somewhat more complex, but enough detail has now been presented to enable the reader to understand spectral representation in the real thing, namely the [i] vowel in *speaker* illustrated at the beginning of this section,to which we now return.

Harmonic spectrum

There are many different types of spectral representation. One type, called a fast Fourier transform or FFT (Johnson 1997: 37–40), is shown in Figure 8.6. This is actually a spectral representation of the three periods of the [i] vowel shown in Figure 8.1. The axes are the same as in the line spectrum of Figure 8.5. Thus frequency is shown horizontally, running from 0 Hz to 5000 Hz, and amplitude is shown vertically and quantified in decibels (dB).

Unlike the line spectrum in Figure 8.5, the spectrum of [i] in Figure 8.6 shows a jagged profile. But important structure can be discerned in this profile. Firstly, the

Speech acoustics

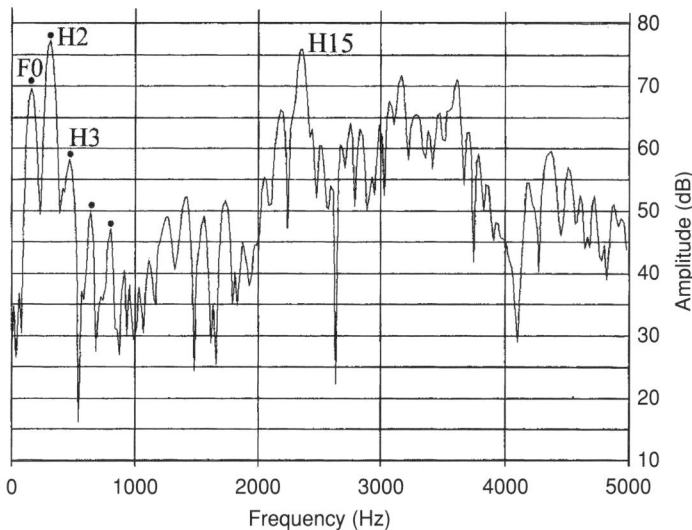

Figure 8.6 FFT Spectrum of [i] vowel

jagged profile can be seen to consist of local spikes of energy with different amplitudes, occurring at equal frequency intervals. Some of these are easier to see than others: the first five have been marked with black dots.

Although the spectral profile for the [i] in Figure 8.6 is more messy than the idealised line spectrum in Figure 8.5, the principle is the same. The local spikes of energy correspond to the sinusoidal components, each with a given frequency and amplitude. Thus the sinusoidal component with the lowest frequency can be seen to have an amplitude of about 70 dB. Its frequency is too hard to specify exactly, given the size of the frequency scale, but it can be appreciated by eyeballing the horizontal frequency scale that it is around 150 Hz.

These sinusoidal components are called *harmonics*, and the one with the lowest frequency is the fundamental frequency (the rate of repetition of the complex wave, and the main correlate of pitch). As explained above, each harmonic has a given frequency and amplitude, and can be thought of as representing a sinusoidal component which, when summed with the other components in the way demonstrated above, results in the time-pressure wave magnified in the bottom panel of Figure 8.1.

The fundamental frequency (marked as F0 in Figure 8.6) is the first harmonic, the next higher in frequency is the second harmonic (marked as H2 in Figure 8.6), the next higher the third (H3), and so on. The harmonics occur at whole number multiples of the fundamental. That is, assuming F0 is 150 Hz, H2 will be at 300 Hz and H3 at 450 Hz, and so on. This even spacing of the harmonics can be seen fairly well in Figure 8.6.

It is worthwhile noting how the spectrum, or frequency-domain representation, of [i] in Figure 8.6 relates to the expanded wave-form, or time-domain representation of [i] in Figure 8.1. The fundamental frequency in Figure 8.6, at about 150 Hz, corresponds to the fundamental frequency (measured at 154 Hz) in Figure 8.1. The second harmonic, at double the fundamental – about 300 Hz in this example – has an amplitude greater than the fundamental, at about 75 dB. It corresponds to the energy visually noted at twice the fundamental in the speech-wave in Figure 8.1. The third frequency component visually extracted from the speech-wave in Figure 8.1 was at about 2.4 kHz, and it can be seen that this corresponds to a relatively stronger, (i.e. amplitudinous) 15th harmonic, marked as H15 in Figure 8.6.

Smoothed spectrum

In addition to the fine harmonic structure in the spectrum, grosser structure can be detected. This can be best appreciated if the jagged structure of the harmonics is smoothed, as is shown in Figure 8.7. The smoothing has been carried out by a rather complicated, but well-established signal processing technique called linear prediction (or LP) analysis (Wakita 1976: 10–20; Johnson 1997: 40–4).

From Figure 8.7 it can be seen that the smoothed spectral envelope from the LP analysis shows five major peaks, these are marked P1 – P5. One peak (P1) is low in frequency, at about the frequency of the second harmonic, and the rest are above 2000 Hz. *The frequencies of the lowest three major peaks are the primary correlates of vowel quality*: they are what makes this vowel sound like an [i], and not a [u] or [a]. It can also be appreciated that the two peaks with the lowest frequencies (P1 and P2) also correspond to the energy noted above at twice the fundamental (at ca. 300 Hz), and ca. 15–16 times the fundamental (at 2.4 kHz).

In an oral vowel like [i], then, acoustic energy is present at the fundamental and at harmonics that occur at whole-number multiples of the fundamental. This harmonic

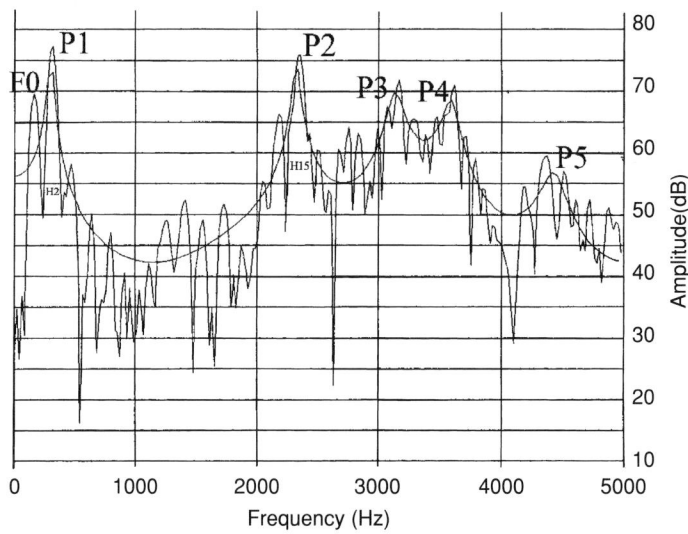

Figure 8.7 Harmonic and linear prediction spectra for [i]. F0 = fundamental frequency; P1 – P5 = first to fifth peak in the LP spectrum; H2, H15 = second and fifteenth harmonic respectively

structure – in particular the lowest harmonic or fundamental frequency – encodes pitch. Furthermore, some of the harmonics will have greater amplitudes than others, and this resolves into a grosser structure of several major peaks. The frequencies of the lowest two or three of these major peaks encode vowel quality. *This acoustic structure not only encodes linguistically relevant pitch and vowel quality, but also carries the imprint of the vocal tract that produced it.* The following sections will explain how this particular structure of harmonics and major spectral peaks is produced, and how it reflects the anatomy of the speaker.

The acoustic theory of speech production

What gives rise to the radiated time–pressure variations of the speech wave? Where do its properties of fundamental frequency, harmonics, and spectral envelope crucial to the signalling of pitch and vowel quality come from? The theory that explains the radiated acoustics in terms of the vocal mechanism that produces them is called the *acoustic theory of speech production*, or *source–filter theory*. It was developed by the Swedish speech scientist Gunnar Fant, and the first full account was given in his 1960 book *Acoustic Theory of Speech Production*. Part of the book tests how the theory works by first predicting, from source–filter theory, the acoustic output of a Russian speaker using estimates of the size and shape of his supralaryngeal vocal tract derived from X-rays, and then comparing the predicted output with the speaker's actual output.

It is worth pointing out that source–filter theory is different from other theories in linguistics. Linguistics, as a hermeneutic science – that is a science of interpreting human behaviour – is full of competing accounts of syntactic, semantic and phonological phenomena (phonemics is one!). Source–filter theory, on the other hand, does not have any competitors and has not as yet been falsified. It is a received theory that is largely responsible for 'raising the field of acoustic phonetics toward the level of a quantitative science' (Stevens 2000: vii). As such, all forensic phoneticians should be expected to know and apply it. It is therefore crucial that its basic ideas are described.

Source–filter theory relates acoustics to production in terms of the interaction of two components: a *source* (or sources) of energy – the energy input into the system – and a *filter*, which modifies that energy. We consider the source first. As before, we restrict the discussion to the most simple case of oral (as opposed to e.g. nasalised) vowels.

Source

In the case of oral vowels which we are considering, the source of acoustic energy is at the larynx, and consists of the flow of air, previously initiated by lung activity, modulated by vocal cord activity. As the air flows through the glottis, vocal cord oscillation is started, whereby the cords come together, stay together for an instant, and then come apart. The cycle is then repeated as long as the aerodynamic and muscular tension conditions for phonation are met. The result of this is a periodic stream of high-velocity jets of air being shot into the supralaryngeal vocal tract.

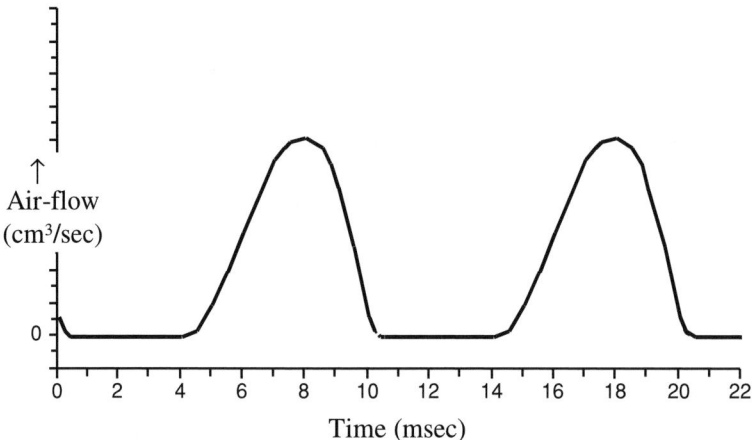

Figure 8.8 Two periods of an idealised glottal volume velocity wave with an F0 of 100 Hz

A typical plot of the volume of air flowing through the glottis over time, the so-called *glottal volume-velocity wave*, is shown in Figure 8.8. The volume velocity is shown vertically and quantified in terms of so-many cubic centimeters (cm^3) of air per second. Time is shown horizontally. It can be seen that the portion of the particular wave shown in Figure 8.8 repeats twice in 20 milliseconds. This means its fundamental frequency is 100 Hz.

The time profile of the wave-form in Figure 8.8 can be explained as follows. The portions of zero air-flow (for example from ca. 0 to 4 msec) are when the cords are closed, allowing no air to flow through the glottis. When the cords come apart, the flow of air increases, reaches a maximum, and then decreases as the cords come together again. Usually the rate of decrease is greater than that of increase, and the most rapid change in flow rate occurs during the closing of the cords (in the figure, this would be at about 9.5 msec and 19.5 msec). The time course of the flow can therefore be divided into three parts: a closed phase, an opening phase and a closing phase. It can be seen from the air-flow profile that the cords are closed for about half of a single glottal cycle.

The air-flow profile in Figure 8.8 is a time-domain representation: it shows how the air-flow changes as a function of time. As explained above, however, the precise energy content of a time-domain wave can be best understood in its spectral, or frequency-domain transformation. The spectrum of the wave in Figure 8.8 is shown in Figure 8.9. This figure shows the energy present in the volume velocity wave-form: the energy input into the system.

Figure 8.9 shows the set of sinusoids, or harmonics, specified by frequency and amplitude that, when summed with the appropriate phase relations, would give the time-domain volume velocity wave-form in Figure 8.8. Figure 8.9 shows that energy is present in the volume velocity wave at many discrete frequencies. Energy is present at the fundamental, i.e 100 Hz, but also at whole-number multiples of the fundamental – in this example at 200 Hz, 300 Hz, 400 Hz, etc. The fundamental, and second and fourth harmonics are marked in the figure.

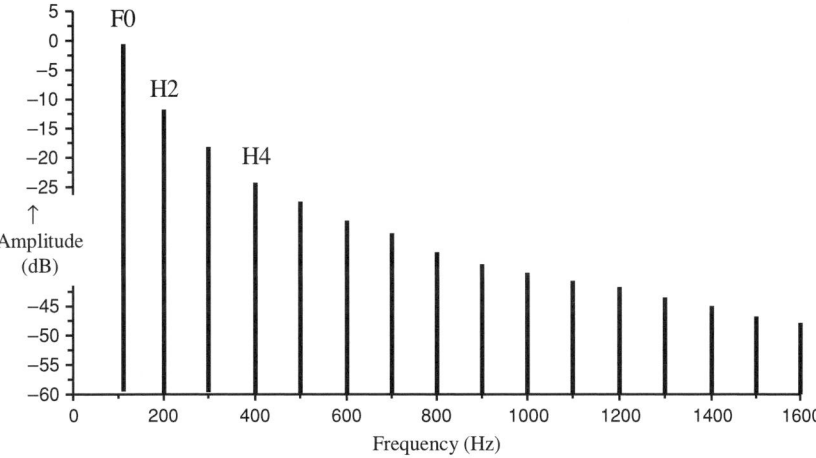

Figure 8.9 Idealised spectrum (up to 1.6 kHz) of the glottal volume velocity wave shown in Figure 8.8. The fundamental (F0) and second and fourth harmonics (H2, H4) are indicated

It can be seen that the spectrum of the glottal source is very rich in harmonics. This is because the more a wave deviates from a sinusoidal shape the more sinusoidal components will be required to approximate it. The complicated time profile of the volume velocity wave deviates considerably more from a sinusoid than, say, the wave form of the radiated speech wave in Figure 8.2, so many more sinusoidal components are needed to specify the energy in the volume velocity wave than the three shown in Figure 8.5 for the wave in Figure 8.2.

It can also be seen that the amount of energy in each harmonic present drops off quite sharply: the accepted representative value for the modal phonation of males is −12 dB per octave. This means that the amplitude of a harmonic will be 12 dB less than that of a harmonic with half its frequency. The difference in amplitude between the fundamental at 100 Hz and H2 at 200 Hz is 12 dB, as is also the difference in amplitude between H2 at 200 Hz and H4 at 400 Hz. This property of the spectrum is called the *spectral slope*: the spectrum in Figure 8.9 has a spectral slope of −12 dB/octave.

Suppose now that the speaker were to increase their fundamental frequency, by increasing the rate of vocal cord vibration, from 100 Hz to 200 Hz. The source spectrum would now have energy present at 200 Hz (the fundamental frequency), and at whole-number multiples of this: 400 Hz (the second harmonic), 600 Hz (the third harmonic), etc. This is shown in Figure 8.10, where it can also be seen that the glottal source spectral slope of −12 dB/octave is preserved.

If the speaker were then to decrease their rate of vocal fold vibration, the fundamental would decrease and the harmonics would concertina together again.

Filter

As just pointed out, Figures 8.9 and 8.10 show the energy content of the volume velocity wave at the glottis. These spectra specify the energy input to the system. This

Forensic Speaker Identification

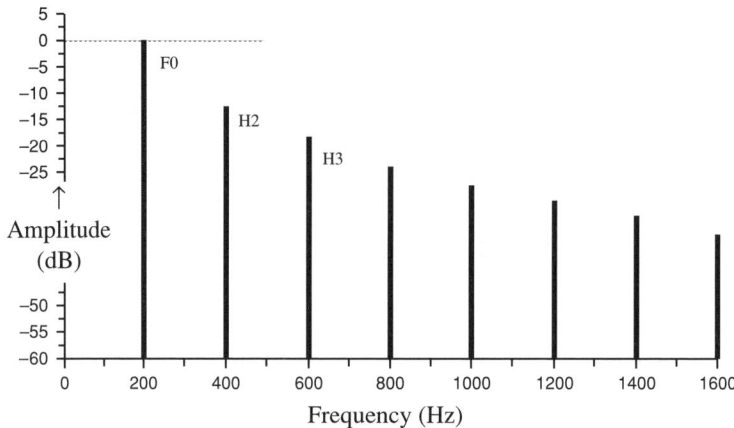

Figure 8.10 Idealised spectrum of glottal source (up to 1.6 kHz) with F0 of 200 Hz

energy is then modified by its passage through the supralaryngeal vocal tract. The contribution of the supralaryngeal vocal tract is to act as an acoustic filter that suppresses energy at certain frequencies and amplifies it at others.

What actually happens? During the production of a vowel, air is being expelled at a fairly constant rate through the vocal tract. (This is usually referred to as a *pulmonic egressive airstream*, because the movement of air is initiated by the lungs, and the direction of the movement of air is outwards). As explained above, this airstream is interrupted by the vibratory action of the vocal cords, so that a sequence of high-velocity jets of air is injected into the supralaryngeal vocal tract. The effect of these high-velocity jets is to cause the air present in the supralaryngeal vocal tract to vibrate. The way the air vibrates – in particular, the frequencies at which it vibrates and the amplitude of those frequencies – is determined by the shape of its container: the supralaryngeal vocal tract. The point in time at which the main response of the supralaryngeal air occurs corresponds to the most rapid change in the glottal flow rate described in Figure 8.8, that is, when the cords are closing. This process has been compared to water from a tap dripping into a bucket full of water. When the drip hits the water it produces a ripple that radiates outwards. At the same time there is a continuous flow of water over the side of the bucket (Daniloff *et al*. 1980: 165).

Again, spectral rather than time-domain representations can be used to make the process – how the source energy is modified by the filter – easier to conceptualise. The way in which the air in the supralaryngeal vocal tract will vibrate, given a particular supralaryngeal vocal tract shape, can again be conveniently shown by a frequency–amplitude spectrum, often called a transfer function. Figure 8.11 shows such a spectrum. This spectrum, or transfer function, represents the acoustic response of the air in the supralaryngeal vocal tract for a schwa [ə] – a vowel like that in the word *heard* – when said by a speaker with a supralaryngeal vocal tract 17.5 cm long. (The length of the supralaryngeal vocal tract means the distance between the glottis and the lips.)

This particular vowel, and this particular supralaryngeal vocal tract length, are chosen because they represent, productionally and acoustically, the least complex case. When a schwa is produced, the shape of the supralaryngeal vocal tract can be

Speech acoustics

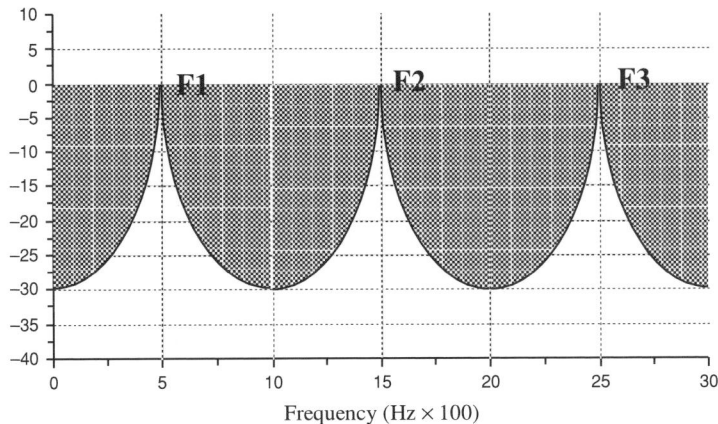

Figure 8.11 Transfer function for schwa said with vocal tract 17.5 cm long

considered to approximate a tube of uniform cross-sectional area. The length of 17.5 cm simply results in acoustical values that are easy to remember and must not be taken to imply that all speakers have supralaryngeal vocal tracts 17.5 cm long, nor that a given speaker's supralaryngeal vocal tract has an invariant length.

Formants

As can be seen from the spectrum in Figure 8.11, exciting the air in a supralaryngeal vocal tract with uniform cross-sectional area gives rise to an acoustic response whereby the air vibrates with maximum amplitude at the frequencies of 500 Hz, 1500 Hz, and 2500 Hz. (Mathematically, these frequencies exist to infinity, but for practical purposes, both phonetically and forensically, only the first few are considered.) It can also be seen that there is minimum energy in-between these frequencies, at 1000 Hz, 2000 Hz, etc.

The frequencies at which there is maximum energy are called *resonant frequencies*. The air in the supralaryngeal vocal tract is vibrating with maximum amplitude at these frequencies. In acoustic phonetics, vocal tract resonances are usually called *formants*. Formants are very commonly used in forensic speaker identification, both in experimental work and case-work (e.g. Greisbach *et al.* 1995; Ingram *et al.* 1996; Kinoshita 2001; LaRiviere 1975). The formant with the lowest frequency is the first formant, or F1. The most important dimension of a formant in acoustic phonetics is the frequency at its point of maximum response – its so-called *centre frequency*. So in Figure 8.11, F1 has a centre frequency of 500 Hz, F2 has a centre frequency of 1500 Hz etc. The reader might like to give, from Figure 8.11, the centre frequency of F3.[1]

In order to specify the overall spectral shape in Figure 8.11, additional formant dimensions are needed. The other dimensions of formants are their (peak) amplitude

---

1 The centre frequency of F3 is at 2500 Hz.

and bandwidth. The *formant amplitude* can be thought of as its height: in Figure 8.11 the amplitude of all formants is 30 dB. (Amplitude is often measured relatively, from a peak value downwards. This has been done in Figure 8.11, thus the peak amplitude of all formants in Figure 8.11 is 0 dB.)

The *formant bandwidth*, which is quantified in Hz, is a measure of the formant's width, and reflects the amount of energy absorption: high energy absorption, for example from compliant, acoustically absorbent vocal tract walls, results in wider formants. The bandwidth for all the formants in Figure 8.11 at 15 dB down from peak is about 200 Hz, or 100 Hz either side of the formants' centre frequency. In other words, for example, the width of F1 at 15 dB down from its peak is from (500 Hz − 100 Hz =) 400 Hz to (500 Hz + 100 Hz =) 600 Hz. This means that in the frequency range between 100 Hz above and below a formant's centre frequency, energy will be passed with at least half that formant's amplitude of 30 dB. Energy present at 500 Hz, for example, will be passed with maximum amplitude (30 dB); energy present at 400 Hz will be passed with 15 dB, half of 30 dB; energy present at 300 Hz, however, will be passed with less than 15 dB, that is less than half the formant's amplitude.

Interaction of source and filter

The spectral envelope in Figure 8.11 is abstract: it shows how the air *would* vibrate, given a supralaryngeal vocal tract in the shape for a schwa (= with uniform cross-sectional area). This abstract representation is often called a *filter* or *transfer function*. Now we need to see what happens when in our *gedankenexperiment* we actually provide some energy from the laryngeal source. Recall that the energy input to the system, for vowels, is the spectrum of the volume velocity wave at the glottis (cf. Figures 8.9 and 8.10 above). Let us assume that we have a source with a fundamental frequency of 100 Hz, as in Figure 8.9. This means that the vocal cords are vibrating and allowing a high-velocity jet of air to be injected into the supralaryngeal vocal tract 100 times a second. Recall also that the supralaryngeal vocal tract is in the shape for schwa, that is, with uniform cross-sectional area.

Figure 8.12 shows what happens when the source is combined with the filter. Visually, it appears as if the shape of the transfer function has been superposed on the harmonically rich spectrum of the source. It can be seen that energy contributed by the source is now present at the fundamental and harmonics. It can also be seen that the overall falling shape of the source spectrum has been modified by the envelope of the supralaryngeal vocal tract response. It no longer falls off in a simple slope, but has the three formant peaks of the transfer function. Thus the amplitude of the fundamental (F0) has been attenuated because it lies far away from the first formant. Its amplitude is now about 10 dB, and low compared to that of the 5th harmonic which, because it lies directly under the peak of the first formant frequency at 500 Hz, is passed with maximum amplitude (30 dB). (Note that amplitude in this figure is not quantified down from peak, as in the previous Figure 8.11.)

It is important to understand that the harmonic structure is independent of the transfer function. During speech, the fundamental frequency moves up and down and the harmonics concertina in and out as the speaker changes their pitch by changing

Speech acoustics

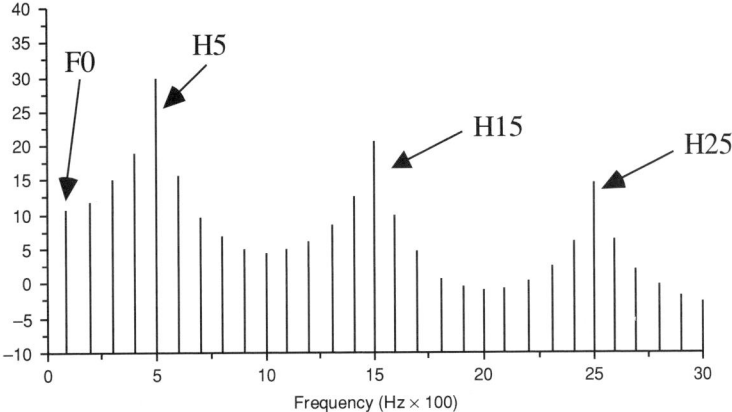

Figure 8.12 Spectrum for schwa on F0 of 100 Hz with vocal tract length of 17.5 cm

the rate of vibration of their vocal cords. The formant centre frequencies move up and down as the speaker changes their vowels by changing the shape of their supralaryngeal vocal tract. A harmonic does not therefore have to coincide with a formant centre frequency. Thus if the transfer function in Figure 8.11 were excited with the 200 Hz source in Figure 8.10, there would be no energy under the first formant transfer function peak at 500 Hz, because the second harmonic would be at 400 Hz and the third at 600 Hz. The fifth harmonic, instead of being directly under the F1 peak, would now be at 1000 Hz, and be maximally attenuated.

One feature of the source that appears preserved in the spectrum in Figure 8.12 is its overall slope. The formant peaks in the transfer function for schwa in Figure 8.11 all have the same amplitude, but the harmonics under these peaks in Figure 8.12 actually decrease in amplitude. The overall slope is not the original one of −12 dB/octave, however, but less: −6 dB/octave. This represents a complication that is often encountered in source–filter representations. The effect of radiation from the vocal tract is to decrease the spectral slope by about 6 dB/octave, since higher frequencies radiate more easily than lower frequencies. This radiation effect is often built in to spectral source–filter models of speech production, as here.

In the previous section it was pointed out that a complex wave can be synthesised from its sinusoidal components. Now, to look at the process once again from the point of view of synthesis, if the sinusoidal components of the harmonic spectrum in Figure 8.12 are summed with the appropriate phase relationships, and converted to a time-domain representation, the resulting wave-form would resemble that for a schwa said with a fundamental frequency of 100 Hz, when it is picked up by a microphone a given distance from the speaker's lips.

Determining the formant centre frequencies for schwa

In the previous section, the first four formant centre frequencies for schwa were quoted as 500 Hz (F1), 1500 Hz (F2), 2500 Hz (F3) and 3500 Hz (F4). Where do these numbers come from? The supralaryngeal vocal tract for schwa approximates a uniform tube.

We introduce a further simplification and assume that the tube is closed at one end – the larynx end. (This is a simplification because, as has been pointed out in reference to Figure 8.8, when the vocal cords are vibrating normally the tube is effectively open at the larynx end about half the time.)

The frequencies at which the air will vibrate with maximum amplitude in a tube of uniform cross-sectional area closed at one end (its resonant frequencies) are a function of the tube's length, and are easy to calculate. The formula is given at (8.3). This formula says that the frequency, in Hz, of any resonance of such a tube is two times the resonance number, minus one, times the speed of sound in centimeters per second divided by four times the length of the tube in centimeters. For example, assuming conventional values of 35 000 cm/sec as the speed of sound, and an average male supralaryngeal vocal tract length of 17.5 cm, the frequency of the first resonance will be $[(2 \times 1) - 1] \times 35\,000/(4 \times 17.5) = 500$ Hz.

> Formula for resonant frequencies F in hertz for tube of uniform cross-sectional area and length l in cm closed at one end. c = speed of sound in cm/sec.

$$F_n = (2n - 1) \times \frac{c}{4l} \qquad (8.3)$$

Thus the lowest frequency at which the air will vibrate with maximum amplitude in a 17.5 cm long supralaryngeal vocal tract of uniform cross-sectional area closed at one end is 500 Hz, and this is the frequency of the first resonance, or first formant F1. Substituting 2 for n in the formula gives 1500 Hz, or the second formant frequency, and the frequencies of F3 and F4 will be 2500 Hz and 3500 Hz, respectively.

The formula at (8.3) shows the resonance frequencies to be a function of l, the length of the supralaryngeal vocal tract (the speed of sound c is assumed to be a constant). It is easy to see that *as the length l of the supralaryngeal vocal tract increases, the resonant frequencies decrease, and vice versa*. Specifically, the resonant frequencies vary with, and are inversely proportional to, the length of the supralaryngeal vocal tract. So a speaker with a short supralaryngeal vocal tract will have higher resonances and a speaker with a longer supralaryngeal vocal tract will have lower resonances. One of the commonest instantiations of this relationship is the difference between males and females. Since females have on average about 20% shorter supralaryngeal vocal tracts than males, their formant frequencies will be on average about 20% higher. However, the same sort of variation will also be found within members of the same sex who have different length supralaryngeal vocal tracts.

For example, moderate positive correlations have been demonstrated between speaker height and vowel formant frequencies for German speakers of both sexes (Greisbach 1999.) The best correlations were found for F3 and F4, and the best vowel was the mid front-rounded [ø], which is fairly similar to schwa. Presumably these correlations reflect the relationship between vocal tract length and supralaryngeal resonances. The moderate degree of correlation observed – for any given height there was a fairly wide range of corresponding formant values – presumably reflects two additional sources of variation: in the relationship between height and vocal tract length, and also between-token variation in the vowels themselves.

Area function

The prediction of the resonant frequencies of a vocal tract from its length is a specific instance of a more general relationship between the shape of the vocal tract and its transfer function. The transfer function of a vocal tract can be predicted from a specification of how its cross-sectional area varies as a function of its length. This is known as its *area function*. Part of the area function of a supralaryngeal vocal tract in the shape for [i] might be that at the mid pharynx its cross-sectional area is 10.5 cm^2, whereas at the middle of the hard palate its cross-sectional area might be 0.65 cm^2. A glance back at the three-dimensional shape for the [i] vowel in Figure 6.8 will help understand this. The calculation of the formant frequencies for schwa is the simplest case because the area function of a supralaryngeal vocal tract in the shape for a schwa is uniform.

If the source spectrum can be fully specified, and the area function is known, the acoustic output can be predicted. The *acoustic output of a vocal tract is thus a unique function of the vocal tract that produced it*. In other words, if we know the dimensions of the vocal tract we can predict its acoustic output. (Or, more accurately, since the vocal tract is usually in motion from one speech sound target to another, if we know the trajectories of the vocal tract over a short stretch of time we can predict the trajectories of the acoustics.) It is in this sense that *the acoustic wave-form carries the imprint of the tract that produced it*.

Given adequate initial assumptions, for example a specification of the supralaryngeal vocal tract length, and radiation characteristics, it is possible to digitally reconstruct a possible supralaryngeal vocal tract area function from an acoustic wave-form. This is called vocal tract modelling. However, it is necessary to point out that *the relationship in source–filter theory is assumed to be unidirectional from articulation to acoustics*. Source–filter theory allows acoustics to be predicted from the area function, for example the formant frequencies from the supralaryngeal vocal tract length. But since different articulations are assumed to be able to give rise to effectively the same area function, it is not possible to infer the articulation from the acoustics. This is because effectively the same acoustic effect can be achieved by different articulations. A common example of this is in the parameter of tongue height, which can be controlled by deliberately lowering the tongue body, or by holding the tongue in a fixed position and lowering the jaw. Moreover, it has long been known that, in a few instances, very similar acoustics can be produced by slightly differing area functions. For example, rounding, or protruding, the lips or lowering the larynx involve different area functions (as well as different articulations), which all have similar lowering effects on formant frequencies (Lieberman and Blumstein 1988: 47–9).

Forensic significance: Vocal tract length and formant frequencies

The relationship just demonstrated between supralaryngeal vocal tract length and formant frequencies in schwa affords the first example of the relationship between supralaryngeal vocal tract anatomy and acoustics that is provided by source–filter theory. It may appear promising for forensic phonetics to be able to establish a function between acoustics and speaker anatomy. However, for several reasons, this relationship must be treated with caution.

Firstly, the human supralaryngeal vocal tract is highly deformable and its length is not invariant for a given speaker. Every human is capable of lengthening or shortening their supralaryngeal vocal tract length, and this happens as a matter of course during speech. Altering the vertical position of the larynx, or protruding the lips, or both, will alter the length of the tract by at least a centimeter. For a demonstration of this *in vivo*, compare the height of my larynx in the X-rays in Figures 6.1 and 6.12. In Figure 6.1, for [u], my vocal cords are at about the height of the inferior border of the body of the 6th cervical vertebra; in Figure 6.12, for [ɑ] they are at about the height of the superior border of the 5th. In me, this is a difference of more than two centimeters.

Since for many speakers a difference of a centimeter constitutes at least 5% of supralaryngeal vocal tract length, the formula at (8.3) predicts that such a difference will consequently increase or decrease the frequency of their schwa vowel formants by the same amount.

Moreover, although speakers certainly differ with respect to resting supralaryngeal vocal tract length, the *range* of differences in the relevant normal population (adult Australian caucasian males, say) is not great. The plasticity of the individual supralaryngeal vocal tract and the narrow range of variation between individuals can therefore result in low ratios of within-speaker to between-speaker variation. Before convicting a suspect with a long supralaryngeal vocal tract of a crime committed by someone with low formant centre frequencies, it is important to be sure that it could not also have been committed by someone with an average length supralaryngeal vocal tract who habitually speaks with a lowered larynx.

Because of the unidirectionality of the relationship between articulation, area function and acoustic output, it is possible to specify, at least in theory, only a range of possible supralaryngeal vocal tract configurations that could have been responsible for a given acoustic output. Such a model would clearly be of use in forensics, since it would be able to say whether the suspect's supralaryngeal vocal tract *could* have produced the observed acoustics, or whether they are excluded. The development of such a model was foreshadowed several years ago, but is not yet a reality.

Illustration with real schwa

It is now appropriate to illustrate the above theory with some more real data, this time from a schwa taken from the second vowel in the word *speaker*, said by the author. (For demonstration purposes, the syllable *-ker* was abnormally stressed, and the vowel was abnormally lengthened.) Figure 8.13 shows its harmonic spectrum, and an LP-smoothed spectral envelope has also been imposed, as was demonstrated above in Figure 8.7 for an [i] vowel.

The characteristic jagged harmonic structure is clear, especially at the lower frequencies. Although it is difficult to read off the exact frequency of the fundamental because of the size of the horizontal scale, its frequency can be easily estimated using the fact that harmonics are whole number multiples of the fundamental. It can be seen that there appear to be nine and a half harmonics within the range from zero to 1000 Hz. Recalling that the harmonic frequencies are at whole-number multiples of the fundamental, this means that the fundamental frequency for this particular schwa is 1000/9.5, or ca. 105 Hz. My vocal cords were thus vibrating at about 105 times per second during its production.

Speech acoustics

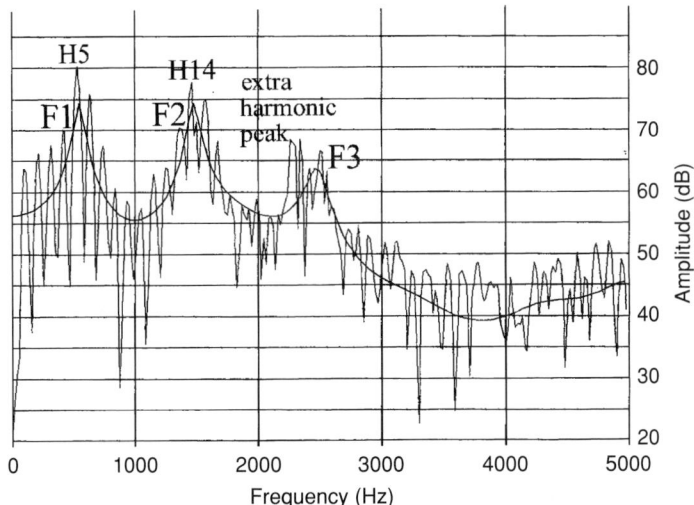

Figure 8.13 Harmonic and linear prediction spectra for real schwa. F1, F2, F3 = first, second and third formants respectively; H5, H14 = fifth and fourteenth harmonics respectively

The overall profile of the LP spectral envelope, with three major formant peaks (F1, F2, F3), is also clear and very similar to that in the idealised schwa spectrum in Figure 8.12. The formants are at slightly different frequencies from the values in the idealised schwa. For example, F1 appears slightly higher than 500 Hz. This can be seen from the fact that F1 is slightly higher than the fifth harmonic (H5), which, since F0 is ca. 105 Hz, is slightly higher than (5 × 105 =) 525 Hz.

Subglottal resonances and antiresonances

Although there is considerable similarity with the idealised schwa spectrum in Figure 8.12, two major differences are evident. There is, firstly, the conspicuous absence of an expected clear fourth formant peak at about 3.5 kHz in the real data. As mentioned above (in the discussion of Figure 8.11), the acoustic theory of speech production predicts a transfer function for schwa that has evenly spaced peaks to infinity, therefore we would expect the amplitude of the harmonics around 3.5 kHz to increase. That there is only a minimal increase is evidence that energy has been absorbed at this frequency by some other effect.

The second difference between the real and ideal data is that, although only one formant peak has been extracted by the LP analysis between 2 and 3 kHz – the third formant at an expected 2.5 kHz – closer inspection of the harmonic structure in this frequency range reveals the presence of two harmonic peaks: one at about 2.5 kHz, and one below it at about 2.3 kHz. (The lower of the two is labelled as the extra harmonic peak.) Source–filter theory predicts that there should only be one peak in the harmonic structure for schwa in this frequency range, namely that evinced by the amplification effect of the third formant. Therefore an additional resonance is occurring from some other source.

217

Both the absence of a fourth formant and the presence of an additional harmonic peak in the real data are in all probability related, and it is important to explain how. As already mentioned, for the calculation of the transfer function of the supralaryngeal vocal tract for a particular vowel like schwa, source–filter theory makes the simplifying assumption that the supralaryngeal vocal tract is closed at one end. In modal phonation, the vocal cords will be together, and the supralaryngeal vocal tract closed off at the larynx end, about 50% of the time. This is for linguistic-phonetic purposes effectively the same as a complete closure.

Under certain conditions, however, the supralaryngeal tract is not effectively closed off, but is to a certain extent continuous with the trachea below it, and this affects the supralaryngeal vocal tract transfer function. These conditions occur primarily in sounds produced when the vocal cords are wide apart. Such sounds, called spread glottis segments, are most commonly the voiceless fricatives [h], [s], [f], etc., and voiceless aspirated stops, like the [p^h], [t^h] and [k^h] at the beginning of the words *pool*, *tool* and *cool*.

The schwa under examination is a vowel, not a voiceless fricative or aspirated stop, and therefore it would be expected that the cords would be effectively closing off the supralaryngeal vocal tract at the larynx. However, excerpted as it was from the word *speaker* [spik^{h}ə], this particular schwa occurred after an aspirated velar stop (moreover one that I strongly aspirated). In the first few centiseconds of such vowels, it is normal for a certain degree of coupling with the trachea to exist, as the cords move from the wide-open configuration for aspiration to the closed configuration for modal phonation, and sometimes air-flow shows the effect even lasts throughout the vowel (Rose 1996c: 595).

The trachea is a tube. It has its own resonant frequencies, one of which (the third) is for males around 2.2 kHz (Stevens 2000: 300). With the supralaryngeal vocal tract coupled to the trachea, the location of the source at the larynx is no longer at the end of the tube, but towards the middle. This gives rise to a phenomenon whereby energy is not only amplified at certain frequencies, but absorbed at others. The frequencies at which energy is absorbed as the result of vocal tract configurations where the source is not located at the larynx are often called *antiresonances* or *zeros*. The effect of coupling with the trachea is therefore actually to introduce pairs of extra resonances and antiresonances, and, since they affect the transfer of source energy, these become part of the transfer function as extra peaks or hollows (Stevens 2000: 196–9; 299–303). (Absorption of energy, and extra resonances, also most typically occur in sounds where the nasal cavity is involved, that is nasal consonants and nasalised vowels. The transfer function in such cases becomes very much more complex; that is the reason why only oral vowels are discussed in this chapter.)

It is thus likely that, since this particular schwa was produced after a strongly aspirated stop, a certain degree of tracheal coupling was present. This resulted in a visible tracheal, or *subglottal resonance* at a frequency just below the third formant, amplifying the harmonics around 2.5 kHz. The lack of expected energy in the region of the fourth formant around 3.5 kHz is also probably due to the presence of a tracheal zero.

It has been remarked (Stevens 2000: 197) that, since the dimensions of the trachea are fixed, subglottal resonances and antiresonances are invariant for a given individual, but can vary between speakers. If this were so, then they would obviously have considerable potential forensic importance, conditional, as always, on the magnitude of between-speaker variation. However, it is not clear how compliant the trachea is under conditions of larynx lowering or raising. Moreover, sounds where substantial subglottal coupling can be expected have relatively low incidence, and a speaker can also always choose to minimise the effect on surrounding sounds. For example, on one occasion a speaker might show substantial coupling to the trachea through most of the word *hello* by maintaining a vocal cord vibratory pattern with a glottal width typical of the /h/ throughout most of the word. On the next occasion, the speaker may change the vibratory pattern immediately after the /h/ to one with normal glottal width, so the effect is minimised, and no coupling is present.

In addition, there are problems with the automated digital extraction, and identification, of subglottal resonances and zeros. Note, for example, that the digital LP analysis in Figure 8.13 failed to pick the extra resonance evident in the schwa harmonic spectrum). Most digital LP furthermore involves so-called all-pole models, which means that they model the shape of a transfer function with resonances only. The forensic power of subglottal resonances and zeros remains to be demonstrated, therefore.

Time domain resynthesis

Finally, let us examine the same schwa from its time-domain wave-form. This is the wave-form that would result if the harmonics in Figure 8.13. were summed with the appropriate phase, and three periods of it are shown in Figure 8.14. Once again, certain relationships to the spectrum can be brought out.

The complex wave-form repeats three times in 0.028 seconds. This means it will repeat 3/0.028 times in one second, and therefore have a fundamental frequency of 107 Hz. This agrees well with the estimated F0 of 105 Hz from the harmonic spectrum in Figure 8.13. Within each fundamental period can be seen five smaller repetitions – a frequency that corresponds to the first formant frequency at about five times the fundamental, or ca. 500 Hz. An even higher frequency component – about 15 times in each fundamental period – corresponds to the second formant frequency at about 1.5 kHz. It can be noted here that the three repeats are not exact – an exact period-to-period repetition never occurs in speech because the conditions are never exactly the same – but the successive periods are similar enough to identify recurrent events.

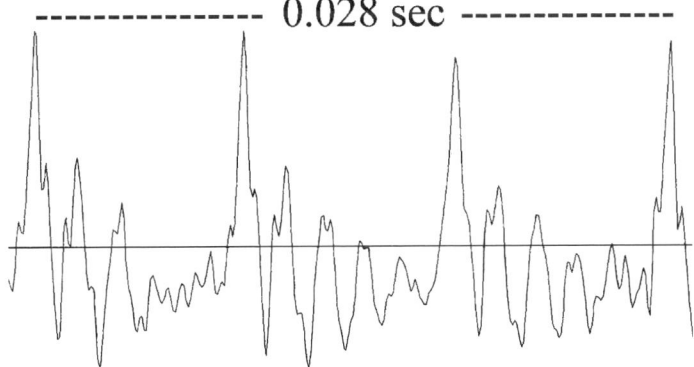

Figure 8.14 Wave-form for the schwa in Figure 8.13

Summary: Source–filter theory

It has been shown how the most important acoustic properties of vowels in the radiated speech wave – their F0 and formant frequencies – arise as an interaction of two independent components: a source and a filter. The source is the energy input to the system, and is associated with vocal cord vibration. The filter is associated with the shape of the supralaryngeal vocal tract, and modifies the source energy. An example was given of how the dimensions of the supralaryngeal vocal tract determine its acoustic output, in this case with the value of formant centre frequencies in schwa. Subglottal resonances and antiresonances were also mentioned as potential forensic-phonetic parameters.

In the independence of fundamental frequency and formant frequencies is seen the acoustic reflection of the most important design feature of speech production, discussed under *the basic dichotomy* in Chapter 6: the independence of vocal cord activity and supralaryngeal activity. This permits the independent control of pitch through fundamental frequency, and of vowel quality (i.e. the production of different vowels) through formant frequencies. It allows us to say different vowels on the same pitch, and say the same vowel with different pitches.

Since vowel quality and pitch are used to signal different linguistic information, it can be appreciated that this allows us to produce an enormously wide range of distinctive speech sounds. It is obvious, for example, that changing the first vowel in the word *filter* to [ɒ] changes the word to *falter*. But saying the word *filter* with high pitch on the first syllable and low on the second conveys something different from saying it with low pitch on the first and high pitch on the second. The difference thus signalled is one of statement (*It's a filter*) vs. question (*Did you say filter?*). For a more drastic example from a tone language (standard Thai), saying the segmental sequence [klai] with a low falling pitch means *near*, but with a high falling pitch means *far*.

The independence of source and filter also allows us to produce different consonants like the alveolar fricative pair [s] and [z]. The filter, or supralaryngeal vocal tract, is the same for both consonants with constriction at the alveolar ridge, but in one ([z]) the vocal cords are vibrating supplying a periodic energy source at the glottis, while in the [s] the cords are not vibrating, and there is no periodic energy source at the glottis. (In source–filter terms, both [s] and [z] also have sources of energy generated by the noise caused by the turbulence as the air flows through the narrow constriction at the alveolar ridge, and hits the teeth. Not being located at the end of the supralaryngeal tract, like the vocal cords, these sources also introduce complications like those from the tracheal resonances, which makes the description of their acoustics more complicated and outside the scope of this chapter.)

Spectrograms

One of the commonest ways of displaying speech data is by spectrograms, so it is important to explain them, and enough speech acoustics has now been covered to do this.

Speech acoustics

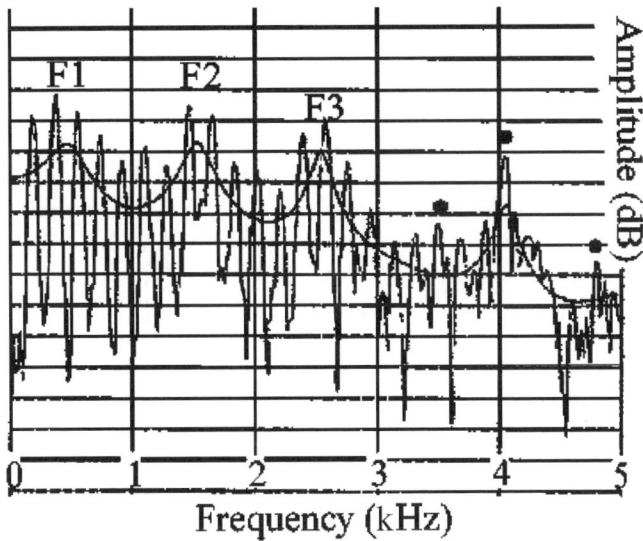

Figure 8.15 Harmonic and linear prediction spectrum for real female schwa

It has been shown how speech acoustics can be represented spectrally as a two-dimensional plot of amplitude against frequency. This shows how much energy is present at what frequencies at a particular instant, and, as explained above, is a unique function of the vocal tract that produced it. As a reminder of this, Figure 8.15 shows the harmonic and LP-smoothed spectrum for the schwa vowel [ə] in the word *bird* said by a female speaker of Cultivated Australian English.

In Figure 8.15 can be seen the usual spiky harmonic structure. The reader should try to identify the 11th harmonic and estimate the fundamental frequency.[2] Major peaks in the harmonic structure can be seen in the 2nd, 8th, 14th and 22nd harmonics, and there are also some minor peaks at the 6th, 19th, and 27th harmonics. The three highest frequency peaks are marked with black dots. There is also a clear zero, or antiresonance (i.e. absorption of energy), at about 4.5 kHz.

The LP-smoothed spectrum can be seen to have resolved the major peaks in the harmonic spectrum, but not the minor ones. The first three major peaks in the harmonic structure coincide with the first three formants (marked F1, F2 and F3). It is likely that the weak harmonic peak at 3.5 kHz, marked with a black dot, actually reflects the fourth formant since we would expect it to be found at that frequency given the frequencies of the first three formants, and the auditory quality of the vowel (the fact that it sounds like a schwa). The strong peak at about 4 kHz possibly reflects a resonance frequency of the speaker's larynx tube. Subglottal factors are in all likelihood responsible for the minor harmonic peak at just above 1 kHz.

---

2 The 11th harmonic can be seen to be just over 2 kHz, which means that the fundamental is at about (2000 [+] Hz /11 =) 185 Hz.

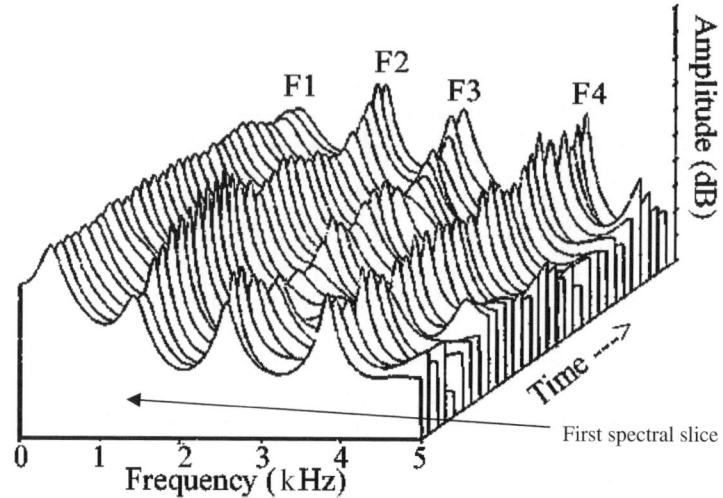

Figure 8.16 Three-dimensional plot of ca. 20 centiseconds of the [ə] in *bird*

Figure 8.15 provided a snap-shot of the speaker's acoustical output over a very short space of time. Figure 8.16 shows what the speaker's acoustic output looks like over a longer stretch of time. It shows about 20 centiseconds of the acoustics of the [ə] vowel in the word *bird* said by our female speaker of Cultivated Australian English. The frequency and amplitude are shown on the horizontal and vertical axes as before. For clarity, only the LP-smoothed spectrum is shown. The third dimension (time) is shown progressing from bottom left to top right. The figure therefore shows what happens to the LP-smoothed frequency–amplitude spectrum of this speaker's schwa over time from bottom left (early in the vowel) to top right (late in the vowel). It should be clear that the shape of the first spectral slice in Figure 8.16, with its four formant peaks, is similar to that in Figure 8.15. It can be appreciated from this figure that the acoustics of speech are in a certain sense three-dimensional, involving the evolution of frequency–amplitude distributions over time.

Of the four well-defined spectral peaks in the first spectral slice, the lowest occurs at a frequency of ca. 440 Hz; the next at about 1.45 kHz; the next at about 2.6 kHz; and the highest at 3.8 kHz. As already pointed out, the first three of these represent the first three formants, and the fourth peak probably represents a resonance connected with another part of the vocal tract. The progression of these peaks can then be followed upwards and rightwards through time, and it can be seen that the smoothed spectrum does not appear to change much. Changes in amplitude in this figure are easier to see than changes in frequency. The amplitude of the peaks changes a little: F2 can be seen to increase in amplitude in the third spectral slice and then decrease again at about the thirteenth, and in about the same period, F3 seems to decrease and then increase. A fifth peak with low amplitude (possibly the fifth formant) is also sporadically visible (at for example the ninth spectral slice).

The spectrum of [ə] in Figure 8.16 does not change much through this short space of time because the schwa in the word *bird* is a steady-state vowel, or monophthong, involving a single target. However, because the vocal tract is constantly changing as it

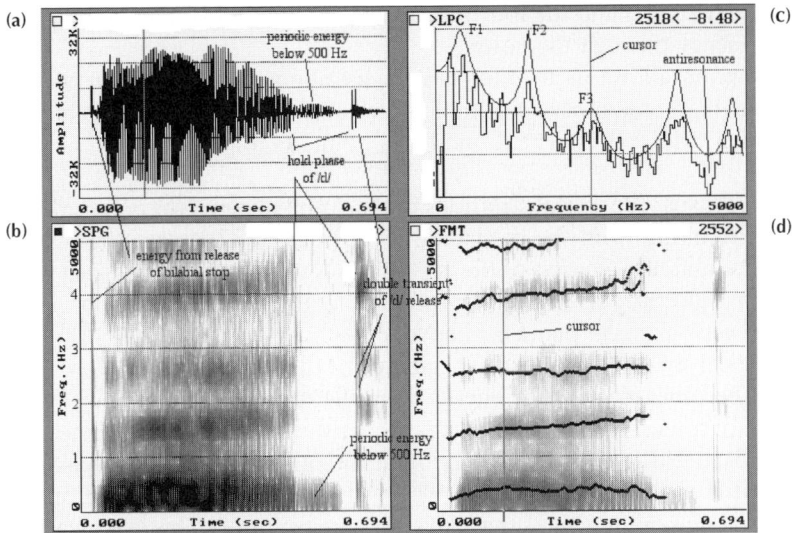

Figure 8.17 Acoustic representations of *bird*. A = wave-form; B = spectrogram; C = harmonic and LP-smoothed spectra; D = spectrogram with formant centre-frequencies tracked by LP analysis

moves from one articulatory target to another, so the frequency–amplitude distributions of its acoustic output change, usually fairly rapidly with time. This gives speech its characteristic acoustic profile. An example showing more dramatic acoustic change, from the word *hello*, will be given at the end of this section.

Now imagine that the three-dimensional plot in Figure 8.16 has been squashed from above by a Monty Python boot and compressed into two dimensions, time and frequency, but with the amplitude still shown by relative darkness, so that the spectral peaks show as dark bands. The result is a *spectrogram: a two dimensional plot of frequency versus time, with amplitude shown by the darkness of the trace*. A spectrogram of the word *bird* from which the three-dimensional plot in Figure 8.16 was also generated is shown in the lower left-hand panel (b) of Figure 8.17. The wave-form of the utterance is above it in the top left panel (a); the contents of the other two panels will be discussed presently.

The spectrogram in panel (b) of Figure 8.17 shows time running from left to right along the horizontal axis, and frequency increasing from 0 Hz to 5000 Hz up the vertical. There are also dotted grid lines at kilohertz intervals. We are now, therefore, looking down on the mountain range of the 3D plot of Figure 8.16, and it is easy to see the spectral peaks – the areas of high amplitude, the formants – as dark bands progressing from left to right. (The lowest one, the first formant, is not quite so easy to see, as this speaker has a lot of high-amplitude energy at low frequencies.) The reader might like at this point to identify the third formant (the third dark band up from the bottom) and estimate its approximate frequency from the grid lines.[3] One thing that *is* easy to see that was not clear from the 3D plot is that the second formant

---

3 The third formant lies between the 2 and 3 kHz grid lines, at approximately 2.6 kHz.

Forensic Speaker Identification

(between 1 and 2 kHz) and the fourth formant (at about 4 kHz) do actually increase a little in frequency through the vowel.

A spectrogram allows us to infer quite a lot about the production of speech. To demonstrate this, some other acoustic features of the spectrogram that reflect aspects of articulation can be commented on at this point, as follows.

- The closely spaced vertical striations in the vowel in panel (b) represent the energy pattern resulting from the individual glottal pulses. These are the points in time where the major excitation of the air in the supralaryngeal vocal tract occurs. The individual glottal pulses can also be seen in the vertical striations in the time-domain wave-form in panel (a) above.
- The vertical striation right at the beginning of the spectrogram, before the formant pattern becomes clear, is the energy from the release of the bilabial stop [p]. The absence of low-frequency energy (that is, between 0 and 500 Hz) before this transient indicates that the cords were not vibrating during the hold phase of the stop and that it is therefore voiceless. The fact that vocal cord vibration starts just after this transient indicates that the stop is unaspirated. This is typical realisation for Australian word-initial stop phonemes /b/, /d/ and /g/.
- The absence of energy above about 500 Hz at the end of the formant pattern indicates the hold phase of the word-final alveolar stop /d/. The presence of periodic energy below 500 Hz at this point shows that the speaker's cords were vibrating during all but the final centiseconds of the hold-phase of this /d/. The low-amplitude periodicity in the hold phase can also be seen in the wave-form in panel (a).
- The two vertical transients almost at the end of spectrogram indicate the energy from the release of the /d/ (these can also be seen in the wave-form in panel (a) above) and the energy is concentrated into formant regions for a very short time after that. The double transient is not typical for alveolars.

Now to return to the formants. Although the dark formant bands are reasonably clear in the spectrogram in panel (b), it is not easy to determine their centre frequency by eye to a useful degree of accuracy, and these frequencies are normally calculated digitally. A very common way of doing this is to use the method of linear prediction analysis, which has already been mentioned in conjunction with the smoothing of the harmonic spectrum. The spectrogram in panel (d) of Figure 8.17 is the same as that in panel (b), but with the formant centre frequencies extracted by linear prediction and plotted as thin black lines over the spectrogram. It can be seen that the linear-predicted values appear to trace through the middle of the formant bands rather well.

Finally, to emphasise the spectral nature of the acoustic structure of speech, both harmonic and LP-smoothed spectra taken at a point early on in the vowel are shown in panel (c) at the top right of Figure 8.17. The point at which the spectra were taken is indicated by the position of the cursor in panel (d). As usual, frequency (from 0 Hz to 5000 Hz) is along the horizontal axis and amplitude (from 0 dB to 80 dB) on the vertical axis.

The smoothed LP spectrum in panel (c) shows five extracted peaks, the first three of which are formants 1 to 3. The cursor in panel (c) has been positioned at the centre frequency of the third formant, which the reader has already estimated by eye to be at about 2600 Hz at this point in time. The actual frequency value at this point, as estimated by the LP analysis, is shown in the top right of the panel as 2518 Hz. With

Speech acoustics

the exception of F1, the peaks in the harmonic spectrum are also clear. Since it was taken from a similar part of the same schwa vowel, the spectrum in panel (c) is similar to that in Figure 8.15. The presence of the speaker's high-frequency antiresonance is very clear as the dip in spectral amplitude at the right end of the panel.

In sum, then, visual inspection of the spectrogram in Figure 8.17 shows that during the production of this schwa the air in the speaker's vocal tract was vibrating with maximum amplitude at averages of about 400 Hz, 1500 Hz, 2600 Hz and 4000 Hz. Since they have been digitally extracted, by the linear prediction analysis, it is easy to specify the centre frequencies of the resonances more accurately. At the cursor position near the beginning of the vowel in panel (d), for example, their values were: 455 Hz (first formant), 1503 Hz (F2), 2518 Hz (F3), 3844 Hz (F4?) and 4777 Hz (F5?).

Spectrograms of other vowels

The previous section has explained a spectrogram as a two-dimensional plot of frequency vs. time, with the amplitude component of the spectrum shown as varying degrees of darkness. It is now possible to see how the formant structure of vowels other than schwa appear in spectrograms. We will examine four more vowels from our female speaker of Cultivated Australian English: her high front unrounded [i], her high central rounded [ʉ], her low back unrounded [ɑ], and her half-close rounded back [o]. Together with schwa, these vowels represent the set of long quasi-monophthongal vowel phonemes of Australian English. It will be shown that *differences between these vowels are mainly to be found in the centre frequencies of the first two or three formants.*

The bottom panels in Figures 8.18 and 8.19 show spectrograms of the Cultivated Australian female speaker's vowels in the words *bead*, *coot*, *cart*, and *board*. These

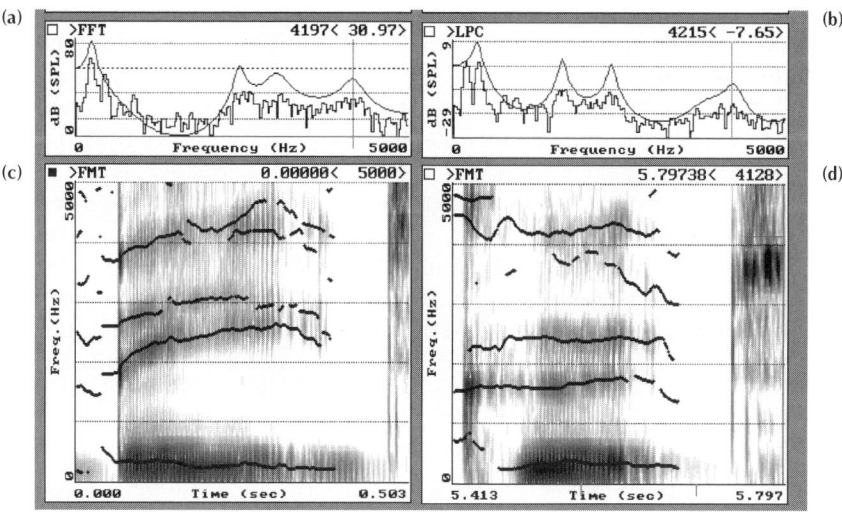

Figure 8.18 Acoustic representations of the vowels [i] (left panel) and [ʉ] (right panel). Top panels show smoothed LP and harmonic spectra; bottom panels show spectrograms with formants tracked with LP analysis

vowels are phonetically very close to [i], [ʉ], [ɑ] and [o]. Harmonic and LP-smoothed spectra, taken at points roughly in the middle of the vowel, are shown in the top panels. The centre frequencies of the vowels' formants have been tracked by LP analysis.

The first thing to see in these four spectrograms is that they differ in formant structure, and indeed *differences in formant structure are the main acoustic correlate of vowel quality*. In each vowel, the quasi-equal spacing of the formants of the speaker's schwa in Figure 8.17 has been perturbed. Sometimes, as in [i] the second formant is higher in frequency than schwa, sometimes, as in [o], lower. Sometimes, as in [o], the first formant is higher in frequency, sometimes, as in [i], lower. These deviations, and the resulting configurations – or *F-patterns* (for formant patterns) – need now to be discussed in greater detail.

F-pattern in [i]

The spectrogram in the bottom left panel of Figure 8.18(c) shows the F-pattern for the word *bead*. The vowel in this particular token sounds very close phonetically to [i], but has a slight schwa onglide, thus: [ᵊi]. It can be seen to have one low resonance, below 500 Hz. This is the first formant (F1). The next highest formants are the second (F2) and third (F3). Both of these can be seen to course upwards in frequency, the second from about 2 kHz, the third just above 2.5 kHz. They reach maximum frequency at about 2.5 kHz (F2) and 3 kHz (F3), and then start to drop in frequency again. F3 has not been well extracted for all of its time course and shows a few breaks and hiccoughs. The speaker's F4 can be seen to course upwards from ca. 3.75 kHz. At about 4.2 kHz there is a break in the trace, after which F4 resumes again at about 4.2 kHz.

There is finally another, dipping, trace just above F4, which rises to just over 4.5 kHz. *Very high-frequency formants are often not extracted well and complicated high-frequency patterns like this are usually difficult to interpret*. It may be the case that there are two resonances between 4 and 5 kHz (F4 and F5), and that the LP analysis has not separated them very well.

Panel (a) above the [ᵊi] spectrogram shows the harmonic and LP smoothed spectra for [ᵊi] taken in about the middle of the vowel. (Frequency increases from 0 Hz at the left to 5.0 kHz at the right). The peaks of the first three formants are easily seen in the LP spectrum, and there is also a fourth peak (at about 4.2 kHz), through which the cursor has been placed.

The most important parts of the F-pattern from the point of view of vowel quality are the closely grouped F2 and F3 at a high frequency relative to the speaker's F2 and F3 in schwa, and the low-frequency F1. The low-frequency F1 and high-frequency F2 (and F3) are what make the vowel sound high front and unrounded, like an [i].

F-pattern in [ʉ]

Panel (d) in the bottom right of Figure 8.18 shows the spectrographic F-pattern for the word *coot*. The vowel in this particular token sounds very close phonetically to a high central rounded vowel [ʉ]. The first three formants can be easily identified. F1 is low, below 500 Hz; F2 lies just above 1.5 kHz; and F3 is at about 2.5 kHz. There is an uninterrupted high-frequency trace at about 4.2–4.3 kHz, and a sporadic trace that

Speech acoustics

meanders between 3 and 4 kHz. It is likely that two formants, F4 and F5, are present here again, but this time it is the upper that has been better extracted.[4]

The LP spectrum in panel (b) above the [ʉ] spectrogram shows four peaks. The lowest three correspond to F1, F2 and F3. The highest peak, at which the cursor has been placed, is actually a combination of two peaks close together, one at about 4.2 kHz, one at 3.75 kHz, whence the peak's lop-sidedness.

The most important parts of the F-pattern from the point of view of vowel quality in [ʉ] are the low frequency of F1, and the average frequency of F2. This is what makes the vowel sound like a [ʉ].

F-pattern in [ɑ]

Panel (c) at the bottom left of Figure 8.19 shows a spectrogram for the word *tart*. The vowel in this particular token sounds very close phonetically to a low back vowel [ɑ]. F1 is at ca. 1 kHz; F2 is about 500 Hz above it, at ca. 1.4 kHz; F3 is at ca. 2.6 kHz; and above this there are two traces, one at about 4 kHz and a broken one just below it at 3.75 kHz. Some energy is also present in this token below 500 Hz, and has been resolved into an LP trace of short duration. This is most likely related to the speaker's breathy mode of phonation, which itself could either be due to the aspirated velar stop [kʰ] preceding the vowel in this word, or reflect a more permanent sex-related speaker characteristic.

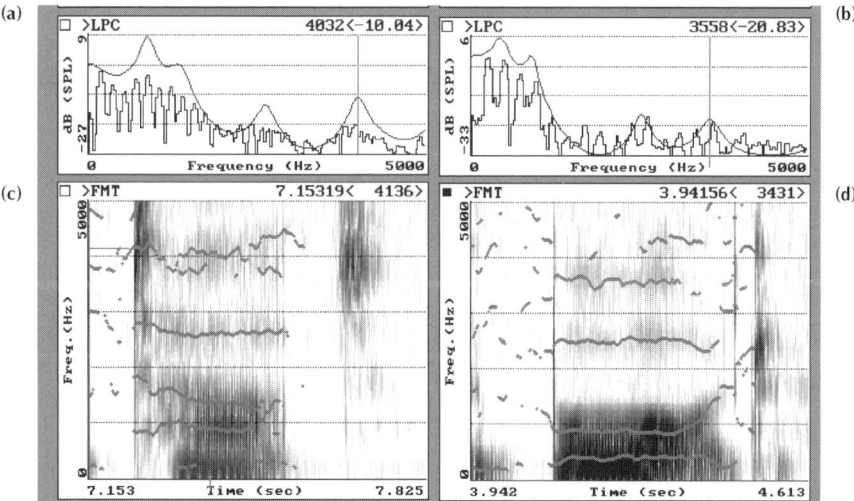

Figure 8.19 Acoustic representations of the vowels [ɑ] (left panel) and [o] (right panel). Top panels show smoothed LP and harmonic spectra; bottom panels show spectrograms with formants tracked with LP analysis

---

4 The drop in frequency of the lower of the two traces possibly identifies it as F4, since constriction at the alveolar ridge (as in the following alveolar stop) can cause such a lowering in F4 (Francis Nolan, personal communication).

The LP spectrum in panel (a) above the [ɑ] spectrogram shows four peaks. The lowest three once again correspond to F1, F2 and F3. The highest peak, at which the cursor has been placed, reflects the single resonance at ca. 4.0 kHz. The harmonic spectrum shows that the fundamental frequency in this particular vowel (the first and lowest harmonic) has in fact the greatest amplitude of any of the harmonics. This is responsible for the low-energy spectral component associated with a breathy phonation type noted above, and it can be seen how the high amplitude of the fundamental actually appears to boost the LP trace in the low-frequency range.

The most important parts of the F-pattern from the point of view of vowel quality in [ɑ] are that F1 and F2 are close together, and that F1 is located high relative to the speaker's schwa F1. This is what makes the vowel sound back and low, like an [ɑ].

F-pattern in [o]

The spectrogram in panel (d) of Figure 8.19 shows the F-pattern for the word *board*. The vowel in this particular token sounds very close phonetically to a mid-high back rounded vowel [o]. The F-pattern for [o] resembles that for [ɑ], with F1 and F2 close together, but located lower in frequency. F1 is at ca. 400 Hz, F2 at ca. 900 Hz. F3 shows as a strong trace at ca. 2.5 kHz, and once again two traces are visible above F3. A strong trace is clear at ca. 3.5 kHz, and towards the end of the vowel a second trace is visible above 4 kHz.

The LP spectrum in panel (b) above the [o] spectrogram shows four peaks, which correspond to the first four formants. The cursor has been placed in the middle of F4, at ca. 3.6 kHz.

The most important parts of the F-pattern from the point of view of vowel quality in [o] are that F1 and F2 are close together, and that F1 is located low relative to the speaker's schwa F1. This is what makes the vowel sound back, high and round, like [o].

Acoustic vowel plots

In the section above, mention was made of how certain features of a vowel's F-pattern are responsible for its auditory-phonetic quality: how they make it sound like a high front rounded vowel, for example, or a low back unrounded vowel. It is now necessary to examine in greater detail how the acoustics relate to phonetic vowel quality. This section will show how vowel quality is encoded in the centre frequencies of the first two or three formants, and how these are commonly plotted to show the main linguistically relevant features of a speaker's vowels. It will also make the point that the F-pattern encodes speaker-specific information too.

The invention of the *sound spectrograph* in the 1940s enabled researchers to actually see the acoustic structure of speech from spectrograms. Soon afterwards it was discovered that if the frequency of a vowel's first formant is plotted against its second formant frequency, and this is done for all of a speaker's vowels, the resulting configuration resembles the position the vowels would have had on the vowel quadrilateral of traditional articulatory phonetics. (This, it will be recalled, is the quadrilateral defined by the features of tongue height and backness, upon which the position of rounded

Table 8.2 The first three formant centre frequencies of single tokens of the Cultivated Australian vowels /iː ʉː aː oː and əː/ spoken by a female

	F1	F2	F3
[i]	250	2475	3046
[ʉ]	369	1656	2357
[ɑ]	868	1371	2652
[o]	435	903	2516
[ə]	455	1503	2542

and unrounded vowels can be specified according to inferred or introspected articulation; see Figures 6.13, 6.14, 6.15.)

The creation of such acoustic vowel plots can be demonstrated with the acoustic data from the female speaker of Cultivated Australian English, spectrograms of whose long vowels have just been presented and described. Centre frequency values for the first three formants in her vowels [i], [ʉ], [ɑ], [o] and [ə] are given for convenience in Table 8.2. These values were taken from LP estimates of the formant centre frequencies of the type demonstrated in the previous section. Formants were sampled in the approximate middle of the vowel.

There are many ways of plotting the formant values, and these ways increase in complexity depending on how closely one wants the resulting position of the vowels to approximate the inferred articulatory or auditory position (Clark and Yallop 1990: 241–3). The simplest way is to plot the frequency of F1 against F2, but this does not show the position of the back vowels as well as it might. The next simplest way is to plot F1 against the difference between F1 and F2, and that is what is shown in Figure 8.20. Other, more sophisticated approaches include using perceptually motivated transforms of the frequency scale, for example the mel scale, and incorporating F3 (Fant 1973: 35–58).

Figure 8.20 shows an acoustic plot for the female speaker's Cultivated Australian long vowels that were shown spectrographically in Figures 8.17, 8.18 and 8.19. In this plot, the position of a vowel in acoustic space, which is often referred to as the *F1–F2 plane*, is determined by its F1 frequency and the difference between its F2 and F1 frequency. Thus, as can be seen in Table 8.2, the F1 and F2 of [i] in mid vowel were 250 Hz and 2475 Hz, and therefore [i] is plotted at the intersection of 250 Hz and (2475 Hz – 250 Hz =) 2225 Hz. The axes of the plot are oriented so that the configuration matches the usual position for articulatory vowel charts, that is with F1 decreasing on the *y* axis and (F2–F1) decreasing on the *x* axis. Other orientations will also be found.

Figure 8.20 shows the speaker's [i] vowel located at the top left of the F1–F2 plane, and her [ɑ] vowel at the bottom right. Two important relationships are evident in Figure 8.20. It can be seen firstly that *the first formant frequency is inversely proportional to the traditional articulatory parameter of vowel height*: [i] is a high vowel, so its F1 is low; [ɑ] is a low vowel, so its F1 is high. Secondly, *the second dimension (F2–F1) is more complicated and corresponds to the articulatory parameters of backness and*

Forensic Speaker Identification

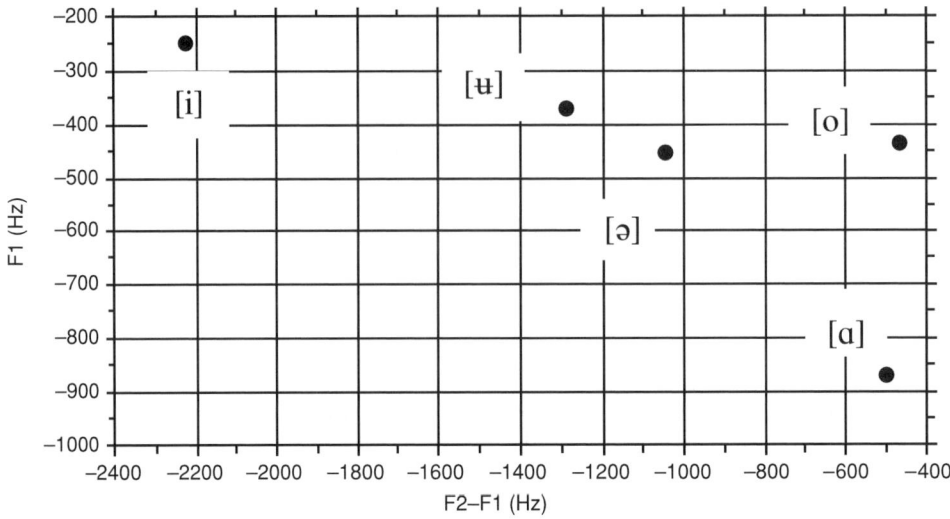

Figure 8.20 Acoustic plot for the five female Cultivated Australian Vowels in Table 8.2

rounding. The backer the vowel, the smaller the difference between F2 and F1, and the more round the vowel, the smaller the difference. The more complicated nature of this F2 relationship means that the three articulatory dimensions of traditional vowel description (height, backness, rounding) map onto two acoustic dimensions: F1 and F2 (or F2–F1).

As mentioned above, displays incorporating F3 information can also be found. One way of doing this is by creating a two-dimensional plot of F1 versus a weighted average of F2 and F3 (called an *effective F2*). This is said to give a better approximation to the auditory position of vowels, and, because F3 encodes rounding for high and mid front vowels, provides a better separation of front rounded from front unrounded vowels in languages that have them, e.g. German, Swedish, French, Shanghai, Cantonese (Fant 1973: 35–58). Another way of showing F3 is simply by making two acoustic plots: one of F1 vs. F2 (or F2–F1), and one of F2 vs. F3 (Clark and Yallop 1990: 241–3). No additional vowel differentiation is gained by doing this for our speaker, since her F3 in four out of five of her vowels is very similar.

Acoustic vowel plots are convenient for demonstrating both between-speaker and within-speaker differences in vowel acoustics, to which we now turn.

Between-speaker variation in vowel acoustics

Given the correlation between vocal tract length and formant frequencies explained above, the best illustration of how big between-speaker differences in vowel acoustics can be will be found between the vowels of a male (long vocal tract) and a child (short vocal tract). The next best demonstration will be between adult speakers of different sexes, and this is what is shown here. In order to maximise the chances of getting large

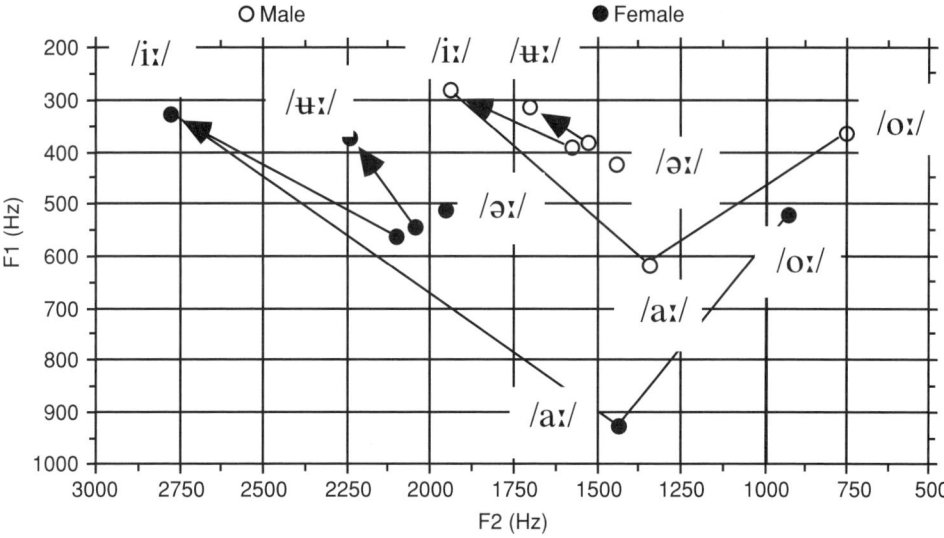

Figure 8.21 Comparison of vowel acoustics in a male and female speaker

differences in formant values, I also chose a fairly tall male and a very short female for comparison. This was because, as shown above, formant frequencies correlate with vocal tract length, and vocal tract length can be assumed to show at least some correlation with height. The data were maximally controlled for vowel quality. Both subjects were speakers of Broad Australian English, and both were recorded saying words containing the five long vowel phonemes /iː ʉː aː əː oː/ in monosyllabic words before an alveolar stop, as in *bead* or *dirt*. Their vowel allophones had very similar auditory features, with the schwa-like first targets in both /iː/ and /ʉː/ (i.e. [əi] and [əʉ]) that characterise Broad Australian. The female had a slightly fronter schwa, however.

Mean values of F1 and F2 were calculated and plotted on an F1–F2 chart, which is shown in Figure 8.21. Both the first and second targets were measured and plotted for the diphthongal allophones of /iː/ and /ʉː/, and the trajectory between them is indicated with arrows. The positions for the peripheral targets in /iː/ /aː/ and /oː/ are joined to facilitate visual comparison.

Figure 8.21 shows plots that are expected from the speakers' sex and the auditory quality of their vowels, with the male vowels lying higher and to the right of the female vowels. Some of the absolute differences in formants are rather big: values for /aː/ differ for example by about 300 Hz in F1, and values for F2 in /iː/ differ by about 900 Hz. This figure, as Figure 3.2 did for Cantonese tonal F0, shows clearly how vowel acoustics encode simultaneously both individual and linguistic information. Formant frequencies reflect the overall length of the vocal tract that produced them, but also the particular vowel that is being produced.

Of course, from a forensic point of view, differences of such magnitude will never be the source of legal controversy because the voices that produced them sound so different that they will never be confused in the first place. A fairer comparison might

Forensic Speaker Identification

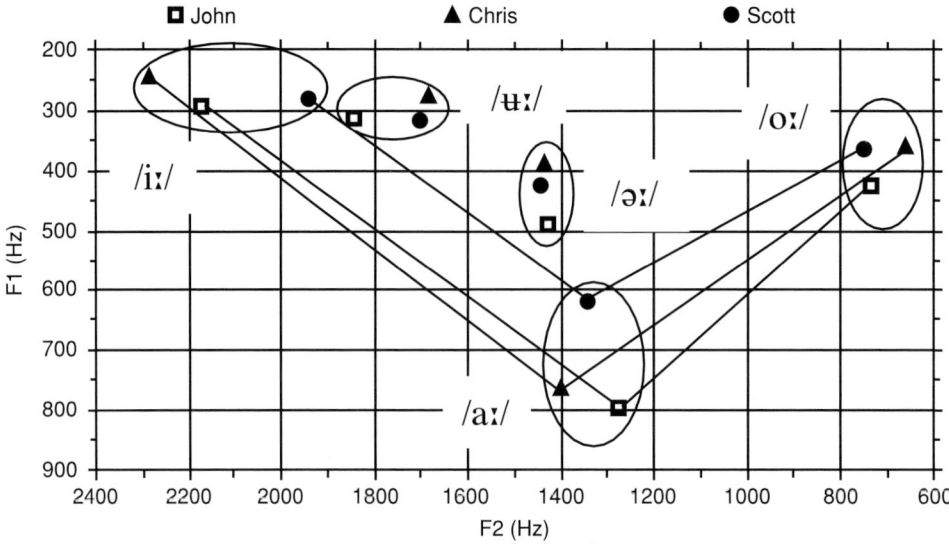

Figure 8.22 Comparison of vowel acoustics in three male speakers from the same family

be between same-sex speakers from the same family, and this is what is shown in Figure 8.22.

Figure 8.22 shows mean vowel acoustics from two brothers and their father (Rose and Simmons 1996) obtained in the same way as in the previous comparison (where one of the brothers, Scott, was actually the male subject). To make it easier to see things, the position of the first target in /iː/ and /ʉː/ has not been plotted, and the same vowels have been enclosed in ovals. Note that the axes are also a little different from those in Figure 8.21.

It can be seen from Figure 8.22 that the differences between the speakers are, as expected, smaller than in the different-sex comparison in Figure 8.21, but that the speakers by no means have the same vowel acoustics. Scott, for example, clearly differs from his relatives in /aː/ and /iː/, and John differs from the other two in his /ʉː/.

Within-speaker variation in vowel acoustics

It probably does not need to be pointed out by this stage of the book that, as with all acoustic parameters in speech, vowel plots are not invariant, even for the same speaker. The plot for the female speaker in Figure 8.20 showed formant estimates taken at a certain position during the vowel. However, it was pointed out above that nowhere is the F-pattern for a vowel totally static, even for a short length of time. Therefore one digitally extracted frame of speech will always differ from the next, at least slightly. If the formants had been sampled at a different point during the vowel, therefore, the vowels would have been shown located at slightly different positions in Figure 8.20. Thus even the same tokens can give slightly different results, depending on where the measurements were made. This is why it is important to specify where F-pattern measurements are made, and to ensure that they are made in comparable positions across samples. (The author usually estimates the F-pattern of a single vowel token using the average of several measurements made around an assumed target position.)

More important factors than sampling point variability contribute to within-speaker variation in vowel acoustics, however, and must therefore be taken into consideration. One is the phonological environment of the vowel; the other is occasion-to-occasion variation. These will now be explained, again with the help of spectrograms and vowel plots.

Phonological environment of vowel

A large part of the phonological structure of words and utterances is understood linguistically as consisting of sequences of discrete segmental units (phonemes or allophones). Thus the word *expert* is understood as being composed of the six segmental phonemes /ɛ k s p ə: t/. When we speak, however, these discrete segments have to be combined smoothly so that our speech is fluent. This means the vocal tract has to achieve a smooth transition from one segmental target to the next. The speed of articulation during normal speech is such, however, that it is seldom possible to attain a segmental target before moving on to the next. This results in what is called *target undershoot* (Daniloff *et al.* 1980: 317; Clark and Yallop 1990: 119), and it can be demonstrated in the following way.

Consider the movement of the tongue body in a sequence of a high central vowel occurring between two velar consonants, as in the Australian English nonsense word *kook* [kʰʉːkʰ] for example. In order to articulate this sequence, the tongue body has to move from its dorso-velar consonantal target [kʰ] down and forwards to its high central vowel target [ʉ] and back again to its velar consonantal target. The round trip is not far, because the vocalic high central position for the body of the tongue is not far from the velum, where the [k]s are made. Because the round trip is not far, there is a possibility that the high central vocalic target will be attained, or nearly so, in the short time available to say this syllable at a normal speed.

Now consider tongue body movement for the same vowel between two post-alveolar consonants (the affricates in the English nonsense word *chooch* [tʃʰʉːtʃʰ] for example). This time the round trip is further, because the post-alveolar place of articulation for the consonants is further from the high central target for the vowel. The tongue body has to move upwards and back from the first consonant [tʃʰ] to the vowel [ʉ], and then front and down again to the second [tʃʰ]. Since the tongue body has to move a greater distance in the same time it is likely that it will not have reached the high central vocalic target before having to head back towards the second post-alveolar target. In this case, the high central vocalic target will have been *undershot*.

It has been shown above that articulatory backness is reflected in the second formant frequency (the backer the tongue body the lower the F2 frequency). Therefore we expect the F2 in /ʉː/ in *chooch*, between two post-alveolar consonants, to have a higher frequency than the F2 in the same /ʉː/ vowel in *kook*, between two velars. This is because (1) post-alveolars and velars differ in the backness of the tongue body: velars are more back than post-alveolars, and (2) the tongue body will not have moved quite as far back for /ʉː/ between the two post-alveolars as for the /ʉː/ between the two velars.

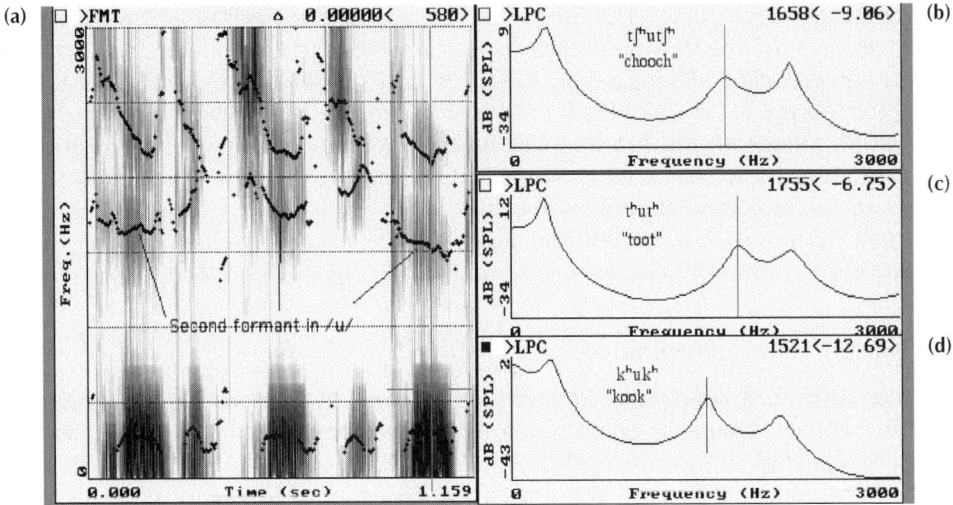

Figure 8.23 Spectrograms of *chooch toot* and *kook*, showing F2 undershoot in /ʉː/

That this is indeed the case is shown in Figure 8.23. This figure shows on the left a spectrogram, with LP traced formants, of three English words *chooch, toot* and *kook*. (I first recorded each separately in a frame sentence *say ___ again*, and then digitally edited them out and combined them so as to fit into a single figure. The /ʉː/ vowel is transcribed with 'u' in the figure.) The second formant has been indicated. The right of the figure shows the LP smoothed spectra of the /ʉː/ vowels in approximately the middle of their duration. Panel (b) at the top shows /ʉː/ between two post-alveolars in *chooch*; panel (c) in the middle shows /ʉː/ between two alveolars in *toot*; panel (d) shows /ʉː/ between two velars in *kook*. The cursor has been placed through the centre frequency of the second formant, and its frequency in Hz is shown at the top right of each panel. Thus F2 in the middle of the /ʉː/ in *chooch* was 1658 Hz. It can be seen that F2 in /ʉː/ between the two post-alveolars is indeed higher than between the two velars, by about 140 Hz.

It can also be seen that the /ʉː/ F2 value in the extra word *toot* is higher even than the value between the post-alveolars. In *toot*, with its alveolar consonants, the tongue body is not actively involved, since alveolar stops are made with the tip of the tongue, not the body. However, the tip of the tongue is attached to the body, so that if it has to be raised in the front of the mouth, it will also pull the tongue body forward to some extent, and thus make it undershoot the /ʉː/ target even more than in the post-alveolar case. This is reflected in the relatively high F2, and is an example of the principle of vocal tract interconnectedness mentioned above. Actually, there is disagreement as to whether the tongue body is supposed to be involved as an active articulator in the post-alveolar [tʃʰ]. Some phonologists claim that post-alveolar affricates are made just with the front, or crown, of the tongue as active articulator, and some that both the body and the crown are involved (Gussenhoven and Jacobs 1998: 199ff.; Roca and Johnson 1999: 110). However, as just pointed out, the front of the tongue is of course attached to the body, so that if the front has to be raised in the front of the mouth, it will also pull the tongue body forward with it.

The demonstrated differences in vocalic F2 conditioned by adjacent consonants mean that *phonemically the same vowel in the same speaker will have different positions in the acoustic vowel plot depending on its phonological environment*, and this therefore contributes to within-speaker variation.

That adjacent segments affect each other is a phonetic commonplace, and quite often the effect can operate at a distance, spanning several segments. Neither is the effect restricted to the influence of consonants on vowels. The reader can experience the opposite effect of a vowel on its preceding consonant by saying the words *key* and *car*, and noting the difference in the place of the initial stop, which is clearly articulated further forward in *key* than in *car*. In the vocabulary of phonemics developed in Chapter 7, the velar stop phoneme /k/ is realised by a fronted velar allophone [k̟] or [c] before a high front vowel [i], and as a cardinal velar [k] before the central vowel [a] (or if you are an American or British English speaker as a backed velar [k̠] before a back vowel [ɑ]). This is because the body of the tongue is front in [i], and pulls the place of articulation of the /k/ forwards.

Forensic significance: Differential effects of phonological enviroment

The forensic significance of this is that, since phonological environment is a source of variation in vowel plots, *it must be controlled in comparing different forensic samples*. From the above example in Figure 8.23 it would be obviously incorrect, for example, to claim exclusion of a suspect on the basis of the F2 difference between his /ʉː/ in *chooch* and the offender's /ʉː/ in *kook*.

It is usually the case that when F-pattern comparisons between samples are made, the investigator knows to adequately control for this factor. However, it should be noted that the ability to do this – to assemble samples comparable with respect to effect of phonological environment – is part of the linguistic knowledge that a forensic phonetician can be expected to bring to bear on the problem.

Occasion-to-occasion variation

Another important, pervasive and well-known source of within-speaker variation can be illustrated with acoustic vowel plots. This is the variation that characterises the speech of someone on different occasions. 'Different occasion' can be construed both as short-term, as when perhaps someone has two telephone conversations in the space of a couple of minutes, or long-term, when the speech activities are separated by days, months or even years (Rose 1999b: 2).

To illustrate both long-term and short-term variation here is some data, from Kinoshita (2001) on the vowels of a single Japanese speaker recorded on two separate occasions. Figure 8.24 shows an acoustic F1–F2 plot for the five short vowel phonemes of Standard Japanese (/i, e, o, a, ɯ/) elicited from the normal speech of one male speaker on several occasions. (The last phoneme /ɯ/ is a high back *unrounded* vowel). The vowels were elicited from natural speech in two sessions separated by a month. In each session there were two repeats, separated by a few minutes. The elicitation method and the corpus were the same for each repeat, so that differences in the data can be taken to reflect long-term and short-term variation. A single symbol in the figure

Forensic Speaker Identification

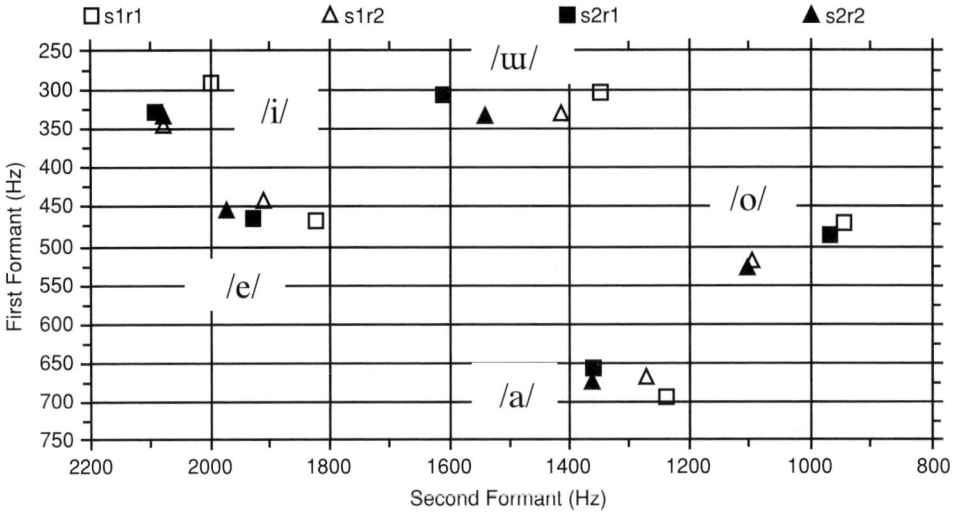

Figure 8.24 Within-speaker long- and short-term variation in the acoustics of Japanese vowels. s = session; r = repeat

represents the mean of five vowel tokens elicited in each repeat. Data from the first session are shown with empty symbols, and data from the second session one month later are shown with filled symbols. The two repeats within a session are shown with different symbols (triangle and square).

It can be seen first of all that the configuration in Figure 8.24 is as expected, given the phonetic composition of the vowel system, and the relationship between F1 and vowel height, and F2 and backness/rounding. Thus high front unrounded /i/ has high F2 and low F1, and low central unrounded /a/ has a high F1 and mid F2. The central position of /ɯ/ in the F2 dimension reflects its unrounded nature. Since F2 also correlates with rounding, and this vowel is unrounded, this vowel has a higher F2 than a corresponding high back rounded [u] vowel.

Figure 8.24 is typical in showing both short-term and long-term variation in F-pattern, even for the tightly controlled material. A small amount of short-term variation is evident, as for example in the /o/ vowel, where in the first session F2 was on average some 100 Hz higher in the second repeat than in the first. Although the same short-term pattern appears in the second repeat of the /o/ vowels, it can be seen that essentially the short-term variation is random. Thus for /ɯ/, F2 in the first repeat is lower than in the second repeat in the first session, but the other way round in the second session.

Long-term variation is clear in several vowels. Thus in /a/, /ɯ/ and /e/, for example, it appears that in the first session the speaker had a lower F2. Since there do not seem to be any long-term differences for the other vowels, the speaker was perhaps positioning their tongue body a little further forwards for these three vowels in the second session. It is normally the case that greater within-speaker variance is to be found in the long rather than the short term, and the Japanese data are also typical in this respect.

Forensic significance: Within-speaker variation in vowel acoustics

The kind of within-speaker variation illustrated here is of importance in forensic phonetics for the following reason. Since the same speaker can differ on different occasions, less confidence attends the estimation of a speaker's characteristic value for a particular parameter from values taken on one occasion than from several. In likelihood ratio terms, *the number of samples involved has an effect, as it should, on the strength of the evidence*. This is because the magnitude of the LR, which reflects the strength of the evidence, depends on the number of samples. The larger the number of samples, the more the likelihood ratio will be shifted away from unity, either positively, indicating greater support for the prosecution hypothesis that the same speaker is involved, or negatively, indicating greater support for the defence hypothesis that different speakers are involved. This means that *conditions are more favourable for forensic-phonetic comparison when several samples are available from both criminal and suspect speech*. This enables a better characterisation and quantification of the distribution of the parameters in both the samples, and a better estimate of the strength of the evidence.

Higher-frequency formants

There is a general assumption that the formants in the higher frequency region (F3 and up) reflect individual characteristics more than those in the lower frequencies (F3 and below). Thus Stevens (1971: 216) mentions that mean F3 is a good indicator of a speaker's vocal tract length, and Ladefoged (1993: 195, 211) mentions that F4 and F5 are 'indicative of a [speaker's] voice quality'.

There are several reasons for the belief in the high speaker-specificity of the higher formants. Firstly, it is assumed that variation in the lower-frequency formants is constrained by the additional function of signalling linguistic differences in vowel quality, and therefore they are less free to directly reflect individual anatomy. It is also assumed that the higher formants often reflect the resonances of relatively fixed smaller cavities in the vocal tract, for example the larynx tube, which are assumed to be relatively unaffected by the gross configurational changes of the vocal tract involved in the production of different vowels. This would contribute to a desirable smaller degree of within-speaker variation. Finally, there is also an empirical basis to these deductive arguments. It is often found that F-ratios (ratio of between-speaker to within-speaker variation) tend to be bigger for the higher formants (e.g. Rose 1999a: 20–3), and consequently speaker recognition is more successful when they are used. Mokhtari and Clermont (1996: 127) for example, were able to show that 'the high spectral regions of vowel sounds contain more speaker-specific information' (p. 127), and Rose and Clermont (2001) showed that F4 outperformed F2 in forensic discrimination of same-speaker and different-speaker *hello* pairs.

Forensic significance: Higher-frequency formants

Despite their promise, it must be borne in mind that the formants in the higher frequency regions are quite often not accessible under real-world conditions owing to poor quality of recordings or telephone transmission. In addition, higher-frequency formants are not always well extracted digitally even under good recording conditions – *vide* the examples for the female Cultivated Australian speaker's vowels in the sections above. It is consequently often difficult even to identify them for certain speakers, which one needs to do in order to be able to compare their frequencies (Rose 1999a: 12, 13; 1999b: 9, 10). Finally, in some specific cases it can be shown that lower formants can also carry high F-ratios, and permit acceptable discrimination rates (Rose 1999a: 22, 35, 36), possibly because they encode between-speaker *linguistic* differences in the realisation of phonemes (this was mentioned in Chapter 7 as one of the ways in which speakers can differ in their phonemic structure).

It is also a sensible working assumption that *no part of the vocal tract is so totally mechanically uncoupled from any other part that its acoustic response will be totally immune from influences from movements of other structures.* For example, resonant frequencies associated with the nasal cavities, which are the most fixed structures in the vocal tract, are still affected by the degree to which the nasal cavities are coupled to the rest of the tract by varying degrees of the soft palate opening, and by different vowels. Even the frequencies of the smaller cavities, for example the larynx tube, will be different depending on whether the tongue body is in a high front position for an [i] vowel, or in a low back position for an [ɑ] vowel.

All formant frequencies reflect the vocal tract that produced them. This is because, as demonstrated above, the frequency of a formant is dependent not only on the configuration of the vocal tract but also its length. Thus, for the lower formants, the acoustics signalling the linguistic information and the acoustics signalling individual information are convolved. The realistic situation, therefore, is that centre frequencies for both high and low formants will be found used as parameters in forensic speaker identification.

A whole-word example

Earlier in this chapter spectrograms were presented showing vowels with quasi steady-state F-patterns. It was pointed out that the acoustics of speech are constantly changing as the vocal tract moves from one articulatory target towards the next. In this section an example is given of a few such changes, in the word *hello* said as a declarative statement by an Australian male speaker from Adelaide. This token consists of the phonemic sequence:

/halaʉ/

(that is, a voiceless glottal fricative /h/, a short low central vowel /a/, a lateral /l/, and the /aʉ/ diphthong).

Phonetically, the short low central vowel /a/ phoneme is realised as a half-open central unrounded vowel [ɐ] (this is a vowel a little higher than fully open [a]). The lateral is uvularised [ɫ], which means that the tongue body was raised towards the uvula in its production, and the diphthong moves from a half low central unrounded target [ɐ] to a high front-of-central rounded target [y]. The second syllable is stressed [']. To implement the statement's declarative intonation, the pitch is low on the first

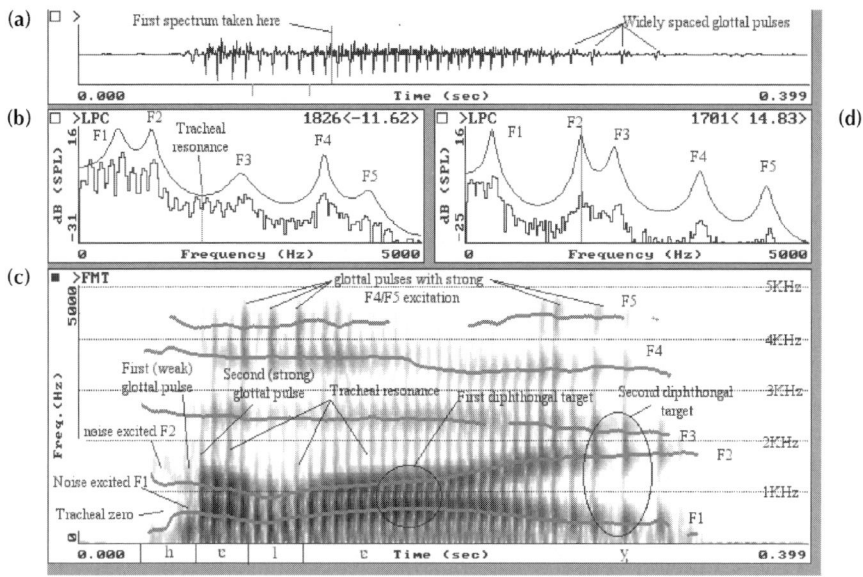

Figure 8.25 Spectrogram, wave-form, and two LP and FFT spectra from the word *hello*. See text for explanation

syllable and high falling on the second [L.HL], and the phonation type becomes creaky at the end of the word [~]. Thus phonetically this particular token of the word is

[hɐˈtɐy L.HL]

Panel (c) in Figure 8.25 shows the spectrogram of this utterance, with formants tracked by LP analysis. Time runs from left to right, and frequency increases vertically, from 0 to 5 kHz. Portions of the spectrogram corresponding to the phonetic segments have been labelled along the bottom (in the panel marked 'Time (sec)'). Along the top of the figure in panel (a) is the wave-form corresponding to the spectrogram; and in panels (b) and (d) below it are smoothed LP and harmonic spectra taken at two points in the *hello*.

I shall now guide the reader, segment by segment, and then suprasegmentally, through some important acoustic features of this *hello*, linking the features where possible to inferred articulations.

Voiceless glottal fricative /h/

This token of *hello* starts with an [h], which is a voiceless glottal fricative. Since the important feature of this sound is its voiceless glottal friction, there is no position unique to the /h/ that the tongue has to assume, unlike, for example, a voiceless alveolar fricative, where the tongue tip has to create a narrow channel near the alveolar ridge. Accordingly, the tongue is free, in /h/, to anticipate the articulatory position for the next segment. Since this must be a vowel in English, the /h/ is actually a voiceless version of the following vowel. In order to produce this [h], therefore, the speaker first put his vocal tract in the position for the following half-open central unrounded vowel [ɐ]. Then, keeping his vocal cords open so that they would not vibrate but let air pass through, a flow of air was initiated.

The turbulent flow of air through the vocal cords acted as a source of energy setting the air in the vocal tract vibrating at its resonant frequencies, in much the same way as the air in a milk bottle can be made to resonate by creating turbulence by blowing across the top. Note that, in terms of source–filter theory, this is an example of an energy source at the glottis, but an aperiodic one, not the periodic glottal source described above. The acoustic result is visible in the first few centiseconds of the word where there are two weak locations of noise-excited energy (look just above the '| h |' in the figure). The energy between 1 and 2 kHz (marked 'noise-excited F2') is the second formant. The energy between 0 and 1 kHz is more complicated and probably initially reflects the first tracheal resonance, and then, after the increase in frequency, the first formant (marked as 'noise-excited F1'). Since the cords are open for the [h], it being a voiceless fricative, the effects of subglottal acoustic coupling (extra resonances and zeros) are expected. It can be seen that the first formant energy has in fact been zeroed out (at the point marked 'tracheal zero'). This then constitutes the /h/.

Short low central vowel /a/

The next segment is the short half-open [ɐ] vowel, which is the realisation of /a/. Since the supralaryngeal vocal tract is already in position for this sound, all the speaker has to do, assuming that the pressure drop across the cords is adequate, is to adduct the cords with the appropriate tension so that they start vibrating. Vocal cord vibration is visible from the 35 or so vertical striations in the spectrogram, which represent the points of major excitation of the air in the vocal tract by the glottal pulses. Typically, the first pulse is not very clear: it has strong low-frequency energy up to about 400 Hz, but weak energy up to about 2.6 kHz. This is a pulse that has occurred as the cords are in the process of coming together and is marked as 'first (weak) glottal pulse' in the figure. All subsequent glottal pulses except perhaps the last two show strong energy up to about 4 kHz.

The four glottal pulses after the initial weak pulse occur when the supralaryngeal tract is open, in the position for the [ɐ] vowel. This segment shows four strong resonances: the lowest, at about 500 Hz is the first formant; the next highest, at about 1 kHz, is F2. F3 is at ca. 2.5 kHz, F4 at ca. 3.7 kHz. All these formants have been well tracked by the LP analysis, and in addition a weaker resonance at ca. 4.3 kHz, has also been picked out. These are all supraglottal resonances. The LP analysis has failed to pick up some energy located around or just below 2 kHz, which is a tracheal resonance, and labelled as such on the figure. This shows that the speaker's cords were still allowing some coupling with his subglottal system during the [ɐ] as a result of the preceding [h], and in fact the tracheal resonance can be seen to persist (as a thickening in each of the vertical striations) well into the second syllable.

Of the first two formants in [ɐ], F1 is relatively high in frequency and F2 is neither high nor low. Given the formant/vowel-feature relationships described in the previous section (F1/height, F2/backness and rounding), this particular combination of values indicates a half-open central vowel. From the LP trace of the first and second formant centre frequencies it can be seen that F1 and F2 are not static during the [ɐ], but are both decreasing in frequency, as the supralaryngeal vocal tract is already moving towards the articulation of the next segment, the uvularised lateral [ɫ].

The lateral /l/

Since the lateral involves mid-line contact between the tip of the tongue and the alveolar ridge, whilst keeping the sides of the tongue low to let the air pass laterally, this involves raising the front of the tongue towards the alveolar ridge while keeping the sides of the tongue down. In addition, the secondary uvularisation of the lateral means that the body of the tongue has to move somewhat upwards and backwards in the direction of the uvula. The movement of the tongue body is reflected in the F1 and F2 trajectories in the spectrogram in panel (c). F1 decreases, indicating raising; F2 decreases, indicting backing.

The hold-phase of the lateral, during which time the mid-line alveolar closure is maintained, is reflected in the slight drop in overall amplitude occurring between the fifth glottal pulse and the tenth. In order to see where the amplitude drop starts, compare the darkness of the fifth and sixth glottal pulses in the spectrogram in panel (c) at the 2 kHz line. A closure of the vocal tract is always associated with a drop in amplitude. If the closure is total, the drop is large. In this case there is not a very great drop because there is not total closure due to the lateral passes. This portion of the spectrogram – from just after the fifth glottal pulse to just after the ninth glottal pulse – can therefore be taken as reflecting the lateral.

The supraglottal area function in laterals is complicated by the fact that the mid-line alveolar closure and the lateral passes effectively create a short cavity as an offshoot to the vocal tract. This cavity, also referred to as a shunt, introduces its own acoustic resonances and zeros. However, none of these are unambiguously clear in the spectrogram in panel (c), where all the F-pattern appears continuous from the preceding and onto the following vowel.

The diphthong /aʉ/

As just mentioned, by the tenth glottal pulse the tongue tip has come away from the alveolar ridge, and the diphthong can be considered to have started. Panel (b) in Figure 8.25 shows the harmonic and LP-smoothed spectra just after the release of the lateral (the point at which the spectrum was taken is marked in the wave-form in panel (a)). Panel (b) shows F1 and F2 close together, indicating proximity of the tongue body to the pharyngeal wall. The harmonic spectrum also shows a low-amplitude amplification at about 1.8 kHz, which reflects the subglottal resonance mentioned above (this is marked 'tracheal resonance' in panel (b)).

As explained in Chapter 6, a diphthong involves two vocalic targets in a single syllable. The first target in this diphthong is probably low central and unrounded, like in the speaker's vowel in *hut*. The second target is high, rounded and front of central, probably the same as in the speaker's vowel in the word *pool*. The high, rounded and front-of-central target in this diphthong is said to be a typical value for Adelaide speakers (Oasa 1986: 83–6). Because of the influence of the height of the second diphthongal target, the low first target is probably undershot so that the vowel is actually more half-open than open. In any case, this diphthong involves the tongue body moving upwards and slightly forwards from a fairly low central position to a front-of-central high position, with the lips moving at the same time from unrounded to rounded.

These movements are reflected nicely in the continuously changing pattern of the first three formants in panel (c). Over about the first third of the diphthong F1 and F2 are parallel and close, and increase slightly in frequency, while F3 remains fairly stable. Then F1 and F2 diverge: F1 decreases in frequency, and F2 increases to meet a slightly decreasing F3.

These changes in the F-pattern are continuous, because the vocal tract moves smoothly from one target to another. Thus there is no place in the time course of the diphthongal F-pattern – no acoustic discontinuity – indicating the boundary between the two *auditorily* discontinuous targets [ɐ] and [y].

Nevertheless it is possible to indicate points in time that can be taken to reflect the diphthongal acoustic targets. The first target, marked in panel (c) by an oval around F1 and F2, can be said to occur at the point where the F1 shows minimum rate of change, that is, at its frequency maximum between the 17th and 19th glottal pulses. Since F1 is inversely proportional to vowel height, this is the point in time when the tongue body is lowest, just after it has descended from the position for the secondary uvularised articulation in the lateral and before it starts to rise towards the second diphthongal target. There is no corresponding minimum rate of change in F2, which at this point can be seen to be steadily increasing. This reflects the fact that at the time that the tongue body has reached its nadir, it is still moving forwards, its trajectory thus describing a kind of a parabola. It can be seen, however, that the rate of change in F2 increases slightly at about the 22nd glottal pulse, which indicates an increased forward movement.

The second acoustic diphthongal target can be assumed to have been reached near the end of the word, where both F1 and F2 approach minimum rates of change. This is during the last three glottal pulses, and is again marked by an oval in panel (c), this time around F1, F2 and F3. At this point F1 is low, indicating a high vowel; F2 is high, indicating a relatively front vowel; and F3 is close to F2, indicating rounding. (In a high front vowel, F3 is associated primarily with the mouth cavity, so changes in its aperture will be reflected strongly in F3.) Note that the idea of a steady-state target is something of a convenient fiction: it is only during the last strong glottal pulse, if at all, that F1 and F2 (and F3) can be said to have zero rates of change. The harmonic and LP smoothed spectra corresponding to the second diphthongal target are shown in panel (d), where the low frequency value of F1 and the high frequency values of F2 and F3 are clear (the cursor has been placed on the LP F2 peak, the frequency value of which can be read as 1701 Hz in the top right corner of panel (d)).

Intonation and stress

In addition to its segmental sounds (its vowels and consonants), this *hello* token is also characterised linguistically by its pitch, which reflects its intonation and stress. Pitch has therefore to be controlled during the word. The pitch on this token was described as low on the first syllable and high falling on the second. Since pitch is the perceptual correlate of the rate of vocal cord vibration, and vocal cord vibration is

Speech acoustics

reflected in the glottal pulses, what cues the pitch changes can be seen in the spacing of the glottal pulses. It can be seen in panel (c) that they are spaced fairly wide on the first vowel, indicating a low rate of vibration. Then the spacing becomes narrower over the first part of the diphthong, indicating a higher rate of vibration, and finally wider again towards the end of the diphthong, indicating a fall in rate.

Creaky voice phonation type

At the end of the word, over the last four periods, it is clear that the spacing between the pulses becomes very big. This is indicative of the change in manner of vocal fold vibration that is audible as a creaky phonation type. This speaker tends to creak quite a lot at the end of utterances, and sometimes also shows an idiosyncratic tendency to deviate from a modal phonatory pattern in the middle of words. This particular token is a very nice example. It can be seen in panel (c) that during the /l/ there are three glottal pulses that have strong excitation of F4 and F5, between about 3.5 and 4.3 kHz. These strong pulses alternate with pulses that show much weaker excitation at these frequencies. (It may be that the alternating weak–strong pattern extends a little either side of the lateral.) This differential excitation in the higher frequencies is indicative of a difference in the glottal volume velocity flow (cf. Figure 8.8). A volume velocity flow with a more discontinuous time domain profile, and thus sharper changes in rate of flow, will have a less steep spectral slope, and therefore excite the higher frequencies more than a less discontinuous profile. Such changes in discontinuity in volume velocity profiles are usually the result of abnormal vocal fold vibration. The cords might come together rapidly in one cycle, but not so rapidly the next, for example. Such abnormalities can be caused by a variety of factors, ranging from the pathological (vocal cord tumour, smoking) through the organic (voice dropping at puberty) to the linguistic (use of different phonation types).

Forensic significance: Vowel acoustics

The descriptions above, of both the individual vowel and *hello* acoustics, are intended to demonstrate two things. Firstly, the acoustics of normal speech can be described quantitatively in considerable – one might say, after the *hello* example, excruciating – detail *with respect to the linguistic and articulatory events that they encode*. Secondly, the descriptions give an idea of the great number of potential acoustic features available for the forensic comparison of speech samples.

It has been shown that acoustic vowel plots relate demonstrably to traditional descriptions like *front rounded vowel*. This relationship between auditory and acoustic F-pattern features is undoubtedly one reason why formants are typically sampled in forensic-phonetic comparisons. In such investigations, a prior auditory analysis is indispensable in order to identify areas of potential relevance that are then investigated and quantified acoustically. F-pattern (and fundamental frequency, to be discussed below) are correlates of auditory, transcribable qualities of speech that make the quantification and evaluation of the auditory analysis possible. In addition, of course, the behaviour of some formants is interpretable, via the acoustic theory of speech production described earlier in this chapter, in an articulatory coherent way. This is of obvious forensic use when one is trying to make inferences from acoustic patterns to the vocal tract that produced them.

This point is mentioned because in automatic speaker recognition – speaker recognition carried out under optimum conditions – formants are not generally used. In automatic speaker recognition, other acoustic parameters, e.g. LP derived cepstral features are employed. Although more powerful than formants – they can discriminate between voices better – it has been assumed that they are more difficult to apply forensically because they do not relate in a straightforward way to articulation. Another reason put forward for the preference of formants over automatic parameters is that they are a lesser of two evils when it comes to explanation to a jury.

Fundamental frequency

In this section is discussed and exemplified what is considered by many to be one of the most important parameters in forensic phonetics: fundamental frequency. Braun (1995: 9) for example, quotes four well-known authorities (French 1990a; Hollien 1990; Künzel 1987 and Nolan 1983) who claim that it is one of the most reliable parameters.

Fundamental frequency is abbreviated F0, and also called 'eff-oh', or 'eff sub-zero'. It is also often referred to by its perceptual correlate, namely *pitch*. This is not a recommended practice, as it is a good idea to keep auditory descriptors separate from acoustic in forensic analysis, and thereby acknowledge the difference between the perceptual and acoustic levels of description (cf. Rose 1989 and Chapter 9).

In Chapter 6 it has been described how, during speech, the vocal cords vibrate as an important part of the production of many sounds. By its *absence* or *presence* this vocal cord vibration may function to distinguish sound segments. This occurs in the contrast between voiced or voiceless sounds, for example [s] vs. [z]. By changes in its *rate*, the vocal cord vibration may function to signal the deliberate changes in pitch that underlie linguistic contrasts in tone, intonation and stress. For example, the difference between the question *yes?* with a rising pitch and the statement *yes*, with a falling, is produced by changing the rate of vocal cord vibration.

F0 is the acoustical correlate of rate of vocal cord vibration, and is directly proportional to it. Apart from the linguistic target involved, one of the main factors that determines the rate of vocal cord vibration, and hence fundamental frequency, is the size of the cords (usually understood as their mass and length). A formula was given above, at (8.3), relating acoustics (formant centre-frequencies for schwa) to an individual's anatomy (vocal tract length) for F-pattern. It is therefore appropriate to show with formulae how F0 can be related to an individual's anatomy. How F0 is related to the mass and length of the cords is briefly described below.

Vocal cord mass and length as determinants of fundamental frequency

The vibratory behaviour of the vocal cords is exceedingly complex, and contains both spring-like and string-like components (Titze 1994: 193). Each cord vibrates like a plucked string (a ribbon is an even more appropriate metaphor) because they stretch, string-like, from the back of the larynx (the arytenoids) to the front (the back of the thyroid cartilage). Each cord behaves like a mass attached to a spring in its medial–lateral movement.

The fundamental frequency of the cords when they behave like a string can be described with the formula at (8.4), which is a version of the formula for the F0 of an ideal string (Titze 1994: 172, 173). In this formula, L_m (quantified in metres) represents the length of the cords, actually the length of the membraneous part; σ_c represents the longitudinal *stress* in the cords, in particular the stress in the so-called cover of the cords (the stress is the tension in the cords divided by the cross-sectional area of vibrating tissue and is quantified in units of pascals (Pa)); and ρ is the *tissue density*, which remains constant at about 1040 kg/m³, and does not play a role in F0 regulation.

Formula for F0 of cords vibrating as a string (Titze 1994: 200)

$$F0 = \frac{1}{2L_m} \times \sqrt{\frac{\sigma_c}{\rho}} \qquad (8.4)$$

This formula makes clear the relation between F0 and vocal cord length (L_m) across different speakers. Since F0 is inversely proportional to the length of the cords, other things being equal, if the speaker's cords are long, the F0 will be low. For example, assuming a tissue density of 1040 kg/m³, and a constant longitudinal stress of 15 000 Pa, a vocal cord with a membraneous length of 1.6 cm (an average value for males) will vibrate with an F0 of 119 Hz. Shorter cords in a different speaker, say with a vocal cord length of 1.5 cm, will produce a slightly higher F0 of 127 Hz. Even shorter cords, say with a length typical for females of 1 cm, will have a corresponding F0 of 190 Hz. (The representative membraneous vocal cord lengths are taken from cadavers (Titze 1994: 179).)

The fundamental frequency of the cords when they behave like a spring can be described with the formula at (8.5), which is a version of the formula for the F0 of an oscillating spring attached to a mass (Titze 1994: 86, 193). In this formula, *m* (in kilograms) represents the effective vibrating mass of cords, and *k* represents their stiffness (in units of newtons per metre).

Formula for F0 of cords vibrating as a spring (Titze 1994: 86, 193)

$$F0 = \frac{1}{2\pi} \times \sqrt{\frac{k}{m}} \qquad (8.5)$$

This formula makes clear the relation between F0 and vocal cord mass (*m*). Since F0 is proportional to the square root of the ratio of stiffness to mass, other things being equal a greater mass will correspond to a lower F0.

It is important to remember that the inverse relationship between cord length and F0 – longer cords means lower F0 – must be thought of as applying to a comparison of the cords of different speakers producing the same linguistic sound (e.g. tonal pitch target) and not to the regulation of F0 to produce different sounds *within a speaker*.

Forensic Speaker Identification

This is because speakers control their F0 mostly by actively stretching the cords to increase their tension: increased tension results in increased F0. When tension is increased actively, by stretching the cords, they of course become longer. Thus F0 *within a speaker* tends to be proportional to cord length: within a speaker, longer cords means higher F0.

Thus longer and more massive vocal cords vibrate at lower frequencies than small cords, other things being equal. Since speakers can differ with respect to vocal cord size – females tend to have shorter, less massive cords than males, for example – parameters associated with the rate of vocal cord vibration are dimensions in which speaker-specific characteristics are encoded, in much the same way as the F-pattern reflects the size and shape of the supralaryngeal vocal tract.

Forensic significance: Fundamental frequency

One reason why F0 is considered important is undoubtedly that measures associated with it have been shown to be among the more successful in speaker recognition (Nolan 1983: 124). Another reason why F0 is considered important forensically is that it conforms to many of the desiderata for forensic-phonetic parameters (see Chapter 3). It is robust, and it can be extracted with relative ease from poor-quality recordings, for example. Moreover, it is not adversely affected by telephone transmission, unlike F1 and the higher formants discussed in the previous sections (Künzel 2001). Since it has to do with voicing, and the majority of speech sounds (e.g. vowels, nasals and laterals) are usually voiced, it is also readily available.

It is important to realise that this does not imply that a speaker's F0 is largely invariant. There is a bewildering number of factors that will affect within-speaker variation in F0. Many of them are summarised, with references, in Braun's very useful (1995) paper, where they are categorised (pp. 11–14) under the three general categories of *technical*, *physiological* and *psychological* factors. Under the technical factors that can, for example, correlate with F0 are sample size and tape speed. Examples of possible physiological factors are race, age, smoking and intoxication. As examples of psychological factors are listed emotional state, and situational factors like the time of day or background noise level in telephone conversations. It is clear from these examples that forensic speech samples will need to be controlled for such variables if they are being compared with respect to F0 (or indeed probably all other parameters).

An example of fundamental frequency in **hello**

It will also be recalled from previous discussion that F0 is acoustically the rate of repetition of the complex periodic speech wave. It is the number of times the pattern of the complex periodic wave repeats per second, and is quantified in hertz (Hz). Panel (a) of Figure 8.26 shows a very short portion, twenty-two thousandths of a second, of the wave-form in the word *hello* said by a male speaker of Australian English. The wave-form of the whole word from which the short portion was magnified is shown in panel (b) below it. The location of the magnified portion in the whole wave is shown within the thin rectangle.

Speech acoustics

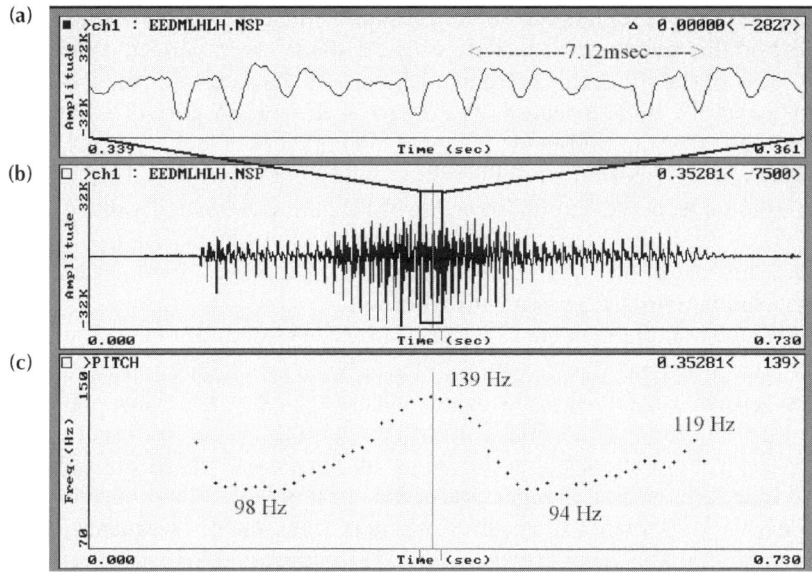

Figure 8.26 Fundamental frequency in one token of the word *hello*

The pattern of the magnified complex wave in panel (a) can be seen to repeat about three times in this stretch of 22 milliseconds, which means that the F0 of this wave at this point is somewhat less than 140 Hz. (If the wave repeats three times in 22 milliseconds, in one second it will repeat (one second divided by 0.022, times 3 =) 136 times, and therefore its average F0 will be 136 Hz.

The particular *hello* token shown in Figure 8.26 was chosen because it was said with a complex intonation in which the pitch changes twice from low to high. This gives the reader the chance to see clearly what the F0 changes look like that signal the pitch changes. In this *hello*, the pitch is low on the first syllable, but on the second it rises and falls and rises again. This pitch pattern can be represented as [L.LHLH], where L and H stand for low and high pitch targets, and the full stop indicates a syllable boundary. If the reader imitates this intonation correctly, with low pitch on the *he-* and rising–falling–rising pitch on the *-llo*, they will sound as if they are conveying the following: 'Hello, I'm trying to attract your attention, but for some reason you are not responding and it's annoying me.'

Panel (c) in Figure 8.26 shows the time course of the F0 for the whole word. The F0 has been automatically extracted from the wave-form by digital means. In panel (c), time runs from left to right, and frequency from bottom to top. The frequency range shown is from 70 Hz to 150 Hz. The F0 has been estimated every 10 milliseconds, and its estimated value is shown by a dot at 10 millisecond intervals. The F0 values at various inflection points of the trace have been put on the figure. Thus the F0 at the beginning of the trace falls to about 98 Hz, rises then to 139 Hz, falls to 94 Hz and then rises again to 119 Hz. These values reflect the rate of vibration of the speaker's vocal cords at various points when he said the word, and the F0 shape largely reflects the [L.LHLH] rise–fall–rise pitch of the *hello*.

Forensic Speaker Identification

The cursor in panel (c) has been placed just after the dot with the highest estimated F0 value, and this point occurs just after the middle of the expanded portion in panel (a). The value of the F0 at this point can be read off the display at the top right-hand corner of panel (c), and this shows the F0 to be 139 Hz. A period of the expanded wave-form just after its mid-point has been identified in panel (a) and measured at 7.12 milliseconds. A period of 7.12 milliseconds corresponds to an F0 of 140 Hz, which is as close as makes no difference to the digitally estimated value of 139 Hz.

Long-term fundamental frequency distributions

The previous section has shown how the F0 can be automatically extracted, displayed, and easily quantified. Given good quality recordings, it is generally easy to display and quantify the time course of the F0 throughout an utterance. In forensic phonetics, however, one is more interested in the *long-term distribution* of the F0, rather than its temporal structure (how it changes over time), since statistical parameters associated with F0 over a long stretch of speech (long-term fundamental frequency, LTF0) are assumed to be useful in speaker recognition. The notion of a long-term distribution is an important one, not just for F0, but for any continuous acoustic-phonetic parameter, and can be conveniently approached via the F0 in the *hello* example just discussed.

In the section above, it was shown how the F0 could be estimated at short intervals and its time course plotted. If the point in time where an F0 value was estimated was ignored, and the individual F0 observations simply collected and plotted, Figure 8.27 would result. Figure 8.27 is a *histogram* of the F0 data in panel (c) of Figure 8.26. It shows the number of times a particular F0 value occurs in the data: the F0's frequency of occurrence. F0 is plotted along the bottom and the number of occurrences ('count') up the left side. It can be seen for example that this *hello* contained one

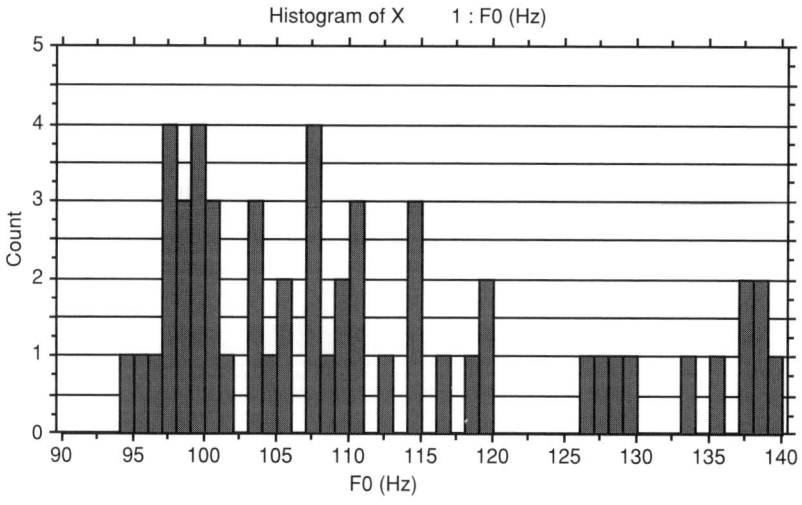

Figure 8.27 Fundamental frequency distribution in the *hello* token of Figure 8.26

occurrence of an F0 value of 94 Hz (this corresponds to the dot marking the lowest F0 observation in Figure 8.26) and one occurrence of an F0 value of 139 Hz (this corresponds to the peak F0 observation). The F0 value of 97 Hz occurred four times, as did also the values of 99 Hz, and 108 Hz. There were no occurrences of F0 values of between 121 Hz and 126 Hz.

Taken together, the pattern in Figure 8.27 constitutes the distribution of the F0 in this particular speaker's one token of the word *hello*. The distribution can be characterised by several important statistical properties that may be forensically useful, that is, have relatively large F-ratios. However, before these are discussed it is better to present a more realistic distribution based on many more F0 observations than are present in the *hello* in Figure 8.27. This is because the values in Figure 8.27 still reflect something linguistic, namely the F0 realising the intonation of the *hello* token. (Had the word been said with a different intonation, for example the calling intonation $_{he-}$ ^{lo}-o, the F0 distribution would have been slightly different, with overall higher values.)

A long-term F0 distribution needs to be built up over a stretch of speech long enough to ensure that all the local linguistic variables responsible for its shape cancel each other out so that the remaining profile, and its statistical properties, characterise the speaker rather than reflect the linguistic content. This is the idea behind all long-term distributions. It must be emphasised, of course, that the long-term characterisation is of the speaker *as they are speaking on that particular occasion*, and is not necessarily valid for all occasions: *no acoustic feature is invariant*.

One obvious question is how much speech you need. The amount of speech required to obtain a characterisation of a speaker depends on the measure used, but there is experimental support for at least 60 seconds' worth of speech (Nolan 1983: 13, 123) for meaningful long-term F0 measurements. There is also evidence that this value may be language-specific. Rose (1991: 241), for example, found that long-term F0 values for seven Chinese dialect speakers stabilised very much earlier than after 60 seconds.

Figure 8.28 shows a long-term F0 distribution. It represents the long-term F0 of one of two male participants in a Cantonese telephone conversation, and is based on 1591 sampled F0 values. Since F0 was sampled every two centiseconds, and about 40% of Cantonese sounds are voiced, the distribution in Figure 8.28 represents about 80 seconds of speech.

In contrast to the profile of the distribution in Figure 8.27, it can be seen that the LTF0 distribution in Figure 8.28 is well filled-out. It is monomodal, that is it has one main peak, and the peak is centered around 145 Hz. The majority of the distribution lies between about 100 Hz and 190 Hz. When this person was speaking on this occasion, most of the time his vocal cords were vibrating at rates between 100 times and 190 times per second. There appear to be very slightly more higher F0 values than lower. In the sections below, we look at some of the important ways in which such a distribution can be described and quantified.

Mean and standard deviation

The most important statistical properties of a distribution like that in Figure 8.28 are those that specify its average value and the spread of the values around this average value. These values are called the *mean* (or *arithmetical mean*) *F0* and *standard deviation*

Forensic Speaker Identification

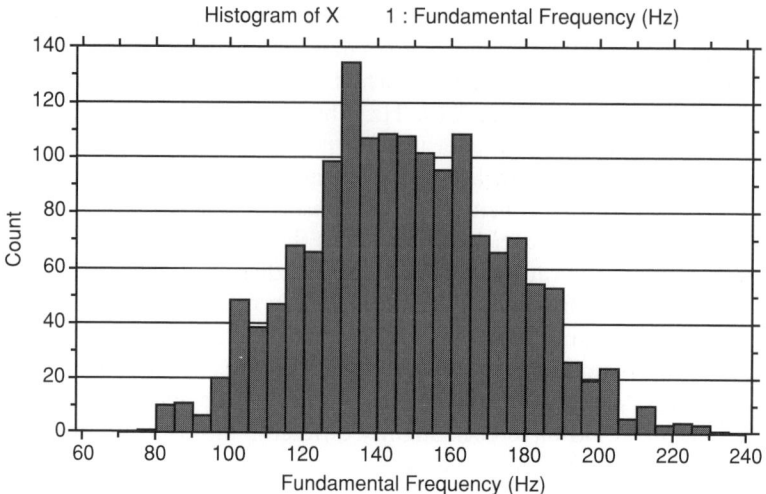

Figure 8.28 Long-term fundamental frequency distribution for a male speaker of Cantonese in a telephone conversation

F0 respectively. It has been claimed that measures based on them are among the more successful long-term measures in speaker recognition. Mean and standard deviation will also be crucial in the demonstration of the use of the likelihood ratio in Chapter 11.

The mean F0 is very commonly cited in forensic case-work. It is, of course, the sum of the F0 observations divided by the number of observations. The mean F0 in Figure 8.28 is 147 Hz. (Mean values are often symbolised with $\bar{x}$ (called x bar), or with a macron, thus $\bar{x}$ = 147 Hz, or $\overline{F0}$ = 147 Hz.)

$$\text{Standard deviation} = \sqrt{\frac{\sum_{i=1}^{n}(x_i - \bar{x})^2}{n-1}} \tag{8.6}$$

Another commonly used long-term parameter is the *standard deviation* (abbreviated *s* or SD, and also called the second moment around the mean). The standard deviation is, as just mentioned, a measure of the spread of values around the mean, and its formula is given at (8.6). It is calculated in a commonsense way by finding the average distance of each observation from the mean. The distance of each observation has to be squared, because otherwise the positive deviations from the mean would be cancelled out by the negative ones. This is the part that is represented by the $(\bar{x} - x_i)^2$ term of the equation. $(\bar{x} - x_i)^2$ represents taking each F0 observation x_i from the mean $\bar{x}$, and squaring the result. These squared deviations from the mean are then summed, represented by the Σ symbol, and then the average squared deviation is found by dividing by the number of observations *n*. (For statistical reasons that are not important here the divisor is actually one less than the number of observation: $n - 1$). Once

Speech acoustics

the average squared deviation is found, the squaring of the deviations is reversed at the end of the process by finding its square-root. Because the procedure involves taking the square root of the mean of the squared differences, it is sometimes referred to as a *root-mean-square* measure.

The standard deviation of the data in Figure 8.28 is 27 Hz. It has been proposed that a measure of an individual speaker's F0 range, or *compass*, is twice the standard deviation above and below their mean (Jassem 1971). The Cantonese speaker's F0 range in Figure 8.28, then, would be from ([147 Hz − [2 × 27 Hz] =) 93 Hz to (147 Hz + [2 × 27 Hz] =) 201 Hz, which is a range of 108 Hz. It is known from statistical theory that, under certain well-defined circumstances (when the distribution is symmetrical or nearly so), a range of two standard deviations around the mean will include about 96% of all observations in a distribution. As can be seen from Figure 8.28, this range, from 93 Hz to 201 Hz, does indeed include most of the speaker's F0 observations.

Skew

It was mentioned above that the F0 distribution in Figure 8.28 seemed to show a very slightly greater number of higher F0 values (say from 200 Hz upwards) than lower (say from 100 Hz downwards). Asymmetries of this type are called skewing. More higher values than lower is called *positive skewing*. Positive skewing is fairly typical in F0 distributions (Jassem *et al.* 1973: 218), and is probably due to the exponential relationship between elongation and tension of the vocalis muscle in the vocal cords (Rose 1991: 239). If the distribution contains more lower values than higher it is *negatively skewed*.

Figure 8.29 shows a long-term F0 distribution with very clear positive skewing. The subject is a young female from Shanghai reading out a prose passage in Shanghai

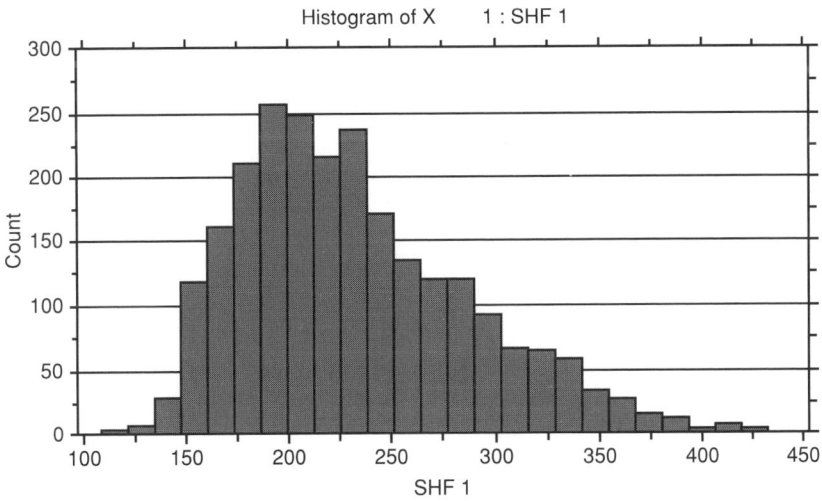

Figure 8.29 A positively skewed long-term F0 distribution

dialect. The distribution is based on about 24 seconds of voiced speech, which represents about 40 seconds of speech overall. It can be seen that the upper part of the F0 distribution has been stretched out markedly, and the distribution is very asymmetrical.

$$\text{skewness} = \frac{\sum_{i=1}^{n}(x_i - \bar{x})^3}{n-1} \bigg/ s^3 \tag{8.7}$$

The statistical measure of skewness is quantified by cubing, rather than squaring the deviations from the mean, and dividing the average cubed distance by the cube of the standard deviation (Jassem et al. 1973: 218). A formula for skewness is given at (8.7). Skewness is also known as the third moment around the mean. A symmetrical distribution has a skewness value of zero. Values greater than zero indicate positive skew, and smaller than zero negative skew (Jassem 1971: 65). The skewness value for the highly asymmetrical distribution in Figure 8.29 is 0.718. This can be compared with the visually much smaller skew for the F0 distribution in Figure 8.28, which is 0.153.

Kurtosis

A third measure for the deviation of a distribution away from the ideal is its degree of peakedness. This is called kurtosis, or fourth moment around the mean, with distributions that are more peaked than normal called *leptokurtic* and less peaked called *platykurtic*. A formula for kurtosis, after Jassem et al. (1973: 218) is given at (8.8), where it can be seen that it is the same as that for skewness except that the distances from the mean are raised to the fourth power, and the average distance is divided by the standard deviation raised to the fourth power. A value of 3 indicates zero kurtosis; greater than 3 indicates leptokurtosis; less than 3 indicates platykurtosis (Jassem 1971: 67).

$$\text{kurtosis} = \frac{\sum_{i=1}^{n}(x_i - \bar{x})^4}{n-1} \bigg/ s^4 \tag{8.8}$$

Modal F0

In many distributions there is a unique F0 value that occurs the most frequently of all. This is called the *mode*. In relatively neutrally skewed distributions the most common F0 observation occurs near the mean. In distributions with positive skew it occurs lower than the mean and in negatively skewed distributions it occurs higher. It looks as though the mode in the distribution in Figure 8.28 is between 130 and 135 Hz, but this is a quirk of the way that the F0 observations have been grouped into frequency bins. The skew in Figure 8.28 is quite small, and the mode is actually very near the mean value of 147 Hz, at 159 Hz. It is, of course, perfectly possible for an F0 distribution to lack a modal value. The distribution in Figure 8.27 is a case in point, where it can be seen that there are three frequency values (97 Hz, 99 Hz, 107 Hz) that each

Speech acoustics

occur four times in the sample, and that there is no one F0 value that occurs most often.

Speakers can differ in their long-term F0 distributions in all of the ways illustrated above: in mean and standard deviation F0; in skewness and kurtosis; and in mode. But, of course, so can the same speaker and, as always, the ratio of between- to within-speaker variation in F0 must be born in mind. This section can be concluded with an illustration of how the same speaker can vary in long-term F0 distribution as a function of one of the variables affecting F0 surprisingly not included in Braun (1995), namely health.

Fundamental frequency and health

Any changes in health that affect the size or shape or organic state of the vocal tract, or its motor control, will alter its acoustic output. Panel (a) of Figure 8.30 shows the author's long-term F0 distribution obtained when reading a passage in good health; panel (b) shows the author's F0 distribution for the same passage read with a bad head cold; and panel (c) shows the distribution for the passage when the author had severe laryngitis. The same scale on the horizontal axis is used to facilitate comparison. It can be seen from Figure 8.30 that laryngitis appears to correlate with extreme differences in F0 distribution. The mean F0 value is 32 Hz lower than in the normal condition; and the mode is 40 Hz lower. For someone with a normal standard deviation of about 15 Hz, these are very big differences. There are also large differences in the two distributions' skew, kurtosis and standard deviation. The F0 distribution associated with the head cold does not show a drastic shift of values down from normal, rather the mean and mode are slightly higher than normal.

Fundamental frequency distributions and probability

Since this section has been concerned with distributions, it will be useful to introduce here another notion central to forensic phonetics that is closely related to distribution. This is that a distribution can not only indicate frequency of occurrence, but also represents probability.

As explained in the discussion of Bayes' theorem in Chapter 4, probability has two interpretations. It can be a quantification of our confidence in a hypothesis, or it can reflect a property of data, calculated from the long-run frequency of an event. It will not harm things here if for the purposes of exposition we amalgamate these two interpretations and treat probability as a measure of confidence in predicting the occurrence (or non-occurrence) of an event.

Probability is expressed as a fraction of 1, or in per cent: a probability of unity for an event, or 100%, means the occurrence of the event is completely certain. A probability of 0, or 0% means that it is completely certain that the event will not happen. A probability of 0.5, or 50%, means that *in the long run* we are confident that an event will happen about half the time. If we were to toss a coin 100 times, for example, we would expect that it would come down heads in about half the tosses: there would be a 50% probability of heads. (If the number of heads deviated substantially from 50% over 100 tosses, we would suspect that the coin was biased.) The phrase *in the long run*

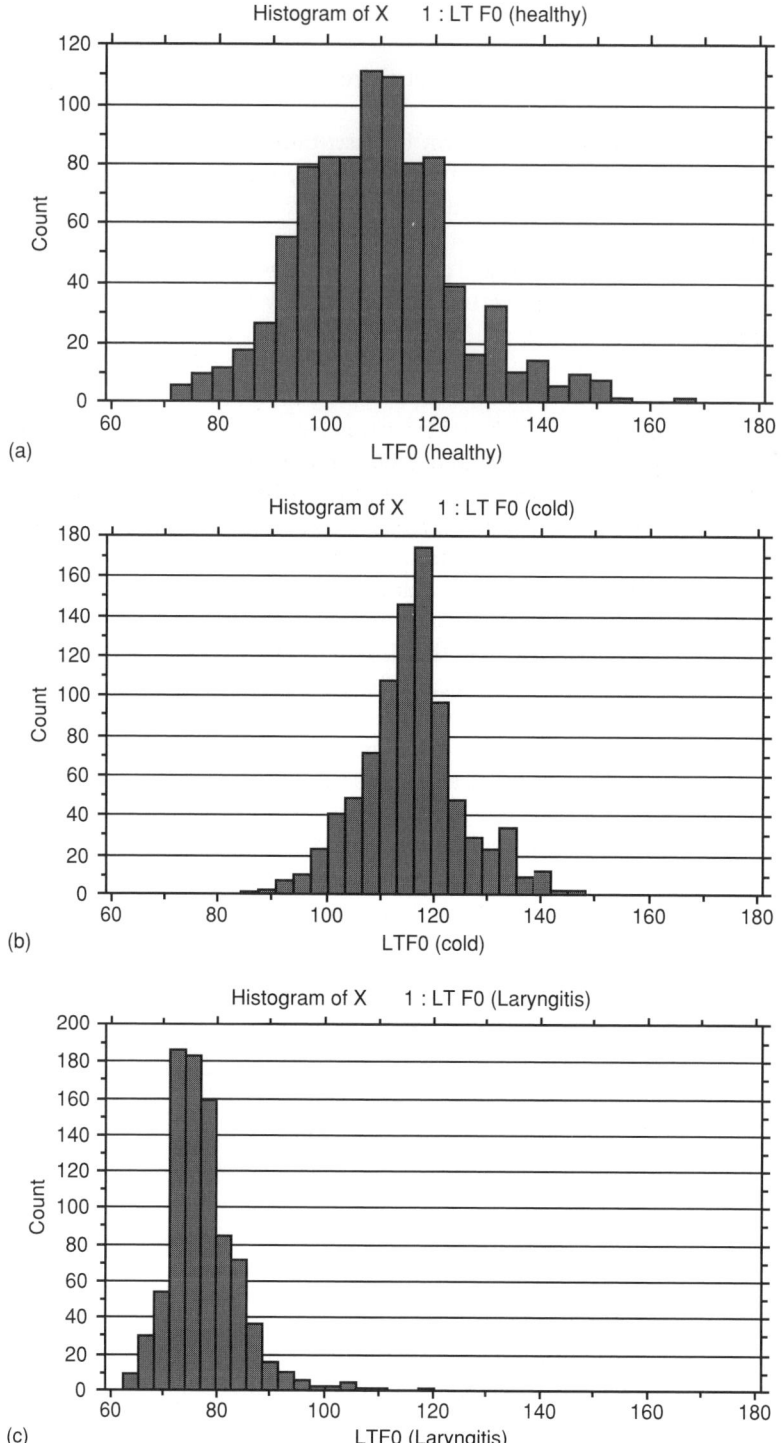

Figure 8.30 Long-term F0 distributions of the same speaker in differing states of health. Top = healthy. Middle = head cold. Bottom = severe laryngitis

Speech acoustics

is emphasised because a distribution will not predict a specific, individual event: what will happen, for example, in the next toss of the coin. It only predicts what will happen if we repeat the experiment, the coin-toss, many times.

An F0 distribution allows us to see the proportion of times a particular F0 value or set of F0 values occurs relative to the total number of F0 values. For example, in the distribution of the 53 observed F0 values in *hello* in Figure 8.27, the value of 97 Hz occurred 4 times, and therefore had a relative frequency of 4/53 which is 0.0755, or between 7% and 8%. Now, this relative frequency can be understood as an expression of the probability of observing an F0 value of 97 Hz at random in the *hello* F0 data. In other words, if we were to put all 53 F0 values from the *hello* in Figure 8.27 into a hat and select an F0 value at random, note the value, put it back, and repeat this 100 times, we would expect to pick the 97 Hz value between seven and eight times out of 100. A slight complication is introduced by the fact that F0 is in fact a continuous variable. This means that we have to express the probability of observing an F0 value (or set of values) *within a given F0 range*. It is sensible to illustrate this using more realistic data from another female Shanghai speaker. Figure 8.31 presents her long-term F0 distribution, based on 2304 observations. As can be seen, F0 values have been grouped into 10 Hz bins.

Suppose we wished for the speaker in Figure 8.31 to estimate the probability of observing an F0 value at random between 220 Hz and 250 Hz. The three bins between 220 Hz and 250 Hz indicated in Figure 8.31 contain in all 114 F0 observations. Since there are 2304 F0 observations in all, the probability of observing an F0 value between 220 Hz and 250 Hz is about 114/2304, or just about 5%. Thus, if we selected F0 values at random from this distribution of this speaker, we would expect to observe F0 values between 220 Hz and 250 Hz about 5% of the time. In contrast to this, we would expect to observe F0 values in a range of the same 30 Hz size between 160 Hz and 190 Hz, where a higher percentage of observations is concentrated, with a greater

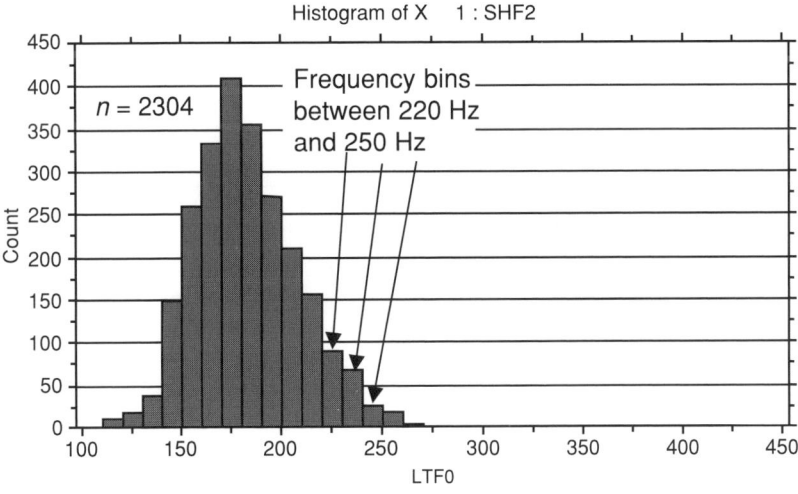

Figure 8.31 Long-term F0 distribution for a female speaker of Shanghai dialect

probability. In Figure 8.31, there are in fact 851 F0 measurements in these three bins; the reader is invited to use this figure to calculate the probability of observing an F0 value at random between 160 Hz and 190 Hz.[5]

A closed-set Bayesian comparison using F0 distributions

A very simple example of the use of such probability observations can now be given using the distributions of two female Shanghai speakers in Figures 8.31 and 8.32. (Figure 8.32 reproduces the data for the speaker in Figure 8.29. The scale on the frequency axis and the grouping of F0 observations into 10 Hz bins in both figures has been made the same to make visual comparison easier.)

Suppose the F0 in a short questioned voice sample was measured and found to lie between 220 Hz and 250 Hz. It is known that the offender is either the speaker in Figure 8.31 or the speaker in Figure 8.32, so this is a closed set comparison. Closed set comparisons were mentioned in the discussion on characterising forensic speaker identification in Chapter 5. There, in the section on categoricality, it was shown that the LR for a parameter in a closed set of two was the probability of observing the values assuming that they had been said by one speaker, divided by the probability of observing them assuming that they had been said by the other.

As just demonstrated, the probability of observing the evidence assuming that the speaker is that in Figure 8.31 is about 5%. It can be appreciated from the long-term F0 distribution for the female Shanghai speaker in Figure 8.32 that the probability of observing an F0 value between 220 Hz and 250 Hz is greater than this. The speaker in

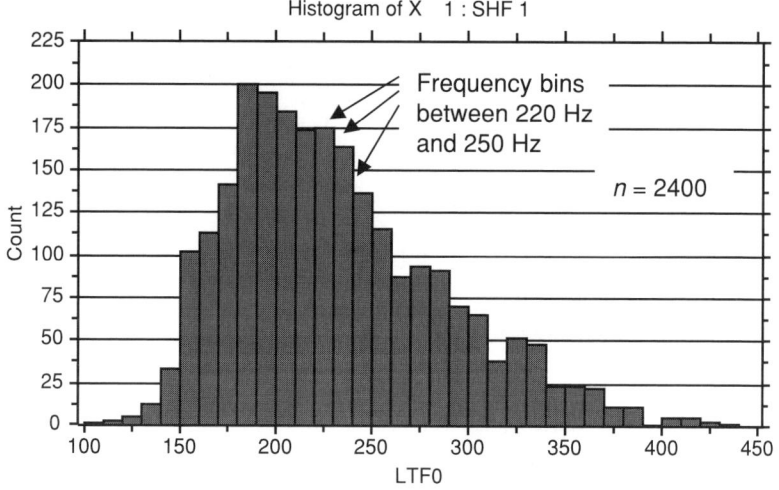

Figure 8.32 Long-term F0 distribution for the Shanghai speaker in Figure 8.29

---

5 The probability is 851/2304 = 0.3694 = 37%.

Speech acoustics

Figure 8.32 has in fact 474 F0 observations in this range, and since there were 2400 observations in all, this is a probability of 19.8%. The ratio of the probabilities for the two speakers is then approximately (20/5 =) 4, which means that one would be four times more likely to observe the questioned F0 values between 220 Hz and 225 Hz if the female in Figure 8.32 were the speaker than if it were the female in Figure 8.31.

The reader hardly needs to be told that this is a contrived example and constitutes a gross oversimplification. In order to have confidence in such a statement, one would need first of all to be sure that the questioned sample was comparable with both suspect samples: this would involve at least being sure that all samples were from telephone conversations; that all conversations had comparable content; and that the interlocutor was comparable in all cases. In addition, it would be necessary to have many more samples than one per suspect in order to have a good idea of the within-speaker variation involved.

Modelling fundamental frequency distributions

An idea of the probability of observing a particular value or range of values at random from a distribution can be obtained, as above, by dividing the number of observations in a given range by the total number of observations. The probability is properly estimated, however, using certain statistical parameters of the distribution together with a formula for the mathematical function that best models the distribution. This will now be demonstrated.

There are several types of distribution, all described by mathematical functions, that various natural phenomena conform to. Perhaps the best known one is the bell-curve, or Gaussian, or 'normal' distribution. Figure 8.33 shows the F0 distribution of Figure 8.31 with a normal curve fitted. It can be seen that there is quite a good match between the observed values of the distribution and the predicted values of the normal curve. The fit, which can of course be quantified, is not perfect – the curve tends to underestimate the actual values on the left, and overestimate them on the right, for example.

If the raw values cannot be modelled well with such a curve, they can often be transformed in such a way that they fit better. For example, distributions with a small degree of skew, like that in Figure 8.31, can often be made less skewed, and more normal, by the simple expedient of taking their square root (this has not been done, however, to keep things less complicated). If the distributions are more complicated, special methods must be used. One method is to model the data with a combination of two or more normal distributions. This is called estimating a *kernel density function* (Aitken 1995: 181–93). Another term is *Gaussian mixture modelling* (GMM) and approaches using forms of GMM, e.g. Meuwly and Drygajlo (2001), Gonzalez-Rodriguez *et al.* (2001), are used in state-of-the-art automatic speaker recognition (Broeders 2001: 8).

To the extent that F0 distributions (or indeed distributions of any other continuous forensic-phonetic parameter) conform to and can be modelled by one of these types, it is possible to specify the probability of observing a particular value or range of F0 values at random from that distribution. The probability of observing a value or range of values from such a distribution is given by the *area* under the curve between

257

Forensic Speaker Identification

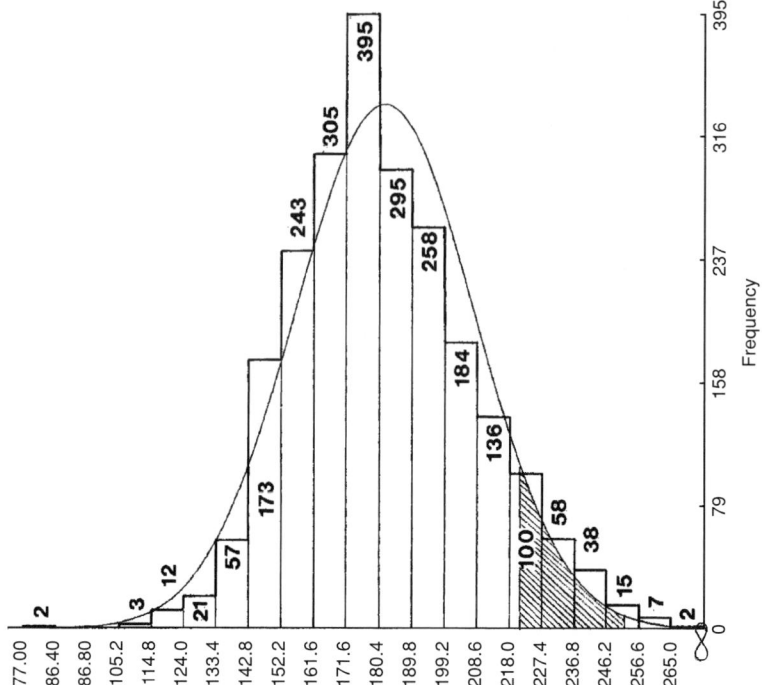

Figure 8.33 F0 distribution of speaker in Figure 8.31 with normal curve fitted. Shading indicates area beneath curve between 220 Hz and 250 Hz. Bold numbers indicate the number of F0 observations in each frequency bin. Width of frequency bins is 9.4 Hz

the limits of the values the probability of which it is required to specify. Thus if we want to know the probability of observing an F0 value somewhere between 220 Hz and 250 Hz in the distribution of Figure 8.32, we need to find the area under the curve between 220 Hz and 250 Hz. This particular region is shown shaded in Figure 8.33, and its area will obviously be its length times a measure of its height. The length is easy to calculate, but since its height varies drastically, some kind of average height measure has to be calculated, and the area has to be estimated using calculus. It is unfortunately beyond the scope of this chapter to illustrate this in detail, but it is relatively easy to show how the height of the curve at various points – its so-called probability density – is found.

$$\text{Probability density } (x) = \frac{n}{s}\left(\frac{1}{\sqrt{2\pi}} e^{-(\bar{x}-x)^2/2s^2}\right) \tag{8.9}$$

The formula for calculating the probability density of a normal curve is given at (8.9). This formula allows us to find the height of the curve, (the probability density), corresponding to any value along the horizontal axis, symbolised as x. The probability density at any point x is derived from three values of the distribution that have already been met: the mean ($\bar{x}$), the standard deviation (s), and the number of observations (n).

How the formula works can be demonstrated with the data in Figure 8.33. This speaker's F0 distribution on this particular occasion had a mean of 181.5 Hz, a standard deviation of 25.6 Hz, and was based on 2304 measurements. Let us assume we wish to calculate the probability density for $x = 220$ Hz. At (8.10) are shown the appropriate figures inserted into the formula. Thus, for example, the reciprocal of the square root of 2 times π is (to three significant figures) 0.399; and e (also to three significant figures) is 2.718. As can be seen, at 220 Hz, the probability density is 11.6.

Probability density (220 Hz) =

$$\frac{2304}{25.6} \times (0.399 \times 2.718^{-(181.5-220)^2/(2 \times 25.6^2)}) = 11.6 \qquad (8.10)$$

Now, given a probability density at 220 Hz of 11.6, the number of frequency observations contained in a frequency bin of interval 9.4 Hz (as is the case in Figure 8.33) centred on 220 Hz is (9.4 × 11.6 =) ca. 109. It can be seen from Figure 8.33 that the frequency bin centred on 223 Hz actually contains 100 observations. Thus the curve predicts the data rather well at this point.

The area under the curve is found by calculus as the definite integral of the function of the normal curve between 220 Hz and 250 Hz. It actually comes out at 0.0626, so the original estimate of 5% (i.e. 0.05) based on the raw histogram data was very close. This is an important result, because it shows the power of statistical models. The F0 distribution of the speaker in Figure 8.31 has been modelled with the formula for a normal curve using only a specification of the distribution's mean, standard deviation and number of observations as input. The model predicted a 6% probability of observing an F0 value between 220 Hz and 250 Hz, which is very close to the reality of 5%.

The above section has used speakers' long-term F0 profiles to show how distributions can be treated as probability statements, which can then be statistically and mathematically quantified in such a way as to allow a simple Bayesian statement (in the contrived example above, the likelihood of observing a range of F0 values in the two suspects of a closed set).

Although F0 has been used in this example, the notion of distribution as probability statement applies to any acoustic speech parameter. Thus it should be clear that it is generally very easy to quantify acoustically the way a speaker is speaking *on a particular occasion*, even if the parameter distributes other than normally, which is quite often the case. The precision and relative ease with which these statements can be made, however, should not obscure the fact that the proper evaluation of between-sample differences must be able to take into account within-speaker variation, that is, how a person speaks on many occasions.

Long-term spectrum

In the previous section, fundamental frequency was used to illustrate the concept of a long-term acoustic parameter. In this section we give an example of another type of long-term measure commonly used in speaker recognition: the *long-term (average) spectrum* (LTAS or LTS). The long-term spectrum is considered to be a powerful parameter in automatic speaker recognition. Hollien (1990: 239) says that 'it is a good cue to a speaker's identity', and that it 'can predict the identity of speakers at very high accuracy levels, especially in the laboratory'. Nolan (1983: 130–5) gives a useful summary and critique of the use of the LTS in speaker recognition experiments.

Many examples have been presented in this chapter of spectra, showing both harmonic and LP extracted formant structure. These have been *short-term* spectra – that is, they are calculated over a very short space of time. This is necessary to capture the acoustic reflexes of individual sounds, like an assumed formant target, which are ephemeral. If, instead of calculating the spectrum over a short period, many spectra are averaged over a long period of speech, say twenty seconds or more, a long-term average spectrum is obtained. The long-term average spectrum shows the average distribution of acoustic energy in the speaker's voice. The idea is the same as with the long-term F0: by taking an average over a long stretch of speech, *acoustic features that are due to individual speech sounds are cancelled out, leaving the overall energy profile of the speaker as they are speaking on that occasion.*

The long-term spectrum thus provides 'a characterisation of the speaker independent of the linguistic content' (Nolan 1983: 130). Its profile will largely reflect the combination of average spectral characteristics of both the laryngeal source and supralaryngeal filter, together with the effect of any subglottal energy in the form of extra resonances and zeros (Nolan 1983: 133), and a small contribution from other non-laryngeal sources.

Figure 8.34 shows the long-term average spectrum of about 30 seconds of monologue from a young Australian male. Panel (a) contains the wave-form. The wave-form is

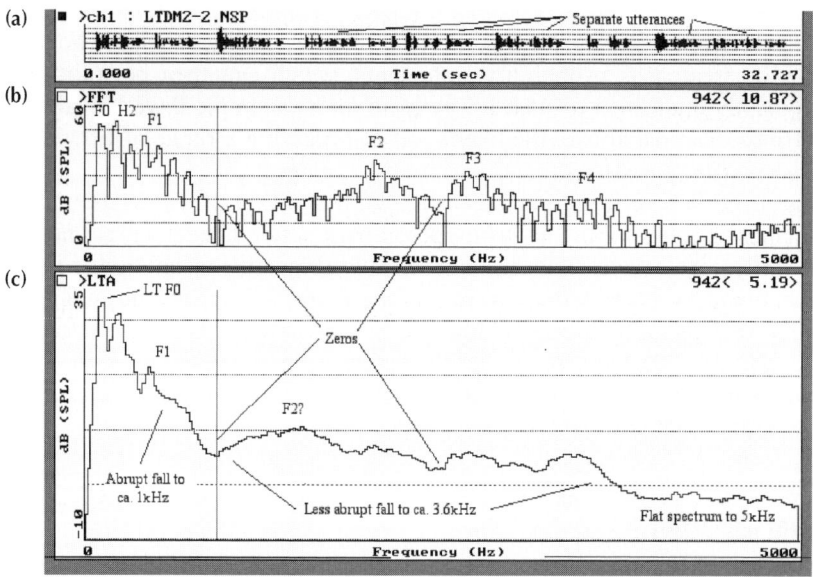

Figure 8.34 Wave-form (A) and long-term average spectrum (C) over 30 seconds of speech. Panel (B) shows a harmonic spectrum from the speaker's [ɛ]

divided into about 11 groups that indicate separate utterances bounded by short pauses. Panel (b) shows the reader what a short-term harmonic spectrum looks like from this speaker. It was taken in the middle of the mid-front vowel [ɛ] in the word *air*, which occurs early on in the monologue. Its harmonic structure is clear (the cursor has been placed on the ninth harmonic), and the approximate location of the first four formants is marked. It will be noticed that the lowest peak in the harmonic spectrum is not the first formant, which is dominated by the high-amplitude fundamental and second harmonic (F0 and H2).

Panel (c) shows the speaker's long-term average spectrum. Its axes are the same as for the short-term spectrum: frequency (increasing along the bottom, in Hz) and amplitude (increasing vertically, in dB). This is the average of some thousands of harmonic spectra similar to that in panel (b) taken throughout the spoken passage. It can be seen that its profile rises abruptly to a low-frequency peak, from which it falls at first fairly abruptly some 30 dB to about 1 kHz, and after that more gently to about 3.6 kHz. There is another short abrupt fall of some 10 dB, after which the spectrum remains flat to 5 kHz.

Some fine detail is evident at the lower-frequency end: the lowest-frequency peak, which is at about 116 Hz, is probably the modal value (i.e. the value which occurs most commonly) for the long-term fundamental frequency (LTF0). The next peak up, at about 228 Hz, is probably the long-term modal second harmonic, since it is about double the F0. Harmonics higher than this have not been separately resolved. The next highest peak, at about 470 Hz, might reflect the long-term modal first formant. The broadband peak at about 1.5 kHz is probably the long-term second formant. Of note are the spectral dips that mark frequencies at which energy has been absorbed. This speaker has a conspicuous zero at about 950 Hz, where the cursor is placed. This zero is also clear in the short-term spectrum. There is another in the middle of the frequency range at about 2.5 kHz, also clear in the short-term spectrum.

Forensic significance: Long-term spectrum

The forensic value of the long-term spectrum, just like everything else, is dependent on the ratio of between-speaker to within-speaker variation it shows. It is clear that speakers can and do differ considerably in their LTS. It is the amount of within-speaker variation that is still not so clear. Nolan (1983) was also able to demonstrate that the LTS did not change much as a function of different supralaryngeal settings, as when he spoke with palatalised as opposed to pharyngealised voice (pp. 143–9). However, it was affected by changes in laryngeal settings, as when he spoke with a creaky as opposed to modal phonation type. Furui *et al.* (1972) have shown that, although a speaker shows a stable LTS 'for two or three days to three weeks', speakers do show significant shifts in LTS over longer periods of time. Hollien (1990: 239) reports that it is 'relatively resistant to the effects of speaker stress', but that 'it does not function well when speakers disguise their voices'.

It is clear that the LTS can be affected by differences in channel transmission (Nakasone and Beck 2001: 3), and various channel-normalising techniques have long been available in automatic speaker recognition for minimising the effect (Furui 1981: 256).

Experts should therefore be open to the possibility of quantifying differences between forensic samples in terms of their long-term spectrum, or of features in the long-term spectrum like local peaks or zeros, providing that the effect of the channel can be controlled for, no differences in phonation type are audible, and there is a short time between recording of both samples. There remains, however, the problem of how forensically to evaluate the inevitable differences between the LTS of different speech samples.

The cepstrum

In Chapter 3 it was pointed out that there are two kinds of acoustic parameters used in speaker recognition: traditional and automatic, the latter being used in commercial speaker identification. The undoubted algorithmic mainstay of automatic speaker recognition is the *cepstrum*, and this section gives a brief non-technical idea of what it is like.

From the mid-1960s to the mid-1970s was a very prolific period in the development of signal processing methods, and the cepstrum was first developed then, as an analytical tool for automatically extracting the fundamental frequency from the speech wave (Noll 1964). The cepstrum very effectively decoupled the parts of the speech wave that were due to the glottal excitation from those that were due to the supralaryngeal response, and the former were used to estimate the F0.

Somewhat later it was demonstrated that the parts of the cepstrum that had to do with the supralaryngeal response could be successfully used for speaker recognition (Atal 1974). Later still its superiority for speech recognition was demonstrated (Davies and Mermelstein 1980). Its use in speaker and speech recognition, therefore, rests primarily on its function as a spectral parameter, not as a fundamental frequency estimator.

Earlier in this chapter, e.g. in Figure 8.7, it was shown how the spiky harmonic spectrum of speech can be smoothed by linear prediction to reveal the formants. The cepstrum smooths too, but more so, and therein lies its power, for the formant structure of the LP smoothed spectrum is still too spiky. A slight difference in the frequency of the formant peaks in two similar LP spectra can result in relatively large overall differences between the spectra that can adversely affect recognition. The cepstrum, however, is more lumpy, and slight differences do not have much effect. It thus tends to exhibit strong immunity to 'noninformation-bearing variabilities' in the speech spectrum (Rabiner and Juang 1993: 169), and hence has greater sensitivity to the distinctive features of speech spectra – the features that are of use in speaker recognition for example (Rose and Clermont 2001: 31). Figure 8.35 helps explain all this.

In Figure 8.35 are shown both cepstrum and LP smoothed spectrum for the same sound, the first target /a/ of the /aʉ/ diphthong in the *hello* in Figure 8.25. A comparison of the LP spectrum in panel (b) of Figure 8.25 with the LP spectrum in Figure 8.35 shows that the spectra are very similar, with F1 and F2 close, indicating a lowish central vowel [ɐ], and F3 to F5 are also clear. The two spectra were taken at points very close together in time in the same word, which explains their similarity.

Speech acoustics

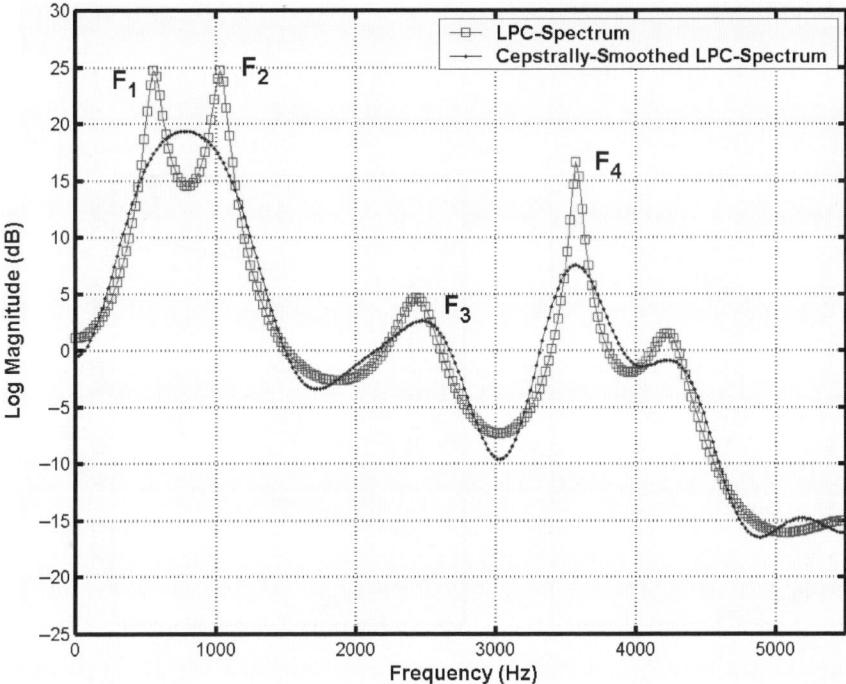

Figure 8.35 LP spectrum (thin line, squares) and cepstrum (heavy line, points) for the first vocalic target in the *hello* in Figure 8.25

Figure 8.35 shows nicely how the cepstrum of this sound relates to its LP smoothed spectrum. The most salient aspect of this relationship is in the lower-frequency region, where F1 and F2 are located. It can be seen that, whereas the LP smoothed spectrum clearly shows the peaks of the first two formants, the cepstrum has smoothed them into a single more rounded, less spiky, spectral peak. The two higher LP peaks of F4 and F5 are also smoothed somewhat by the cepstrum. It turns out that this is just the right amount of smoothing to give powerful results in automatic recognition. Furui and Matsui (1994: 1466), for example, report a 98.9% verification rate for a speaker pool of 20 males and 10 females, using individual vowels in automatic verification experiments. However, it is more often used to model long-term spectra, like the long-term average spectra exemplified in the previous section, where it is equally as powerful (e.g. Furui *et al.* 1972; Furui 1981).

The cepstrum is composed of *cepstral coefficients*, the spectra of which can be summed in the same way as the sine waves in Figures 8.3 and 8.4 to give the cepstrum's final smoothed envelope. The cepstrum in Figure 8.35 is actually made up from 14 cepstral coefficients. Some of these coefficients' spectra, and how they can be added to gradually approximate the final cepstral envelope, are shown in the four panels of Figure 8.36. Panel (a) shows the spectral envelope of the full cepstrum already shown in Figure 8.35. This is marked as $C_{1..14}$. Together with it is the spectrum of the first cepstral coefficient, marked C_1. It can be seen how the simple sigmoid shape of the spectrum of the first cepstral coefficient provides a first approximation to the complex

263

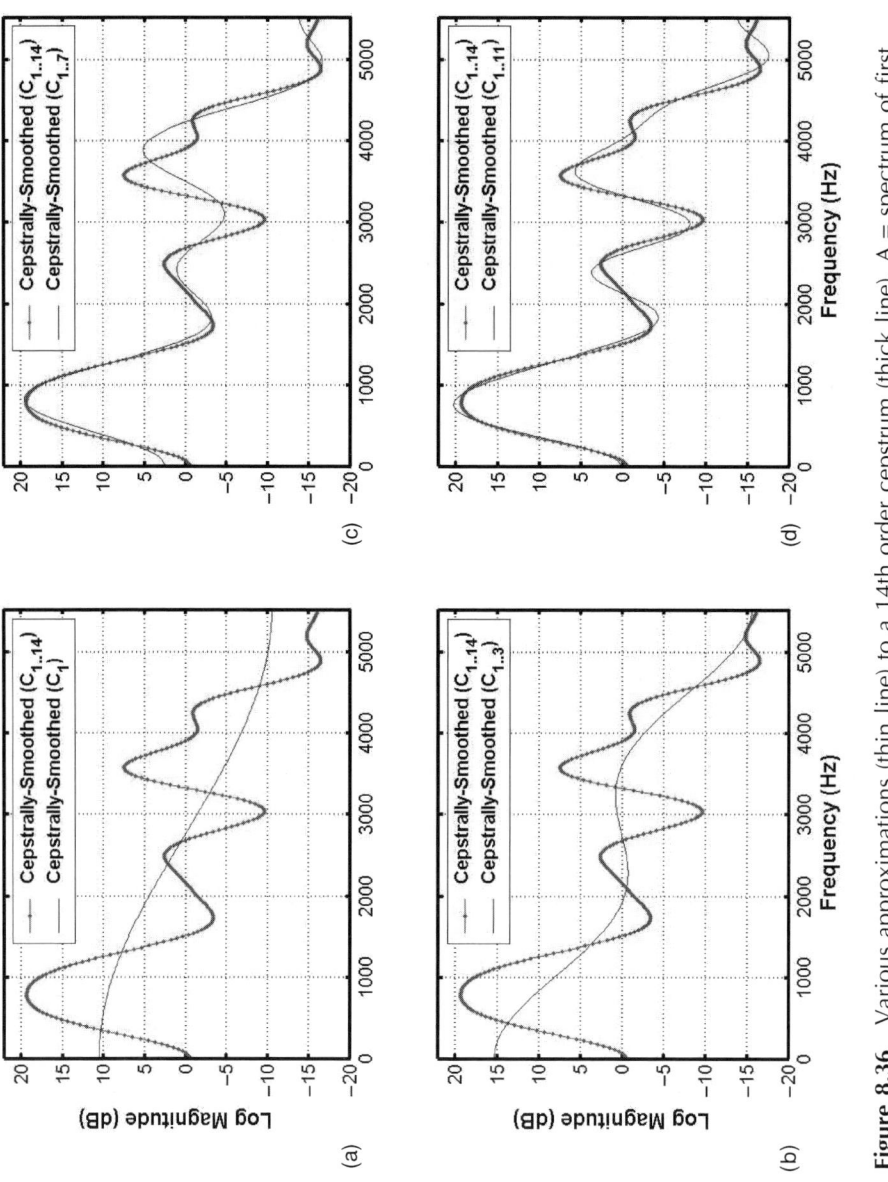

Figure 8.36 Various approximations (thin line) to a 14th order cepstrum (thick line). A = spectrum of first cepstral coefficent, B = sum of spectra of first three CCs, C = sum of spectra of first 7 CCs, D = sum of spectra of first 11 CCs

shape of the full cepstrum. Panel (b) shows the spectrum of the first three cepstral coefficients ($C_{1..3}$) added together. A slightly better approximation to the full envelope can be seen. An even better fit results if you add the spectra of the first seven coefficients ($C_{1..7}$), as in panel (c); and when you sum the spectra of the first 11 of the 14 coefficients, as in panel (d), a very close approximation is achieved.

Forensic significance: Cepstrum

In Chapter 3 it was mentioned that automatic parameters like the cepstral coefficients, unlike traditional acoustic parameters such as formant centre frequencies, do not relate in any straightforward way to the linguistic properties of speech sounds. It is not clear for example what, in terms of speech production or perception, a single cepstral coefficient or a combination of coefficients could correspond to. This is somewhat restrictive from a forensic-phonetic point of view. To be more specific, suppose two speech samples had vowels with audibly very similar phonetic quality, say, or suppose that the samples differed in that one had a slightly retracted allophone for a particular vowel. It would be very difficult to know which particular cepstral coefficient, or set of coefficients, to measure in order to quantify the similarity or difference. Indeed this might be a silly thing to do since, as can be understood from Figure 8.36 a single cepstral coefficient contributes to all parts of the overall envelope. To be sure, certain relationships have been pointed out between cepstral coefficients and other constructs, but they remain too tenuous to be used as equivalent measures in forensics. For example, it has long been known that the first cepstral coefficient relates to spectral slope (mentioned earlier in this chapter); and it has been pointed out that the third mel-weighted cepstral coefficient correlates to F2, while vowel height is reflected to a certain extent in the second cepstral coefficient (Clermont and Itahashi 1999, 2000). Given the tenuous nature of these relationships, it is best to continue to exploit the cepstral coefficients for their power as mathematical abstractions rather than as ways of quantifying the results of auditory analysis.

That said, given its proven power in automatic speaker recognition, the forensic potential of the cepstrum has certainly not yet been fully realised, and its use as a forensic short-term parameter is only now beginning to be explored. Rose and Clermont (2001), for example, have shown that the cepstrum performs better than single formants in discriminating forensically realistic same-speaker and different-speaker *hello*s on the basis of one segment. Aside from its actual performance, the cepstrum is also more easily extracted than the F-pattern, with its inevitable problems of identification and tracking of the higher formants (Rose and Clermont 2001: 31). It is possible, therefore, to envisage that forensic samples could be compared with respect to the cepstra of their segments as well as, or instead of, the F-pattern. This would be of particular use in segments with high speaker-specificity but elusive F-patterns, like nasals.

Differential software performance

It is appropriate at the end of a chapter on speech acoustics to give some attention to the ways they are extracted and quantified. There are many different commercially

Forensic Speaker Identification

available software packages for the acoustic analysis of speech, and they are not totally comparable in their analysis options – in the digital algorithms that they implement in the analysis, for example. Because of this, *they will not necessarily give exactly the same numerical results when analysing the same speech token*. Schiller and Köster (1995) for example compared the performance of four analysis systems in the extraction of F0 under various conditions. They found that all four systems produced similar but not identical results, and some parameters were more variable than others.

To illustrate this with formant data, we will see what happens when we measure the same thing using two different software packages. The F-pattern of the same *hello* as described in the section 'A whole-word example' above, and shown in Figure 8.25, was analysed with a different software package. The second analysis was also performed with linear prediction, using the same analysis settings (called sampling frequency, filter order, pre-emphasis, and window).

Figure 8.37 shows the results of the second analysis. The wave-form is shown in panel (a), with the creaky portion at the end nice and clear. In panel (b) are shown the amplitude and F0, and in panel (c) is a spectrogram, with gridlines at kilohertz intervals, and with frequency increasing from 0 to 5 kHz. The short vertical striations are not glottal pulses, but show peaks of spectral energy at each frame analysed (with the vertical height representing their bandwidth).

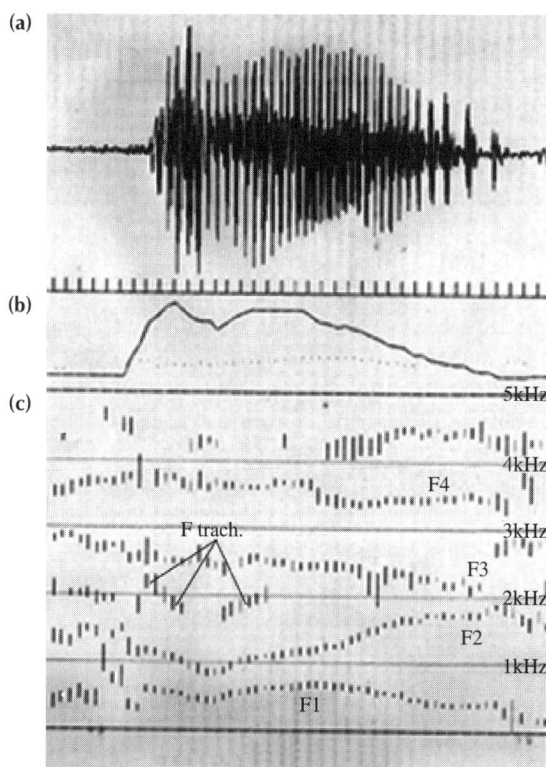

Figure 8.37 A spectrographic representation of *hello* produced with different software

Table 8.3 Estimates of formant centre-frequencies (Hz) in a token of *hello* using two different software packages

	F1	F2	F3	F4	F5
Software A	385	1703	2220	3404	4350
Software B	397	1716	2265	3431	4446
Δ(B − A)	12	13	45	27	96

When compared visually with Figure 8.25, Figure 8.37 shows essentially the same F-pattern features. The parallel, then diverging pattern of F1 and F2 can be seen easily, for example, as well as the sudden slight drop in the F4 frequency mid-way through the word. One clear difference from the pattern in Figure 8.25 is that the tracheal resonance associated with the initial /h/ has been well extracted. It is marked 'F trach.', and can be seen lying between F2 and F3 in the vicinity of the 2 kHz line, lasting into the second syllable.

In order to make a quantified comparison, the F-pattern was estimated at the same well-defined point in time in both analyses − at the third glottal pulse from the end. The values for the centre-frequencies of the five formants at this sampling point as estimated by both software packages are shown in Table 8.3, and the difference between them is also shown. It can be seen that, as expected, they are not the same across the two software packages. Although the differences are not all that big − they are about the magnitude expected for average short-term within-speaker variation for example (Rose 1999b: 17) − one worrying fact is that the differences are all the same sign (B is consistently estimating higher than A). This means that the differences between the results from the two software packages might not cancel each other out in the long run.

Although the differential performance of software packages is common knowledge, there have been very few studies quantifying it, especially on less than ideal quality recordings of the type encountered in forensics. As a result, we do not know how the different software packages available perform. Because of these differences, *it is mandatory to carry out comparison of questioned and suspect material on the same equipment, with exactly the same settings*. It is sometimes the case that analysis results from another laboratory require to be checked. Another consequence of these software differences and their unknown magnitudes is that *complete replication of the analysis will not be possible unless the same software is available*. Both these considerations highlight the need to adhere to the accepted scientific procedure of being totally explicit with regard to the digital options and settings used in analysis. This will be of use should it be necessary to repeat the analysis, or have it replicated by another laboratory.

Chapter summary

This chapter has explained a little of the theory of how speech acoustics are produced, and applied this in detail to the description of two forensically important acoustic parameters: vowel F-pattern and long-term fundamental frequency. Vocalic F-pattern, usually comprising the centre-frequencies of the first two or three formants, simultaneously encodes linguistic information on the particular vowel and individual information on the dimensions of the vocal tract that produced it. Long-term F0, with its primary dimensions of mean and standard deviation, tries to abstract away from linguistically relevant F0 values and reflects individual information ultimately related to the size of the speaker's vocal cords. Two other important constructs – the long-term spectrum and the cepstrum – were also briefly exemplified.

Speech acoustics reflect the vocal tract that produced them, and can be extracted and quantified with relative ease using currently available signal-processing software. Distributions of acoustic parameters like F0 can then be successfully modelled statistically to enable quantified comparison between speech samples, and spectral features can be modelled with the cepstrum.

Speech acoustics only reflect how a speaker is speaking on a particular occasion, however. Just like any other parameter, they are not invariant and show both within-speaker as well as between-speaker variation. Successful comparison requires well-controlled data from different occasions numerous enough to characterise the speaker of both questioned and suspect samples.

9

Speech perception

The raw acoustic energy reaching the listener from a speaker undergoes much processing and many transformations before it is ultimately heard as speech being spoken by a particular individual. Two main stages in the processing are usually distinguished. The first, sometimes referred to as the *organo-receptive* stage, is where acoustic energy is transformed into mechanical energy at the ear drum and transmitted further through the auditory ossicles to the fluid-filled cochlea. Here it becomes hydraulic energy, before being transformed once more into mechanical energy as the fluid movement sets up a shearing force in hairs embedded in cells in the basilar membrane of the cochlea (Daniloff *et al.* 1980: 368–410).

One probably thinks of the vibration of the ear-drum in response to the rapid fluctuations in incoming air pressure as the first part of this organo-receptive sequence. But modification of the acoustic information begins before this. The outer ear – the part you can see – differentially affects the intensity of certain frequency components and thereby helps in locating where the sound is coming from, and the acoustic signal is also amplified by passage through the ear tube (Warren 1999: 5, 6). The dimensions of the ear tube determine that the range of frequencies in which this amplification occurs lies between about 2.0 kHz and 5.5 kHz (Warren 1999: 6) with maximum amplification between 3.0 kHz and 4.0 kHz (Daniloff *et al.* 1980: 373; Stevens 2000: 20; Warren 1999: 6). This amplification makes a difference: it has long been known that it corresponds to, and is at least partially responsible for, the frequency range of greatest auditory sensitivity (Daniloff *et al.* 1980: 375).

Why should humans be evolutionarily tuned to this frequency range by the length of their ear canal? It cannot have too much to do with the perception of speech sounds *per se*, because the most important acoustic information for them lies in the range occupied by the first two formants and the fundamental: from about 100 Hz to 3.0 kHz. This is similar to the range in 'bionic ear' use, for example (Clark 1998). It is tempting to speculate that this range of enhanced frequency sensitivity may have had something to do with responding to voices rather than speech.

The second, or *neuro-receptive*, stage of speech perception begins with the response of the nerve cells to the shearing force in the basilar membrane of the cochlea, which results in patterns of electrochemical energy being transmitted up the auditory nerves (Daniloff *et al.* 1980: 410–34). These neural signals are modified in complex ways at several points in their transmission before arriving at the auditory cortex of the brain and being decoded (Stevens 2000: 212–13). The primary outcome of this decoding is a double percept involving the previously described separation into linguistic information (speech) and information about the speaker.

As with speech production and speech acoustics, speech perception constitutes a vast area of scholarship in its own right, and cannot be covered in any detail. A very simple summary may be found in Crystal (1987: 142–8). More advanced treatments are Studdert-Kennedy (1976), Delgutte (1997), Moore (1997), McQueen and Cutler (1997), and Warren (1999: 155–88). There are, however, aspects of the way humans perceive speech that are of importance for forensic phonetics and that therefore require discussion: *speech-specific perception*; the *expectation effect*, and *perceptual integration*.

Speech-specific perception

One of the important things to happen to the incoming acoustic signals is that the brain must decide whether they represent speech or some other sound. This appears to be a relatively easy task, since the acoustic properties of speech are very different from those of other sounds, involving as they do rapidly modulated frequencies that reflect the continuous changes in vocal cord and supralaryngeal activity necessary to produce fluent speech. Some of these frequency modulations, for example changes in formant frequency, were illustrated in the previous chapter. The result of this decision is that one either hears the sequence of vowels and consonants that is speech or one hears other sounds: it is not possible to hear something in-between, or to hear speech as a sequence of pops, hisses and buzzes.

Although both speech and non-speech sounds are percepts resulting from rapid fluctuations in air pressure, then, speech is perceived in a different manner from other sounds. Speech is perceived *as if it has been produced by a vocal tract* (Studdert-Kennedy 1974: 2351). This is what is known as speech-specific perception. Some examples will make this clear.

Speech-specific perception of fundamental frequency

As mentioned in the previous chapter, the acoustical correlate of pitch is fundamental frequency (F0). Other things being equal, a vowel with a low F0 will be perceived as having a lower pitch than one with a higher F0. What happens if other things are not equal?

Suppose one of the vowels is a [i] (a vowel said with a high tongue position) and one is [ɑ] (a vowel said with a low tongue position), but they are both said with the same F0. If they have the same F0, one would expect that they would have the same pitch, but listeners will actually tend to hear the [ɑ] as having a higher pitch than the [i] (Hombert 1978: 98–102). Why? This is an example of speech-specific perception, and can be explained in the following way. In speech, high vowels tend to have slightly higher F0 than low vowels. This in turn is because of mechanical coupling between the tongue and the vocal cords: raising the tongue, as for high vowels, imparts some kind of tension on the cords, thus increasing their rate of vibration. This phenomenon is very well known and is called intrinsic vowel F0. Now, the perceptual mechanism 'knows' to interpret F0 as pitch, but it also 'knows' that high vowels are intrinsically associated with slightly higher F0. Therefore it compensates by perceptually scaling the F0 on high vowels down (and the F0 on low vowels up) in calculating the appropriate pitch. Whence the lower pitch percept of the [i] relative to the [ɑ], despite their same F0. The perceptual mechanism is thus behaving in its response to a physical frequency (the F0) as if it 'knows' that the frequency has been produced by a vocal tract.

Perceptual compensation for sexually different F0

Suppose, once again, the pitch of two vowels is being compared. Let us make them the same ([i]) given the perceptual complexities associated with different vowel quality just described. But let us imagine them being produced one by a male and the other by a female. Males have overall lower F0 than females because they have larger cords, so the chances are that the male [i] vowel will have a lower F0 than the female [i]. Will this F0 difference be perceived as a pitch difference? That depends on the *linguistic value of the pitch.*

Imagine the male and female are the Cantonese speakers C and B whose tonal fundamental frequency was shown in Figure 3.2 in Chapter 3. If the pitch of both male and female [i] vowels is encoding the same linguistic tone, say the Cantonese low tone, then you will be able to hear that they both have the same linguistic pitch, despite the difference in F0 (which is shown in Figure 3.2 to be about 60 Hz).

What is happening here, then, is that the perception of F0 as pitch is being mediated by the brain's knowledge of what size of vocal tract the F0 is being produced by. There are many cues in the speech acoustics that allow the perceptual mechanism to infer the dimensions of the vocal tract that produced them. A particular F0 is then perceived accordingly, using something like the normalisation process illustrated in Figure 3.3. If it is perceptually inferred that the cords that have produced a particular F0 value are relatively small, then the pitch percept of that F0 is scaled down; if the inference is that the cords are relatively big, the pitch percept is scaled up. The perceptual mechanism is thus perceiving the F0 *as if it has been produced by a vocal tract* with specific dimensions.

Other examples of speech-specific perceptual compensation

There have been three other instances of speech-specific processing in this book. The first occurred with the relationship between the perception of vowel nasality and degree of soft-palate lowering described in Chapter 6. A certain amount of acoustic nasalisation on low vowels will not be perceived as nasalisation because the perceptual mechanism 'knows' that low vowels are usually produced with the soft palate somewhat open.

The second instance has to do with the perception of amplitude. Amplitude is the acoustic correlate of loudness, but amplitude also varies as a function of other things, in particular the openness of the vocal tract. The speech-wave in Figure 8.1 in Chapter 8 is a case in point, where the difference in amplitude between the [i] and [ə] wave-forms was commented on. The amplitude difference between these two vowels has to do with the difference between the vowels themselves. Vowels produced with a more open vocal tract have inherently greater amplitude, thus [ə], being produced with a more open tract, is more amplitudinous than [i], other things being equal. The perceptual mechanism 'knows' about this intrinsic relationship, and compensates for it in its perception of loudness. In this case, therefore, you would not perceive the [ə] in the wave-form in Figure 8.1 as being louder than the [i]. The [ə] would have to have a much bigger amplitude than the [i] for it to be heard as louder.

The third instance of perceptual compensation in this book occurred with the *chooch–toot–kook* demonstration of the effect of phonological environment on vowel formant frequencies in Chapter 8. The perceptual mechanism makes use of the fact that the F2 of /ʉː/ will be intrinsically affected, as demonstrated, by the place of articulation of surrounding consonants. This enables the same vowel, i.e. the central vowel /ʉː/, to be heard for all three vowels with different F2 values. If the conditioning environment is removed, that is if the vowels in *chooch, toot* and *kook* are digitally excised, then differences in backness correlating with the F2 differences become audible: the vowel taken from *kook*, especially, sounds like a central [ʉ], whereas that taken from *chooch* sounds clearly fronter: [y].

This kind of perceptual processing is pervasive. Indeed, because the acoustics of speech are unique functions of the dimensions of the vocal tract that produced them, and vocal tracts differ in their dimensions, perceiving the acoustics in terms of the vocal tract that produced them is vital for speech perception to function at all. Speech-specific perception enables the perceptual mechanism to make linguistic sense of all kinds of intrinsic variation in speech acoustics, for example the difference in formant frequencies and F0 associated with different vocal tract sizes, not just associated with different sexes, but different individuals.

Top-down and bottom-up processing

Linguistic structure is generally characterised by a degree of redundancy, and speech is no exception. There are many cues as to the linguistic message in the speech signal, both from within the speech signal itself and from the ambient non-linguistic context. For example, suppose that in a recording the last word was for some reason obscured in the utterance:

Look at those cute ____.

A lot can be inferred about the missing word. Because of its syntactic position, the missing word will be a noun or pronoun; because it is preceded by the plural demonstrative *those* it will be a plural, and is likely to end in either [s], [z] or [əz]: the three common sounds, or allomorphs, that encode the meaning of the *plural* morpheme in English. Because of the meaning of *cute*, the missing word is likely to refer to something that is animate, small and liked by the speaker. If the speaker's predilections are known, then the identity of the things that they consider cute can be guessed at. And if the speaker is actually looking at two baby elephants at the zoo then the reference of the word, and its identity, is clear.

Speech perception relies heavily on redundancies of this kind. *What we hear is therefore not necessarily always actually contained in the acoustic signal impinging on our ear-drum but can be actually supplied by our brain from its knowledge of the linguistic structure of the message, or the non-linguistic context.* The distinction between the two types of information and the processing that mediates them is often characterised as *bottom-up*, that is where information is actually present in the signal, versus *top-down*, that is where information is supplied by the brain.

Top-down processing at work can be easily demonstrated by so-called *phonemic restoration* experiments, where segment-sized chunks are digitally excised from a stretch of speech that is then repeatedly played against increasing levels of background noise. At a certain level of noise the listener becomes aware that the gaps in the signal that were previously audible are no longer there. It is as if the brain has seamlessly restored the missing speech sounds. This effect can be experienced on the CD accompanying R. M. Warren's 1999 book *Auditory Perception*.

Forensic significance: Speech-specific perception

Perceptual compensation and the existence of both top-down and bottom-up processing are good examples of the complexity of the relationship between acoustic and perceptual categories in speech. Knowledge of this relationship is essential for the interpretation of the complex acoustic patterns of sounds in forensic samples, for the forensic phonetician is a perceiver as well as transcriber and analyst.

Expectation effect

The power of both the linguistic and non-linguistic information in cueing a percept is considerable. One of the ways in which the non-linguistic information can affect judgement is the *expectation effect*, and this applies to voices as well as speech sounds.

As is self-explanatory from the title of Ladefoged's 1978 paper, 'Expectation affects identification by listening', we hear who we expect to hear. This undoubtedly contributes to our performance in recognising telephone voices whose owners are familiar to us, since they are by definition those who are likely to telephone us.

Forensic significance: Expectation effect

The forensic relevance of the expectation effect is clear. If it is true that we are more likely to hear whom we expect to hear, then care has to be taken with voice samples tendered for analysis. If the police identify the samples for an expert that they think come from the same speaker, then the expert becomes open to possible influence by the expectation effect (Gruber and Poza 1995: section 62). A better strategy is simply for the expert to be provided with unidentified samples. In that way, too, a degree of confidence in the results of the analysis, and also the method, can be established. This is because the police will presumably know the identity of some of the samples against which the performance of the expert can be checked. More importantly, *it will be possible in this way to estimate a likelihood ratio in a particular case for the expert's performance in correctly identifying a particular voice.* (This will be the ratio of the probability of the expert concluding that two samples are from a particular speaker, when the police know that the two samples are indeed from that speaker, to the probability of the expert concluding that two samples are from that particular speaker when it is known that they are not.) In any case, the expert will normally want to be told as little as possible about the background details of the case to avoid bias (Gruber and Poza 1995: section 62).

This also applies to cases involving content identification. For obvious reasons, an expert should not be told what the interested party thinks is in a disputed message, but should be allowed to come to their own conclusions. If after careful protracted listening the expert cannot hear what is claimed, this is an indication that the information might not be present in the signal itself. In such cases, it is still necessary to do an acoustic examination, however. This is because there might be acoustic information that is, for whatever reason, not being resolved by the expert's ear, which will help decide the matter An example of this is French (1990b), where careful acoustic analysis helped determine whether a disputed utterance was more likely to be *You can inject those things* or *You can't inject those things*. As with speaker identification, the appropriate formulation should still use a likelihood ratio. This will be the ratio of the probabilities of observing the acoustics under the competing hypotheses of what has been said.

Perceptual integration

Every stage in the perceptual process is characterised by modification and loss of information originally present in the acoustics. Thus there is a many-to-few relationship between successive stages. This progressive reduction contributes to efficiency, because not every piece of information present in the acoustics is necessary for speech perception. One example of this loss of information is in the way acoustic frequencies are perceptually integrated.

As explained in considerable detail in Chapter 8, vowel quality is encoded in the central frequencies of the first two or three formants. The graphical analogue of vocalic perception was demonstrated by plotting vowels according to the centre frequencies of their first two formants. This separates them out (into a configuration similar to that of the conventional vowel quadrilateral).

However, it is clear that the perceptual mechanism does not work by identifying all the *individual* formants in the transformed acoustics. This is because the frequency resolution of the perceptual mechanism decreases with increasing frequency. The higher up in the frequency range, the less able the perceptual mechanism is to separate out frequencies close together. To illustrate this, consider two low frequencies separated by 50 Hz within the range of the male fundamental – say 100 Hz and 150 Hz. These can be perceptually discriminated with ease. However, the same 50 Hz difference in the vicinity of the frequency of the second formant in an [i] vowel (at ca. 2000 Hz) would be far more difficult to hear apart, and would in fact be perceptually integrated into a single frequency response.

Speech perception

Forensic significance: Perceptual integration

Perceptual integration is forensically relevant because it allows for two samples to have the same or similar auditory-phonetic quality, and thus be suggestive of provenance from the same speaker, and yet be different in their acoustic structure, which clearly indicates elimination (Nolan 1994: 337–41). 'In principle . . . the ear may be inherently ill-equipped to pick up some differences between speakers which show up clearly in an acoustic analysis' (Nolan 1994: 341).

Such a case is described in Nolan's (1990) paper 'The Limitations of Auditory-Phonetic Speaker Identification'. In this case, involving telephone fraud, two prosecution experts had, on the basis of auditory analysis, identified the questioned voice as coming from the suspect with a very high degree of certainty. Acoustic analysis, however, revealed large and consistent differences between the questioned and suspect samples. For example, for the vowel in the word *that*, one sample showed a strong F3 at 2.5 kHz, but the other showed weak F3 and F4 close together bracketing this frequency. These differences were taken to indicate likely provenance from different vocal tracts and therefore that it was unlikely that the same speaker was involved. What might have been happening is that the single strong F3 frequency in one sample was perceptually equivalent to the two weak F3 and F4 frequencies in the other, thus making the two samples sound similar even though their acoustics were different.

This paper is important because it shows the indispensability of acoustic analysis in forensic-phonetic investigation. For this reason, *it is obviously important not to rely on auditory judgements alone in forensic speaker identification.*

Chapter summary

This chapter has looked briefly at aspects of speech perception that are relevant for forensic phonetics. Highly complex relationships exist between the acoustics of speech and how they are perceived, and knowledge of these relationships must inform forensic-phonetic analysis. In particular, the phenomenon of perceptual integration is a cogent argument for the indispensability of acoustic analysis, and care must be taken to avoid expectation effects.

10

What is a voice?

The aim of this book has been to explain what is involved in forensically discriminating speakers by their voice. Although it is obviously a central notion in this undertaking, it has not yet been possible to say what a voice actually is, because the necessary concepts had first to be developed in the previous chapters. So, what is a voice? The aim of this chapter is to answer that question by presenting a model of the voice and explaining in some detail the components of this model. This model is very nearly the same as that described in Chapter 2 of Nolan's 1983 book *The Phonetic Bases of Speaker Recognition*, and section 2 of his 1996 *Course in Forensic Phonetics*. Considerable use has also been made of concepts in several important articles by Laver and colleagues, all conveniently reprinted in Laver's 1991a collection *The Gift of Speech*.

The meaning of 'voice'

Perhaps the normal response to the question *what is a voice?* is to try to say what *voice* means. The woeful OED gloss of 'voice: – Sound formed in larynx etc. and uttered by mouth, . . .' is one good example of the semantic inadequacy of dictionary definitions (Wierzbicka 1996: 258–82). Another shortcoming is that we cannot assume that the meaning is invariant across languages. After all, many languages, unlike English or French, German, Russian or Japanese, lack a separate non-polysemous word for voice. In Modern Standard Chinese, for example, shēngyin is polysemous: it can correspond to either *sound* or *voice*. In languages like this the meaning of voice is not separately lexicalised, but is signalled with sounds that, depending on the language, also mean *noise*, *neck*, *language*, etc. One should not accord too much significance to whether a language lexicalises a concept like voice or not: the American Indian language Blackfoot, for example, outdoes most other languages in this regard. It is said to have a verb which specifically means *to recognise the voice of someone*. Let us stick with English, therefore.

One key semantic component of the English word *voice*, conspicuously absent from the dictionary definition, must surely be the link with an individual. An acceptable paraphrase of the meaning of the word *voice* might therefore be *vocalisations (i.e. sound produced by a vocal tract)* when thought of as made by a specific individual and recognisable as such.

However, this is still of little forensic use, since, although it correctly highlights both the production and perceptual aspects of a voice, it fails to bring out just exactly how a speaker and their voice are related. A better way of characterising a voice for forensic purposes is semiotically, that is, in terms of kinds of information conveyed.

Within this semiotic approach there are three important things to be discussed:

- The distinction between voice quality and phonetic quality
- Tone of voice
- The model of a voice

It is possible, at a stretch, to characterise the first two as considering a voice primarily from the point of view (point of hearing, really) of the listener, and the last (i.e. the model) from the point of view of the speaker. To a certain extent the voice model presupposes the voice quality/phonetic quality distinction, so this distinction will be covered first, together with tone of voice. The terms linguistic, extralinguistic and paralinguistic (or paraphonological) are often used to qualify features functioning to signal phonetic quality, voice quality and tone of voice respectively (Laver and Trudgill 1991: 239–40).

Voice quality and phonetic quality

As already mentioned, when we hear someone talking, we are primarily aware of two things: what is being said, and characteristics of the person saying it. The aspects of the voice that correspond most closely to these two types of judgements on the content and source of the voice are termed *phonetic quality* and *voice quality* respectively (Laver 1991b: 187–8; Laver 1991c).

Phonetic quality

Phonetic quality refers to those aspects of the sound of a voice that signal linguistic – in particular phonological – information (Laver 1991d: 158). In the more technical terms explained in Chapters 6 and 7, phonetic quality constitutes the fully specified realisations, or allophones, of linguistic units like vowel and consonant phonemes. For example, the phonetic quality of the /aː/ vowel phoneme in the word *cart* as said by an Australian English speaker might be described as long low central and unrounded, and transcribed as [aː]. Some of these phonetic features indicate that it is the linguistic unit, or phoneme, /aː/ that is being signalled, and not some other phoneme. For example, the length of the vowel marks it as a realisation of the phoneme /aː/ rather than /a/ (the vowel phoneme in the word *hut*), and the word as *cart* not *cut*.

Linguistic information is being conveyed here, since the choice of a different word is being signalled. Other phonetic features of the vowel [aː] are simply characteristic of the range of possible allophones for this phoneme in Australian English. For example, in another speaker, the same vowel phoneme /aː/ might have a phonetic quality described as long low and back of central ([ḁː]). Both [aː] and [ḁː] are acceptable realisations of the phoneme /aː/ in Australian English. (This example demonstrates once again that speaker identity can also be signalled by linguistic as well as non-linguistic features, as for example in the choice of allophone.)

Phonetic quality is not confined to segmental sounds like consonants and vowels, but is also predicated of suprasegmental linguistic categories like intonation, tone, stress and rhythm. Thus the stress difference between *INsult* and *inSULT* also constitutes an aspect of phonetic quality, as it signals the linguistic difference between a verb and a noun, as do those aspects of sound that make the following two sentences different: *When danger threatens, your children call the police* versus *When danger threatens your children, call the police* (Ladefoged 2001: 14, 15). The linguistic difference being signalled in this example is the location of the boundary between the syntactic constituents of the sentence, and it is being signalled by intonational pitch (rising pitch, a so-called boundary tone, on *threatens* in the first sentence; rising pitch on *children* in the second). Many phoneticians would also extend the notion of phonetic quality to those features of sounds that characterise different languages or dialects and make one language/dialect different from another.

Voice quality

Voice quality is what one can hear when the phonetic quality is removed, as for example when someone can be heard speaking behind a door but what they are actually saying is not audible.

Voice quality is usually understood to have two components: an *organic* component and a *setting* component (Laver 1991e; 1991b: 188). The organic component refers to aspects of the sound that are determined by the particular speaker's vocal tract anatomy and physiology, such as their vocal tract length or the volume of their nasal cavity, and which they have no control over. A speaker's anatomical endowment typically imposes limits to the range of vocal features; thus a good example of an anatomically determined feature would be the upper and lower limits of a speaker's fundamental frequency range. (Recall that fundamental frequency is an acoustic parameter determined by the rate of vibration of the vocal cords, and this in turn is a function of the size and mass of the vocal cords themselves, which is an anatomical feature. So there are high fundamental frequencies above the range of an average male speaker, and low F0 values below the range of an average female speaker.)

The second component of voice quality, often called the setting or *articulatory setting*, refers to habitual muscular settings that an individual adopts when they speak. A speaker may habitually speak with slightly rounded lips, for example, or nasalisation, or a low pitch range. Since these setting features are deliberately adopted, they differ from the first component in being under a speaker's control.

The components of an individual's articulatory setting (for example lip rounding) are conceived of as deviations from an idealised neutral configuration of the vocal tract. For example, a speaker might speak with the body of their tongue shifted slightly backwards and upwards from a neutral position, resulting in what is described as uvularised voice (the deviation being in the direction of the uvula). This would mean that all sound segments susceptible of being influenced by the setting would be articulated further back and slightly higher than normal. The initial stop in the word *cart*, for example, might be articulated further back in the mouth towards the uvular place ([qʰ]) instead of at the normal velar, or back of velar, place ([kʰ]), and the final alveolar stop [t] might be slightly uvularised (that is, articulated with the tongue body backed and raised at the same time as making contact at the alveolar ridge ([t̪])).

The vowel in this example, too, might be articulated further back as [ɑ̱ː], thus illustrating an important point. This is that *part of the nature of a segment – in this case the backness of the vowel – can often be either the result of a quasi-permanent articulatory setting, or simply a lawful allophonic realisation of the phoneme in question* (it was pointed out above that the Australian English phoneme /aː/ could be realised either as [aː] or [ɑ̱ː]). Which one it is – allophone or setting – is shown by the adjacent segments: a pronunciation of [kʰɑ̱ːt] for *cart* indicates that the backness of the /aː/ is a phonetic feature; a pronunciation of [qʰɑ̱ːt] indicates that the backness is part of a deliberate voice quality setting. This example shows that whether a particular feature is an exponent of phonetic quality or voice quality depends on how long it lasts: the features as exponents of linguistic segments are momentary; features as realisations of settings are quasi-permanent.

Chapter 6 of this book showed how phonetic quality can be described componentially. For example, the phonetic quality of the initial sound [f] in *far* can be specified as a voiceless labio-dental fricative. The controllable aspect of voice quality can also be described componentially. Thus a voice might be described as 'soft denasalised and velarised', or 'deep and whispery'. The conceptual framework and details for such a description are set out in J. M. D. Laver's (1980) *The Phonetic Description of Voice Quality*, which is also accompanied by recordings illustrating the different types of settings. Since this is not considered part of traditional auditory-phonetic training, the number of phoneticians proficient in this approach is much smaller than those proficient in describing phonetic quality.

The difference between voice quality and phonetic quality can also be illustrated from the point of view of their different roles in the perception of phonetic features. It has been pointed out that voice quality provides the necessary background against which the figure of the phonetic quality has to be evaluated (Laver 1991c). For example, a speaker's linguistic pitch – the pitch that signals the difference between a high tone and a low tone in a tone language – can only be evaluated correctly against the background of their overall pitch range. This was actually demonstrated with the Cantonese tone example in Chapter 3 (Figures 3.6, 3.7), where it was shown how the linguistic import of a particular fundamental frequency value – what linguistic tone it was signalling – depended on the speaker's fundamental frequency range, that is aspects of their voice quality. *This relativism applies to nearly all acoustic features*, a further example of which from this book is formant frequencies. It was pointed out in Chapter 8 how vocalic F1 correlates to perceived vowel height, but a particular value for a first formant frequency in a vowel, for example, needs to be evaluated against the speaker's F1 range before it can be decided whether it is signalling a high vowel or a low vowel (cf. Figure 8.21).

Sometimes it is possible to be able to hear both the voice quality and phonetic quality aspects of speech. Thus if one listens to a male and a female speaker of a tone language saying a word with a high falling tone, it is possible to hear that the phonetic pitch (signalling the tone as high falling) is the same. It is also possible to pay attention to the voice quality pitch and hear that, despite the phonetic quality identity, the

What is a voice?

female has a different, higher voice quality pitch than the male. This is not possible with vocalic quality, outside of specialised techniques like overtone singing, however. It is possible, for example, to hear that female and male are both saying the same vowel with the same phonetic quality, but it is not possible to hear the accompanying voice quality difference in formant frequencies other than as one of sex.

Forensic significance: Phonetic quality and voice quality

The independence of voice quality and phonetic quality means that forensic speech samples can differ in four ways with respect to them. Samples can agree in values for the two qualities, i.e. have the same (or similar) voice quality and phonetic quality, or different voice and phonetic qualities; or they can disagree in values: have same voice quality but different phonetic quality, or different voice quality but same phonetic quality. In naive voice discrimination, voice quality is probably the dominant factor. When two samples from the same language are auditorily very close in voice quality, naive listeners tend to form the opinion that they come from the same speaker, irrespective of whether they contain phonetic (or phonemic) differences (as long as the phonetic differences are not too big). For example, the following two *hello* tokens differ in the backness of their offglide in the /aʉ/ diphthong. The first has a backer offglide [ʊ] and the second a more front offglide [ʉ]. They also differ in the vowel phoneme in the first syllable: /e/ (allophone: [ɛ]) vs. /a/ (allophone: [ɐ]).

    [hɛˈlʌʊ L.HL]   [hɐˈlɐʉ L.HL]

If the two *hello*s are from different speakers with very similar voice quality (very similar upper formant frequencies, almost identical F0, similar phonation type and spectral slope, for example), they will be judged by the majority of naive listeners to be from the same speaker. Likewise, when samples differ in voice quality, naive judgements will also tend to infer different speakers. Big between-sample phonetic differences can override the voice quality differences, however. For example, the two *hello* tokens below show considerable phonetic differences (absence vs. presence of /h/; different first vowel phoneme; different intonational pitch; different diphthong.)

    [hɐˈlɐy L.HL]   [ɛˈlɒʊ H.H!]

Even if these two *hello*s are from the same speaker and have very similar voice quality (very similar upper formant frequencies, similar F0 range, similar phonation type and spectral slope, for example), they will be adjudged by the majority of naive listeners to be from different speakers. The above examples are from the author's and S. Duncan's 1995 paper 'Naive auditory identification and discrimination of similar voices by familiar listeners' (pp. 5–7).

This illustrates once again the absolute necessity for an informed analysis from a forensic phonetician. It would be natural, for example, for a suspect to be singled out by the police because they have the same voice quality as the offender. A phonetician is then required to determine whether or not the samples are also similar in phonetic quality.

Tone of voice (1)

It is conceivable that, although we cannot hear what our hypothetical post-portal speaker is actually saying, we can hear something about how they are saying it. They may sound angry, for example, or whingeing. The sound features that communicate this information constitute what is called tone of voice, and *tone of voice* is therefore one of the main ways in which we verbally signal temporary emotional states.

It will perhaps come as no surprise that tone of voice shares the same dimensions of sound as phonetic quality and voice quality. During an armed robbery, for example, the robber threatened: 'Where's yer fucken' till?' and 'I'll blow yer fucken' head off' with *fucken'* pronounced with uvularised setting, and a long vowel, as [fa̱ːqəʀ] ([f] and [ʀ] indicate uvularised [f] and [n]). *Fucken'* would normally be pronounced [fakən] with a short vowel, non-uvularised [n] and [f], a velar stop [k] and short vowel. In the threatening pronunciation, then, we note the same long back central vowel [a̱ː] used to demonstrate the phonetic and voice quality settings above.

Since phonetic quality, voice quality and tone of voice are all realised in the same dimensions, the question arises of how we perceive the differences. It is assumed that difference between phonetic quality, voice quality and tone of voice features lies primarily in how long the features are maintained. Tone of voice features are maintained for a duration intermediate between the quasi-permanent voice quality and the momentary phonetic quality features: for as long as the particular attitude is being conveyed.

Forensic significance: Tone of voice

The obvious forensic relevance of all this is that one needs to have access to a considerable amount of a voice in order to stand a chance of sorting the features out. Unless this is the case, misperceptions may occur. Here is a commonly cited example. If one hears a voice in a short sample of speech to have a whispery phonation type, this could signal a deliberate conspiratorial tone of voice on the part of the speaker. But it could also signal voice quality – either because the speaker had severe laryngitis (in which case it would be a quasi-permanent *organic* voice quality feature), or because they had a whispery voice quality *setting*.

Since not all offenders are native speakers of English, the role of phonetic features must also not be forgotten in this context: there are languages in which whispery voice, being part of the realisation of linguistic sound units, is a phonetic feature. This means that *the correct interpretation of any sound feature, in particular its selection as a speaker-specific variable, must also depend on linguistic knowledge of the phonology of the language in question*. As an example of this, the author is aware of a case involving Vietnamese in which identity was claimed in part on the presence of creaky voice in questioned and suspect samples. However, in (at least Northern) Vietnamese creaky voice is phonemic: it is used as a quasi-tonal feature to distinguish words. Thus this identity claim was roughly analogous to saying that questioned and suspect samples were the same because they both said [b] (or any speech sound) a lot!

In its code of conduct, the International Association for Forensic Phonetics (IAFP) enjoins caution on the part of experts undertaking work in languages other than their own. It avoids specifying just what precautionary measures are to be taken, however.

What is a voice?

Such work is probably not feasible unless the expert cooperates with a linguistically non-naive native-speaker, and can themself demonstrate a thorough knowledge of the phonology and sociolinguistics of the language in question. This concludes the discussion of the phonetic quality/voice quality distinction, and tone of voice. In the next section a model for the voice is presented.

The need for a model

The sections above have pointed out that when we speak, a lot of information is signalled. This information is informative: it makes the receiver aware of something of which they were not previously aware (Nolan 1983: 31). We obviously think first of all about the information in the linguistic message that we intend to convey. But there are many other types of information, some of them intended, some not.

The different types of information in speech are encoded in an extremely complex way. One aspect of this complexity that has already been demonstrated in Chapter 3 with different speakers' F0 traces, and in Chapter 8 with different speakers' acoustic vowel plots, is that the different types of information in speech are not separately and discretely partitioned, or encoded in separate bits of the message. There is not one frequency band, for example, that signals the speaker's health, or one that signals emotion; the phonetic quality is not a frequency-modulated as opposed to an amplitude-modulated voice quality. Such things are typically encoded in the same acoustic parameter. *Unless the details of this encoding are understood, it is not possible to interpret the inevitable variation between forensic samples.*

Let us take once again the example of average pitch to illustrate this. As already explained, pitch reflects the size of a speaker's vocal cords; but it also encodes linguistic differences like that between statement and question, and differences in emotion, and differences in health (recall Figure 8.30 illustrating differences in F0 correlating with health). Unless we understand the details of this encoding, it is not possible to interpret the inevitable pitch variation between samples. An observed difference in pitch might reflect one speaker speaking differently on two occasions (with a preponderance of questions on one occasion, and statements on the other), or two different speakers with different-sized cords speaking in the same way.

The principle involved here is this. Two samples from the same speaker taken under comparable (i.e. totally controlled, as in automatic speaker verification) circumstances are likely to be similar and favour correct discrimination as a same-speaker pair. In the same way, two comparable samples from different speakers are also likely to be correctly discriminated as a different-speaker pair. But non-comparability of samples can lead to incorrect discrimination. It can make two samples from different speakers more similar, thus resulting in evaluation as a same-speaker pair, or it can amplify the difference between two samples from the same speaker, thus favouring evaluation as a different-speaker pair.

This means that in order to understand how these speaker-specific bits of information are encoded in the speech signal, it is necessary to understand what the different types of information in speech are; what the different components of the voice are; and what the relationship is between the information and the components. To these questions we now turn.

Voice as 'choice' and 'constraint'

When we speak it is often because we have information to communicate. However, this information has to be processed through two channels: most obviously, the message has to be implemented by a speaker's individual vocal tract. But the message has to be given linguistic form too, and both these channels affect the form of the message. The result of passing information we want to convey through these channels is the voice.

When we want to communicate something in speech, we have to make choices within our linguistic system. For example, when we want to signal the word 'back' as against 'bag' we choose the phoneme /k/ instead of /g/ after /bæ/. When we want to signal our assumption that the hearer can identify the thing we are talking about, we select the definite article 'the' instead of the indefinite 'a' (*the book* vs. *a book*). But these choices have to be processed through our individual vocal tracts to convert them into speech, and therefore are constrained by the physical properties of the individual's vocal tract. This leads to Nolan's characterisation of the voice as *the interaction of constraints and choices in communicating information*. A speaker's voice is the

> ... interaction of *constraints* imposed by the physical properties of the vocal tract, and *choices* which a speaker makes in achieving communicative goals through the resources provided by the various components of his or her linguistic system. [My emphasis]
>
> Nolan (1997: 749)

A model which makes explicit what is involved in this interaction is shown in Figure 10.1. The model has been simplified somewhat from that in Nolan (1996). This can be regarded as the picture of the components of a voice. It can be seen that the model consists of four main parts, the connections between which are symbolised by fat arrows. Two of these parts are inputs, and two are mechanisms. The two inputs are labelled *communicative intent*, and *intrinsic indexical factors*; and the two mechanisms are labelled *linguistic mechanism* and *vocal mechanism*. The communicative intent maps onto the linguistic mechanism, and the intrinsic indexical factors map onto the vocal mechanism. The vocal mechanism accepts two inputs, from the intrinsic indexical factors and the linguistic mechanism. There is also a picture of a speech wave coming from the vocal mechanism. This represents the final physical, acoustic output of the interaction. This output can be thought of as both the thing that a listener – perhaps best thought of as the forensic phonetician – responds to, and the acoustic raw material that is analysed by the forensic phonetician.

The idea being represented by the configuration in Figure 10.1, then, is that a speaker's voice results from the two inputs (communicative intent and intrinsic indexical factors) processed through the two mechanisms (linguistic and vocal). The sections below describe the components, and their sub-components, in greater detail. Although perhaps the logical place to start is the intended message, it is easier to start with the part that, because it has already been covered, requires the least explanation.

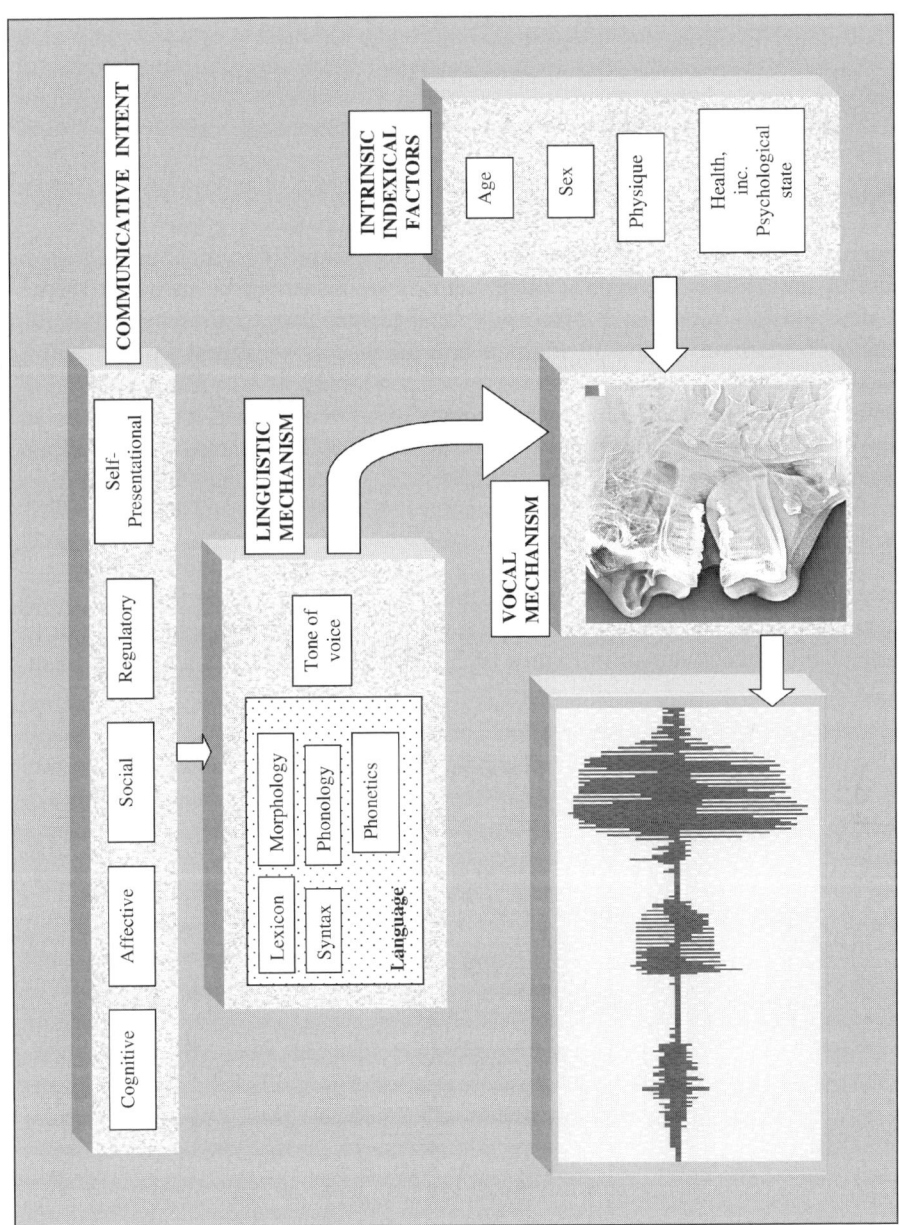

Figure 10.1 A voice model

Forensic Speaker Identification

Vocal mechanism

What is meant by the vocal mechanism is easy to understand. It is the anatomical ensemble of the vocal tract organs that are used in the production of speech and other vocalisations. Its main components have already been described in Chapter 6, and this has allowed me to represent it simply as an X-ray of the vocal tract. The vocal mechanism plays an obviously important role in the voice because the acoustic output is a unique function of the tract that produced it.

Linguistic mechanism

The linguistic mechanism can be understood to refer primarily to those aspects of the structure of the speaker's language other than meaning. It can be seen that the linguistic mechanism contains a stippled box labelled *language* to roughly reflect this. The linguistic mechanism also contains a box labelled *tone of voice*, which will be explained in due course.

Language is, of course, an immensely complex thing in itself, and it will help understanding of this section to describe how linguistic structure is conceived in the discipline concerned with its study, namely linguistics.

Linguistic structure

As conceived in linguistics, language is a complex multilayered code that links sound and meaning by a set of abstract rules and learnt forms. A very simple model for the structure of this code is shown in Figure 10.2. As can be seen, it has five components:

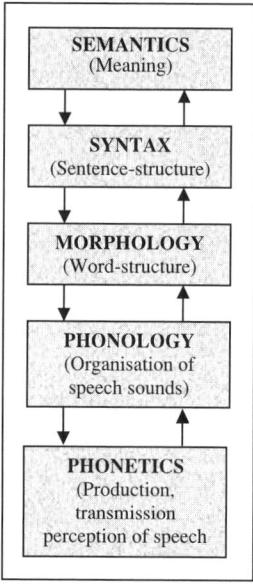

Figure 10.2 Analytic components of Language structure

semantics, *syntax*, *morphology*, *phonology* and *phonetics*. These components, apart from semantics, also appear in the linguistic mechanism's *language* box in Figure 10.1. *Semantics* has to do with the meanings conveyed in language; *syntax* with how words are combined into sentences. *Morphology* is concerned with the structure of words, and *phonetics* and *phonology* encompass aspects of speech sounds. In linguistics, all this structure is termed the *grammar* of a language, and thus *grammar* has a wider meaning than is normally understood. The voice's linguistic mechanism in Figure 10.1 can therefore be properly understood to comprise, in addition to the *tone of voice*, a large part of the speaker's grammar.

With one proviso described below, in addition to indicating the main components of linguistic structure, Figure 10.2 can be understood as representing the suite of processes involved when a speaker communicates a specific linguistic message verbally to a listener. This is sometimes called 'the speech chain', and is symbolised by the bidirectional arrows. Thus the speaker has a meaning they want to convey, and the meaning is expressed in syntactic, morphological and, ultimately, acoustic-phonetic form (this is what the downwards arrows imply).

It is this acoustic-phonetic form that reaches a listener's ears and that they decode, via their naturally acquired knowledge of the phonetics, morphology, syntax and semantics of their native language, to reconstruct the meaning of the original message (the implication of the upwards-pointing arrows). Figure 10.2 thus represents the linking of speaker's to listener's meaning via sound by showing semantics and phonetics peripherally, joined by the remaining three components of the linguistic code.

These five modules of linguistic structure traditionally constitute the core of any linguistics programme, and are the major categories in terms of which the grammar of a previously undescribed language is described in descriptive linguistics. Since they are also part of the voice, and since they may be referred to in forensic-phonetic reports, it is important to provide a brief characterisation of each.

Semantics

One of the main differences in the way linguists view the structure of language has to do with the place of meaning: specifically, whether it is primary or not. The view described here will simply assume that it is. That is, as described above, a speaker has meanings they want to communicate, and these are given syntactic, morphological, phonological and phonetic structure. This is why semantics is placed at the top of the model in Figure 10.2.

Semantic structure comprises firstly the set of meanings that are available for encoding in language in general, and the meanings that have to be encoded in a specific language. For example, all languages allow us to refer to objects, and to their location in space. In order to illustrate this, let us assume that I wish to communicate the location of an object to you, for example: 'The book is over there'. In English there are certain semantic aspects that I do not have to encode. I do not have to refer to the book's location being uphill or downhill from the speaker (over there uphill from me vs. over there downhill from me); nor to whether the location is near you, or away from you; nor whether the object is visible to me, or you; nor to the source of my knowledge about the book's location, and my consequent belief in its truth

(I know it's over there because I put it there; because I can see it; because someone told me it's over there; because there are other reasons for my believing it's over there). All these are semantic categories that have to be encoded in some languages. But if I am an English speaker, I *do* have to encode whether I am talking about a single book or more than one (book vs. books); and whether I assume that you know what book I am talking about (the book vs. a book). These categories are not obligatory in some other languages. These differences give rise, at least in part, to the linguistic aphorism that 'languages differ not in what they can say but in what they must say'.

Three types of meaning are encoded in language, and hence there are typically three subdivisions of linguistic semantics. The first is *lexical semantics*, or the meanings of words. An apposite lexical semantic question might be *What does the word* voice *mean?*

The second type of meaning is *structural semantics*, or the meaning of grammatical structures. As an example of structural meaning, take the two sentences *The man killed the burglar*, and *The burglar killed the man*. The two sentences clearly mean something different, yet they have the same words, so their semantic difference cannot be a lexical one. The difference in meaning derives from the meaning associated with the sentences' syntactic structure: in this case the difference in structural position between the noun coming in front of the verb and the noun coming after. Pre-verbal position, at least with this verb in this form, is associated with a semantic role called *agent*: that is the person who prototypically does the action indicated by the verb. Post-verbal position encodes the semantic role *patient*. This is prototypically the person who is affected by the action of the verb (in these two sentences the degree of affectedness is extreme, with the patient undergoing a considerable, indeed irreversible, change of state).

The third kind of linguistic meaning, *pragmatic meaning*, has to do with the effect of extralinguistic context on how an utterance is understood. An example of this is the understanding of the meaning 'Please give me a bite' from the observation *Mm! that looks yummy*, the form of which utterance linguistically contains no actual request to carry out the action. As an additional example, pragmatics has to be able to explain how the sentence *That's very clever* can be understood in two completely opposite ways, depending on the context.

Although semantics is clearly part of linguistic structure, meaning in the voice model is probably best thought of as a part of one of the inputs to the system (the part labelled communicative intent, which is the meaning that the speaker intends to convey), rather than as a part of linguistic structure. Moreover, attempting to talk about the intent to convey a meaning separately from the meaning itself involves one in unwanted philosophical arguments about whether meaning can exist separately from language. This is why there is no box labelled *semantics* within the linguistic mechanism component in Figure 10.1.

Syntax

Syntax functions as a framework on which to hang the structural and pragmatic meanings. Obviously, linguistic meanings have to be conveyed in sequences of words. However, words are not simply strung together linearly, like beads on a string, to convey a meaning. They are hierarchically combined into longer units like phrases, clauses and sentences, and it is this hierarchical structure that syntax describes.

Syntactic structure is described in terms of *constituents*, which are words that behave syntactically as a single group. Thus in the sentence *The exceedingly ferocious dog bit the man* the four words *the exceedingly ferocious dog* form one constituent (called a *noun phrase*) for three main reasons. Firstly, the group of words has a particular internal structure, expressed in terms of word class, typical of noun phrase constituents. It consists of an article (the), an adverb (exceedingly), an adjective (ferocious), and a noun. Secondly, the group can be substituted by a smaller item, for example the pronoun *it*, and still yield a grammatical sentence (*It bit the man*). Thirdly, their constituent status is shown by the fact that they can be moved *as a group* to form, for example, the related passive sentence: *the man was bitten by* **the exceedingly ferocious dog**.

The hierarchical combination of syntactic constituents like noun phrases into higher-order constituents is shown by the fact that the noun phrase *the exceedingly ferocious dog* forms part of the prepositional phrase *by the exceedingly ferocious dog*. The resulting structure can be expressed by a syntactic rule like 'prepositional phrase = preposition plus noun phrase'. This is one example of what is meant by the structure of the linguistic code being rule-governed.

Morphology

The smallest meaningful unit in a language is called a morpheme, and words may consist of one or more morphemes. The single word *books*, for example, consists of a morpheme meaning 'book' and a morpheme meaning 'plural'. The average number of morphemes per word is one of the main ways in which languages differ. Vietnamese has on average very few morphemes per word; English has on average somewhat more.

The different types of morphemes and the ways they combine to form words are the subject of morphology. The reader might like to consider how many morphemes are present in the word *oversimplification*. It consists of four morphemes: a basic adjectival root morpheme *simple*; a suffix *-ify* that functions to change an adjective (simple) into a verb (simplify); a prefix *over-* that attaches onto a verb or adjective (cf. *oversubscribe, overzealous*); and a suffix *-ation* that functions to change a verb (oversimplify) into an abstract noun. It is not clear whether the -c- in *oversimplification* is a part of the morpheme *-ify*, or the morpheme *-ation*.

Phonology

Phonology deals with the functional organisation of speech sounds. One aspect of phonology central to forensic phonetics, namely phonemics, has already been described in detail in Chapter 7. It will be recalled that phonemics describes what the distinctive sounds, or *phonemes*, of a language are, what the structure of words is in terms of phonemes, and how the phonemes are realised, as allophones.

It is one of the interesting structural features of human language that its meaningful units (i.e. morphemes) are not signalled by invariant sounds. An example of this is the English 'plural' morpheme described in the section of Chapter 7 called

Morphophonemics, where it was shown how the plural morpheme is sometimes realised as an [s], as in *cats*; sometimes as a [z] as in *dogs*; and sometimes as [əz] as in *horses*. Which of these forms (or *allomorphs*) is chosen is predictable, and depends on the last sound in the noun. This is another example of the predominantly rule-governed nature of the linguistic code. The reader might also have noted that the morpheme -*ifi* in the *oversimplification* example above has two different sound realisations or allomorphs: at the end of the word, e.g. in *oversimplify* it is /ɪfaɪ/; in the middle it is /ɪfɪ/. The area of linguistic structure that is concerned with relationships between the morphemes (meaningful units) and their allomorphs (realisations in sound) is called *morphophonemics*. It is usually considered as another aspect of phonology.

Phonetics

As will be recalled from the detailed exposition in Chapters 6, 8 and 9, phonetics deals with the actual production, acoustic realisation and perception of speech sounds.

Lexicon

There is one thing that is, strictly speaking, still absent from the sequence of components in Figure 10.2 if it is to be understood as representing the speech chain. The five components just described – semantics, syntax, morphology, phonology, phonetics – have to do with linguistic structure, but obviously a message will not mean much if it is pure structure. Not much is conveyed for example by the structural description '[[[article] + [[noun] + [plural]]]', compared to, say, *the criminals*. A language must therefore also contain a repository of words, and this is called a *lexicon*. Linguists disagree on whether the lexicon contains words or morphemes, or a mixture of both; and on where in the speech chain the words/morphemes are actually inserted into the linguistic structure. A box labelled *lexicon* can also be found alongside the structural components in the *language* box in Figure 10.1.

Having outlined what is understood by linguistic structure, and how it fits into the voice model, we can now return to the other voice model components.

Tone of voice (2)

The term *tone of voice* has already been encountered in the discussion on phonetic quality and voice quality, but from the point of view of how a voice *sounds*. In order to *sound* angry, sad, respectful, threatening, conciliatory, intimate... even neutral, the speaker must intend to do so. Somewhat confusingly, *tone of voice* is also the term used for the component that accounts for the realisation of these intended emotions in the voice model. It is situated, alongside *language*, in the *linguistic mechanism* box.

Tone of voice is allocated a position with linguistic structure within the linguistic mechanism because it displays some important features typical of linguistic structure. This is why it is often termed a *paralinguistic* feature. Tone of voice features are, like linguistic features, to a large extent arbitrary and convention-governed, and they are realised in the same acoustic dimensions as the purely linguistic features. For example, an angry tone of voice is conveyed partly by using a different pitch range, and pitch is also put to a myriad of linguistic uses (to signal tones, stress, intonation, etc.). As mentioned above, tone of voice features differ from linguistic features in being maintained for longer stretches. A speaker might use a high pitch to signal anger for as long as they desire to sound angry, whereas the use of high pitch to signal a high tone in a tone language lasts only momentarily, for the duration of the linguistic unit (syllable, word) carrying the high tone.

Communicative intent

We now turn to the main input to the system, namely all the information that a speaker intends to convey. This is termed *communicative intent* (Nolan 1983: 35) and is represented by the large box along the top of Figure 10.1. For the speaker to intend something, it has to be the result of a choice (whence the characterisation above of voice as *choice* vs. constraint). What sorts of things can and do speakers deliberately encode in their voices?

The first that springs to mind is the linguistic message itself: a proposition (an utterance with a truth value) perhaps, or a question, or a command. However, speakers also deliberately express emotion; convey social information; express self-image; and regulate conversation with their interlocutor(s), and these also constitute different components of communicative intent. The communicative intent box thus contains five smaller boxes, which refer to these five possible different types of information. These different types will now be described.

Cognitive intent

Probably the first type of information that one thinks of in speech is its 'basic meaning'. This is called cognitive information, and refers to meaning, differences in which are conveyed by a particular choice and arrangement of words (Nolan 1983: 61–2). For example in the utterance *The snake ate the wide-mouthed frog* the cognitive information contains the meaning of (at least) the facts that two animals are referred to; that the speaker assumes that the hearer knows which snake and which frog are being singled out; that eating is involved, and that the snake is the agent and the frog the patient. Another way of thinking of this type of meaning is that it is the type of meaning that is conveyed in writing (for languages that do, in fact, have a written form).

Since we are dealing with linguistic meaning, changes in cognitive content will have consequences for all the components of linguistic structure. A change in the cognitive meaning of an utterance will be represented in its linguistic semantic structure, and result in a change in the selection of a word and/or syntax, and this in turn will cause utterances to be phonologically and phonetically non-equivalent. For example, a different choice of words can signal cognitive intent (*The dog ate the frog* vs. *The snake ate the frog*).

Differences in cognitive intent are, of course, the main reason for forensic samples having different words and syntax and therefore different sounds. It is not uncommon for forensic samples to contain the same words, however, and a certain amount of research has been done on acoustic and auditory within-speaker and between-speaker variation in commonly occurring words like *hello* (Rose 1999a, 1999b), *yes* (Simmons 1997) and *okay* (Elliott 2001). Sometimes even the same phrase occurs in different samples. As explained above, the occurrence of the same word or phrase in different samples furnishes the highest degree of comparability in sounds, because they are in the same phonological environment. Often, however, the forensic phonetician will have to look for equivalence at the level of sounds in comparable phonological environments, that is, environments that are assumed to have minimal effect on the parameter being compared. For example the [iː] vowels in *pea* and *fee* occur to be sure in different environments: after /p/ and /f/. But the difference between the two will not affect the vocalic F-pattern in mid-vowel.

Affective intent

We can choose to signal an emotional state when we speak. Affective intent refers to the attitudes and feelings a speaker wishes to convey in the short term (Nolan 1983: 62). The reader should try saying the same sentence in a friendly, then angry tone of voice and note what changes occur. One change will almost certainly be that the angry utterance will be louder, and perhaps the overall pitch will be higher. So another way in which speech samples can differ is in affective intent.

How different emotions are actually signalled in speech is very complicated. They can be signalled linguistically, for example: we may choose different words and/or different syntax (*Could you please give me the money?/Hand over the fucking cash!*). These would be examples of mapping between affective intent, and lexicon and syntax.

More commonly, perhaps, different emotions are signalled linguistically in sound. This occurs primarily by the control of intonational pitch. For example, *It won't hurt*, with a low rising pitch on *hurt*, sounds friendly. Putting a falling pitch on *hurt* changes the emotion to petulance, with the implication: (It won't hurt) so stop complaining! This is a linguistic matter because the choice of a different intonational pattern – falling vs. rising pitch, for example – is conceived of as a discrete choice between two phonologically different sounds. In this case, then, the mapping would be between affect and phonology.

We can also signal differences in emotion non-discretely, by for example altering our pitch range. *Yes*, said with a pitch falling from high in the speaker's pitch range to low signals more enthusiasm than a *yes* said on a narrower pitch range, with a pitch falling from the middle of speaker's pitch range to low. In these cases, there is a more direct relationship between the actual realisation and the degree of emotion signalled, with the degree of involvement reflected in the size of the pitch fall, or the width of the range. Emotion is also commonly signalled in sound by phonation type – the way our vocal cords vibrate. For example, in some varieties of British English, using breathy voice can convey sympathy, using creaky voice at the end of an utterance can signal boredom. These relationships are not usually conceived of as primarily linguistic; and so the mapping would be between the affective intent and the *tone of voice* box.

In Figure 10.1, the mapping between the communicative intent and the linguistic mechanism is represented by a single fat arrow. However, enough examples of the many-to-many mapping between the subparts of communicative intent and the linguistic mechanism have already been given to show that it is more complex than a single arrow can represent.

Forensic significance: Affect

The forensic importance of understanding and specifying the complex mapping in a voice becomes clearer with the discussion of affective intent. Phonation-type features like breathy or creaky voice, or pitch features like wide pitch range, which are used to signal affective intent, are also often assumed to be speaker-specific, that is, parameters along which speakers differ. Therefore, it is necessary that, if identification or elimination is being claimed on the basis of these features, it must be clear that *the affective communicative intent is comparable across samples*. Otherwise there is no guarantee that the affective mapping will not confound identification or elimination based on those properties. For example, it is no good maintaining that two samples differing in pitch height come from different speakers if there are other indications that the voice in one sample is angry and the voice in the other is not. Conversely, it is no good maintaining that two samples differing in pitch height come from the same speaker unless there are indications that the voice in one is angry but not in the other. This reasoning of course applies to all cases below where a phonetic feature signals two or more different things, which is the norm in speech.

Social intent

Speakers are primates. They interact socially in complex ways. Part of this social interaction is played out in language, and is responsible for both between-speaker and within-speaker variation.

It is often assumed that the primary function of language is to convey cognitive information. However, a very important function of language is to signal aspects of individual identity, in particular our membership of a particular group within a language community. This group can be socioeconomically defined (for example, the speaker is signalling themself as working class, middle class, upper class), or regionally defined (the speaker is signalling their provenance from Melbourne, from Brisbane), or ethnically defined (the speaker is signalling their status as Vietnamese immigrant, Greek immigrant). The idea here, then, is that speakers typically choose to signal their membership of social, ethnic or regional groups by manipulating aspects of linguistic structure. That is part of what is meant by the social intent sub-part of communicative intent (Nolan 1983: 63–8).

As the result of sociolinguistic research over the past decades, correlations have been determined between linguistic features and socioeconomic and ethnic groups of this kind. An enormous amount of this work has been done, was indeed pioneered, by the American linguist William Labov (mentioned in conjunction with the Prinzivalli case above) and his students. One well-known study in the Australian context is Horvath's 1985 book on the sociolects of Sydney. The way speech changes as a function of geographical location has long been studied in the linguistic discipline of dialectology (Chambers and Trudgill 1980).

Being able to correlate linguistic features with geographical and social background seems *prima facie* a good thing for speaker identification, since a statement of the kind 'The speech of both samples is typical of a working-class male of Italian extract from Sydney' would clearly narrow down the range of suspects. There are indeed well-known cases where dialectological information has been used forensically to good effect. But not in Australia. The dialectological side depends firstly on a high degree of dialectological differentiation in the language community. In Australia, regional dialects are only just beginning to emerge in some areas.

More importantly, however, it must be remembered that social *intent* means that we are working with choices, which means that the content is not invariable. This can be seen most clearly if we take another dimension into consideration.

In addition to the unambiguously social dimensions outlined above, it is important to understand that we convey our interpretation of various social aspects of a situation in the way we speak. One of these aspects is the perceived formality of the situation: this is often known as stylistic variation. Another is the perceived status relationship of the speaker to the interlocutor, or others present, or even the people the speaker is referring to. This is encoded lexically, in the choice of words, and grammatically, in the syntactic constructions and endings, in Japanese (Inoue 1979: 281–96). Another might be the perceived, or known, group membership or kinship relationship of the interlocutor. In most Australian Aboriginal languages, for example, a form of language differing in lexicon (the so-called *mother-in-law language*) has to be used by a man when speaking to, or in the presence of, his in-laws (Dixon 1983: 163–93).

Forensic significance: Social intent

All this means that people choose to speak differently depending on how they perceive the speech situation. A well-known and easy to understand example of this is the use of either an alveolar nasal /n/ or a velar nasal /ŋ/ in verbal forms like *playing*, *singing*, *doing*, etc. to reflect formality. I am more likely to say to one of my mates *Where're ya livin'*, with an alveolar /n/ in *living*, than *Where are you living*, with the standard velar nasal. The latter pronunciation, however, I would be far more likely to use when giving evidence in court. Another example of informal speech would be substitution of a /d/ for a /t/ in the middle of a word: saying /sɪdi/ for *city* instead of /sɪti/.

This illustrates the well-known fact that speakers control a range of styles that they use on different occasions: there are no single-style speakers, and once again we note yet another aspect of within-speaker variation.

To complicate the picture still further, sociolinguistic research has shown that the stylistic variation interacts with the socioeconomic/ethnic/regional classification. Thus a working-class speaker will exhibit different linguistic behaviour as a function of perceived formality from an upper-class speaker.

A well-known aspect of social interaction involves the phenomena of convergence and divergence. If you want to establish a rapport with your interlocutor you will signal this in various ways: adopting the same posture, for example. Conversely, it is also possible to signal distance. Convergence and divergence also occur in speech, and can be manifested in many ways. Moving towards your speaker's pitch range, for example, or using the same allophones for some phonemes, or using the same expressions. The consequences of convergence and divergence in forensic-phonetic comparison are not difficult to imagine. It is useful if both parties use the same expressions, because that makes for a high degree of phonological comparability. However, convergence can also give rise to more within-speaker variation. The intercepted voice of the offender converging with his co-conspirators may differ considerably from his voice when diverging from those of the interrogating police officers.

The kinds of social information mentioned above are primarily encoded linguistically, hence the connection between the *social intent* box and the *language* box. Although a lot of the information will be in the speaker's accent, i.e their phonology and phonetics, some will be signalled in their lexical choices. The reader might like to reflect on the sort of social situation in which they would use, or expect to hear, the following: *toilet, lavatory, dunny, shithouse, bog, loo, gents', ladies', powder-room, little girls' room*. Some languages also encode aspects of perceived status relationships in their morphosyntax. As mentioned above, Japanese is a well-known example.

It is clear from the above that attention must be given to potential between-sample differences in social intent. Ideally, samples should be comparable from the point of view of factors related to social intent. If not, knowledge of how social structure is encoded and manipulated in the language in question is necessary to interpret the inevitable between-sample differences properly.

Regulatory intent

We all talk to ourselves (or the computer, or dog) from time to time, but most forensic-phonetic casework involves verbal interaction, usually conversation, with other humans. Conversation is not haphazard: it is controlled and structured, and the conventions underlying conversational interaction in a particular culture are part of the linguistic competence of all speakers who participate in that culture. In traditional Australian aboriginal societies, for example, in contrast to Anglo-Australian culture, it is not normal to elicit information by direct questions. (The obvious implications of this for aboriginal witnesses in court have often been pointed out.) The sub-discipline of linguistics that investigates how speakers manage conversations is called Conversation Analysis (or CA).

Regulatory intent has to do with the conventional things you deliberately do to participate in a conversation in your culture (Nolan 1983: 71). For example, one particular structural component that you will adhere to if you are a native participant in an Anglo conversation is the so-called question/answer/close sequence. Given the following exchange between two speakers A and B, for example, it is possible to say who says what: *Where yer goin? Footy Okay*. (A questions: *where yer goin'*. B answers: *footy*. A closes, (and simultaneously anticipates a new topic of conversation): *okay*.

Conversation involves turn-taking. Conversational participants need to signal when they do not want to be interrupted, and when they have finished saying what they wanted to say and are prepared to yield the role of speaker to their interlocutor. This is done in many ways, but in speech by controlling such things as the rate of utterance, pitch and loudness level, or choosing specific intonation patterns.

Forensic significance: Regulatory intent

Knowledge of the dynamics of conversational interaction is important in forensic phonetics for several reasons. The first point has been made many times before. The phonological cues that speakers use to regulate conversations must be understood to ensure comparability between samples. If utterance-finality is signalled by certain phonetic effects, like increasing vowel duration, choice of intonation, or phonation type, there may be problems in comparing sounds at the end of utterances in one sample with those in the middle in another.

Another point, screamingly obvious perhaps only after it has been made, is that *the comparison of forensic samples is crucially dependent on correct attribution of stretches of speech to a particular speaker*. If I do not know when one speaker has stopped speaking and the other taken over, the consequences are dire: I run the risk of treating samples from two different speakers as coming from the same speaker, and constructing a meaningless statistical profile for a single speaker from a compound of two. Some forensic-phonetic tasks are concerned precisely with this problem – determining how many speakers are present in a given, usually incriminating, situation.

Usually it is clear from differences in voice quality when the speaker has changed, but not always. Sometimes it is useful to know how conversations are structured. For example, telephone conversations in Anglo culture very commonly start with the phonee picking up the telephone and immediately saying *hello*. (In conversation analysis this is known as a *summons* – from the receiver to the caller to identify themself.) To which the phoner then appropriately responds (*Hello / It's me / Oh hi /* etc.). This is partly how we know, given two *hello*s with similar voice quality at the start of an intercepted phone conversation, that they probably constituted a summons–response sequence, and therefore the first was probably said by the receiver and the second by the caller.

Self-presentational intent

Richard Oakapple sings in the Gilbert and Sullivan opera *Ruddigore*: 'If you wish in the world to advance / Your merits you're bound to enhance / You must stir it and stump it and blow your own trumpet / Or trust me you haven't a chance'. This reminds us that speakers can deliberately use their voice to project an image to others (Nolan 1983: 69). This starts early. It is known that little boys and girls, although they are too young to show differences in vocal tract dimensions associated with peripubertal sexual dimorphism, nevertheless choose to exploit the plasticity of their vocal tract in order to sound like (grown-up) males and females. Little boys have been shown to use lower F0 values, and little girls higher.

By their voice, speakers can project themselves as, for example, feminine, confident, extrovert, macho, diffident, shy. To the extent that this self-image changes with the context, and one might very well encounter such a change between the way a suspect speaks with his mates and the way he speaks when being interviewed by the police, we will find within-speaker variation.

Intrinsic indexical factors

The last component of the voice model to be discussed is the other input alongside communicative intent: the vertical box on the right of Figure 10.1 labelled *intrinsic indexical factors*. This term obviously denotes the intersection of indexical and intrinsic information, so these will be clarified first.

Indexical information

It is a semiotic commonplace that a message usually carries some information about the mechanism that produced it. It is generally accepted that we can tell a lot about a person from their voice. Their sex, for example, or, within certain limits, their age or state of health; perhaps also where they come from. Sometimes it is possible to actually recognise the individual themself. The most important thing for the purposes of this book is, of course, the information about the identity of the speaker. Information about the characteristics of a speaker is called *indexical*, the idea being that a particular feature or constellation of features is *indicative* of a particular characteristic, like sex, or age (Laver and Trudgill 1991: 237ff; Laver 1991d).

Intrinsic features

The terms *extrinsic* and *intrinsic* are often used to distinguish between vocal features that are under the speaker's control and those that are not. The former are called extrinsic, the latter intrinsic (Laver 1991c: 163ff). It was mentioned above that this distinction also underlies the difference between organic voice quality and voice quality settings.

Of the indexical features that characterise a speaker, then, some are extrinsic. For example, the use of an elaborately differentiated colour terminology (teal, burgundy, charcoal, cerise, . . .) will usually mark the speaker as female. This of course represents lexical choice, and is extrinsic. Another extrinsic indexical feature is the use of sociolinguistic variables (like [n] vs. [ŋ] in *doin'*, *fucken'*, etc.).

Many indexical features, however, and those that are forensically important, are intrinsic: the speaker has no control over them. For example, we can usually tell the sex of an adult speaker (even when they are not talking about colours) and this is because certain sex-related features of a speaker's anatomy shape the acoustic signal willy-nilly. An insightful metaphor, that of *imprinting*, has been proposed by Laver for the way this information appears in the acoustics: the intrinsic indexical information is said to be *imprinted* on the speech signal by the vocal tract that produced it.

The information about a speaker that is revealed non-volitionally is primarily biological, and shown in the *intrinsic indexical* block on the right-hand side of Figure 10.1. As can be seen, it consists of age, sex, physique and health. Health here is broadly construed as a parameter in which deviations from normal function can be located, and can be equally taken to include the effect of intoxication, stress and psychosis as well as a cold, hence it has been marked to include psychological state.

The presence of the first two, or perhaps three, features – age, sex and physique – will come as no surprise. However, the voice also carries a considerable variety of information about its speaker's organic state: whether the speaker is in robust health, is pre-menstrual, is deaf, has a cold, is suicidal (France *et al.* 2000), is drunk, or is just plain tired (Laver 1991d: 154–5; Laver and Trudgill 1991). All these intrinsic indexical features can contribute to a speaker's organic voice quality.

The imprinting of various types of indexical information in the speech wave from the vocal mechanism occurs for the following reason. The acoustic output from a speaker is determined uniquely by the size, shape and condition of their vocal tract, and the size, shape and condition of the vocal tract differ depending on the speaker's age, sex, physique, psychological state and state of health (plus the particular sound they are making of course!). If there are between-sex differences in vocal tract size, which there are, then this size difference will be reflected in the speech-wave, and is available for interpretation by listeners.

The most important anatomical ways that individual vocal tracts differ are in the overall length of the vocal tract, the size and length of the vocal cords, the relative proportions of the oral and pharyngeal cavities, and the dimensions of the nasal cavity and associated sinuses. They can also differ in ways that affect the compliancy of the tract walls, that is, to what extent they absorb acoustic energy. Tense vocal tract walls with minimum fat and mucosal covering are non-compliant and reflect relatively more acoustic energy and are thus associated with narrow formant bandwidths. Compliant tract walls are acoustically absorbent and are associated with wider formant bandwidths. Differences in the tenseness and mucosal covering of the vocal cords will also be reflected in their vibratory behaviour and thus their contribution to acoustic output.

Age, sex and the vocal tract

Beck (1997: 261–83) presents a summary of how the vocal tract changes during life. From birth to puberty, the vocal tract increases considerably in overall size and changes in shape, with the biggest changes taking place in the first five years. However, as with other aspects of growth, for example in height, there is no substantial difference between the sexes in the growth of the vocal tract in this period. At puberty, the length of the vocal tract will be about equally divided between the oral cavity and the pharynx.

During adolescence, there is in both sexes a rapid increase in overall vocal tract size, again reflecting general growth patterns in body size. But there are also differential changes between the sexes. Firstly, the increase in overall size is more marked in males, resulting in overall longer vocal tracts. At maturity, males have a vocal tract that is assumed on average to be about 20% longer than females. Because of the relationship, explained above, between overall supralaryngeal vocal tract length and average formant centre frequencies, males will have overall lower formants than females. The larynx in males also descends further than in females, and this results in a disproportionately longer pharynx. In adult males, the pharynx is assumed to be about 20% longer than in the adult female, which means the ratio of pharynx length to oral cavity length is greater in males. Since the pharynx acts as a separate acoustic resonator in some vowels, the frequencies that it contributes to the speech signal will be relatively lower in males, for the same linguistic sound, than in females.

Finally, both the larynx increases in size and the vocal cords increase in size and length, so that in adult caucasians male cords are on average longer than female cords (values quoted are 2.5–2.3 cm for males and 1.7 cm for females). Because post-pubertal males have larger and more massive vocal cords than post-pubertal females, adult male cords vibrate at lower rates, and are associated with lower fundamental frequencies for the same linguistic sounds. Typical F0 differences associated with sex were shown in the *long-term fundamental frequency* section of Chapter 8.

Degenerative processes accompanying ageing are responsible for changes in vocal tract shape from maturity to senescence. Males tend to show more marked effects than females. Muscular and bone atrophy, and changes in mucosal layers, are responsible for overall changes in the shape of the supralaryngeal vocal tract. For example, loss of bone from the lower jaw, accompanied perhaps by loss of teeth, will change the shape of the oral cavity. Changes in the mucosal layers of the vocal cords, together with calcification of parts of the laryngeal cartilages (visible, alas, in the X-rays of the author's vocal tract in Chapter 6) will also affect the fundamental frequency.

Since speakers' vocal tract dimensions tend to correlate directly with their overall body size, as quantified perhaps in their height and weight, larger speakers will tend to have longer vocal tracts, and larger vocal cords, with correspondingly lower F-patterns, and lower overall F0 values. It is not clear whether the *physique* dimension can be distinguished clearly from *sex*, since sexual differences are also manifested in physique as understood here. But for analytic convenience, the *physique* dimension is perhaps best construed as operating within-sex. That is, it is perhaps better to recognise sex-determined variation first, and then differences due to different size within a sex. This is because we might be able to recognise a small male from a large female.

Health and the vocal tract

Any changes in health that affect the size or shape or organic state of the vocal tract, or its motor control, will alter its acoustic output, thus contributing to within-speaker variation. A speaker's state of health is thus also imprinted on their acoustic output.

These intrinsic health-related changes can range from temporary (e.g. a head cold), to periodic (effects of menstrual cycle), to chronic (vocal fold polyp), to permanent (effects of surgery, congenital stutter) and can have the usual consequences of making two different speakers more similar, or the same speaker more different in certain parameters. Very little of forensic use is known about the interaction of health and phonetics (but cf. Braun 1995: 14–18), and it is clearly an area where more research is needed.

An instance of the common health factors that affect the acoustic output is a temporary head cold, which might cause inflammation and swelling in the nasal cavities or sinuses, thus altering their volume and compliance and resonance characteristics. Inflammation and swelling associated with laryngitis might make it painful to stretch the cords too much, and this will temporarily alter a speaker's fundamental frequency values, restricting their range of vibratory values, and making it uncomfortable to reach high F0 target values. In Chapter 8 examples were given of how different states of health affected the values of the author's long-term F0 distribution. Peri-menopausal hormonal changes are also known to affect the cords' mucosal layer and thus their vibratory characteristics.

Dentition, too, falls under the heading of health factors affecting acoustic output. To the extent that the front teeth are involved in the formation of a sound, a speaker's dentition will have an effect on the distribution of energy in those sounds. This will have the greatest effect in sibilant fricatives like [s] [ʃ] [z] and [ʒ] for example. Thus a speaker using different dentures, or none, will output different spectra for these sounds. Accommodatory changes in tongue body movement associated with different dentures will also affect the resonance pattern of vowels.

Probably the most important cases from a forensic point of view where state of health affects speech concern intoxication by alcohol and/or drugs. These can affect both the speaker's motor control and the auditory and proprioceptive feedback mechanisms that are necessary for fluent speech. This can result in all sorts of errors in the execution of the complex articulatory plan for the correct realisation of linguistic sounds. The tongue might not achieve closure for a [d], for example, and some kind of [z]-like fricative might result, or there may be local changes in the rate and continuity of speech. Other factors that interfere with normal motor control and feedback are stress and fatigue.

Forensic significance: Intrinsic indexical factors

It is assumed that all the factors just described apply generally to all humans, although there are undoubtedly small differences in physique within different racial groups. Some individual differences in vocal tract size and shape are to be attributed therefore to the fact that the individuals differ in age and sex. But of course there is considerable variation holding both of these factors constant: middle-aged males come in all sizes, and it has been said that there is the same amount of difference between individuals on the inside as on the out. Such remaining between-speaker differences are assumed to arise from the interplay of genetic and environmental factors. The genes determine an individual's maximum growth potential, and the environment – for example an individual's nutrition, exposure to disease, etc. – determines the extent to which the potential is realised (Beck 1997: 283–91). Differences of this type account therefore for between-speaker differences in physique, and in the short term, differences in health.

We all differ in genotype, and during our lives we are all exposed to different environmental factors; but vocal tracts are intended to perform a function, and obviously cannot be allowed to vary without limit. In reality, therefore, the majority of same-sex mature adults in a same-language population will have vocal tracts of very similar dimensions. Therefore, the chances for differentiating speakers on acoustic features that are assumed to reflect between-individual differences in physique must also be assumed to be limited. Two samples from different speakers who have similar anatomical endowment will also have similar organic voice quality acoustic features. Two samples from different speakers with very different anatomical endowment, like a male and female, or an adult and a child, will sound so different that they are not liable to be forensically confused in the first place.

Finally, it must never be forgotten that indexical features as imprinted by the vocal tract are not the only source of between-speaker differences. It is perfectly possible for two individuals to have as similar vocal tracts as possible, for example identical twins, and yet still differ clearly and consistently in the linguistic use to which they put those tracts (Nolan and Oh 1996). This is the main reason why it is important always to examine samples for *linguistic* differences, and the reason why some practitioners believe that it is often easier to demonstrate similarity or difference with respect to these rather than acoustics.

Limits to variation, not absolutes

The most important fact to be understood about the mapping of intrinsic indexical features onto the vocal mechanism is that *the indexical factors are not associated with absolute acoustic values, but impose limits to a speaker's variation*:

The vocal apparatus itself is perhaps the most obvious source of phonetic differences between speakers. What is essential to recognise, however, is that it does not determine particular acoustic characteristics of a person's speech, but merely the range within which variation in a particular parameter is constrained to take place. Thus it is certainly true to say that the dimensions of a person's vocal tract, or length and mass of his vocal folds, will in some sense 'determine' his formant frequencies and fundamental frequency, respectively, and may even define optimum values for him in these parameters . . . ; but the *plasticity* of the vocal tract is such that this scope for variation in these parameters is considerable . . .
. . .
There is, in fact, no acoustic feature which escapes the plasticity of the vocal tract.

Nolan (1983: 59)

The anatomy and physiology of a speaker determine the width of the potential range of operation for any voice quality feature, and the long-term habitual settings of the larynx and the vocal tract restrict this feature to a more limited range of operation.

Laver (1991d: 148)

If intrinsic indexical factors were linked to absolute values, a person of a given height, for example, would produce vowels with a specific resonance pattern, and this would of course facilitate both identification and exclusion. *Rather, indexical features determine ranges of acoustic values*. This means that it will be possible for a speaker, given their individual anatomical endowment, to produce a *wide range of values* for a particular feature, say fundamental frequency, depending on their communicative intent. This range will then overlap to a greater or lesser extent with other speakers' fundamental frequency ranges (in the way demonstrated in the chapters outlining the basic ideas).

Furthermore, it is possible to extend ranges somewhat by implementing other changes. Thus a male speaker might go below his normal F0 range by lowering his larynx and becoming creaky; and go above his normal F0 range by changing to a falsetto phonation type. There are limits to this, however. Fundamental frequency values outside this extended range are not achievable, although the higher F0 values would easily be within the normal range of a female speaker with her smaller cords.

It also seems plausible that intrinsic factors might also determine a speaker's default value for a parameter – the value at which they are most comfortable, given their anatomy. However, the evidence from articulatory settings shows that this is not the case for all speakers, and there is also evidence from cross-linguistic studies that different languages select different default values for a given parameter. Both Gfroerer and Wagner (1995: 46–7) and Braun (1995: 14), for example, demonstrate higher mean long-term F0 values for Turkish than German speakers. Since this is unlikely to be referrable to genetically caused variation in cord size, someone – either the Turks or the Germans or both – is selecting values that are not primarily determined by their anatomy. There are other examples cited in the literature of apparently culturally selected LTF0 values (e.g. Braun 1995: 14).

Thus anatomical endowment imposes limits to the variation a speaker can command, but within these limits there is considerable scope for variation as a function of communicative intent, and this will result in overlap with other speakers.

Chapter summary

This section has described what a voice is from a semiotic perspective, that is, in terms of the information it conveys. It was motivated by the necessity to understand what can underlie variation in a single speaker's vocalisations in order to correctly evaluate differences between forensic voice samples. It has shown that variations in a speaker's output are a function of two things: their communicative intent (itself a combination of what they want to convey and the situation in which they are speaking), and the dimensions and condition of their individual vocal tract (which impose limits, but not absolute values, to the ranges of phonetic features their language makes use of).

The point has been made elsewhere (Nolan 1983), and it is worth repeating here, that if the internal composition of a voice appears complex, that is because it is. The voice is complex because there are many things that humans choose to communicate; because the linguistic mechanism used to encode these things is immensely complex; because the mapping between the linguistic mechanism and the communicative intent is complex; because the vocal tract used to implement the complex message involves an enormous number of degrees of freedom, and finally because individual vocal tracts differ in complex ways. All these complexities must be understood if we are to be able to accurately estimate whether differences between forensic samples are between-speaker or within-speaker.

11

The likelihood ratio revisited: A demonstration of the method

In this chapter, we return to the likelihood ratio (LR) introduced in Chapter 4, and give some examples of its application.

In the section of Chapter 8 on fundamental frequency, a hypothetical example was given of the use of the likelihood ratio to determine the probability that an F0 value had come from one of two Shanghai females. The author was recently involved in a similar closed set case, mentioned at the beginning of the book, where the question was how well incriminating speech samples could be attributed to one of two similar-sounding brothers. However, forensic-phonetic comparison is almost never as simple as this. Usually, one is concerned with an open set evaluation of the difference in selected acoustic and linguistic parameters between questioned and suspect speech samples. *This means saying whether the observed differences between suspect and questioned samples are more likely to be typical of between-speaker or within-speaker differences.*

In order to do this quantitatively, likelihood ratios (LR) must be calculated. The concept of the LR was introduced in Chapter 4 in the Basic Ideas part of the book. It will be recalled that it is the ratio of two probabilities – the probability of observing the evidence assuming the prosecution hypothesis that both questioned and suspect samples come from the same speaker, $p(E \mid H_p)$; and the probability of observing the evidence assuming the defence hypothesis that the samples come from different speakers, $p(E \mid H_d)$.

LRs must be calculated for each of the parameters in terms of which the samples are being compared, and then an Overall Likelihood Ratio derived by combining these separate LRs. Thus if the samples are being compared with respect to parameters of mean F0 and mean F2, for example, a LR must be calculated separately for both mean F0 and mean F2, and the Overall LR is then derived from both LRs.

The outcome of such a calculation is a statement to the effect that one would be *n* times more likely to observe the difference between suspect and questioned speech samples were they from the same speaker than from different speakers, where *n* is the

Forensic Speaker Identification

LR. The Overall LR from the forensic-phonetic data can then be combined with the prior odds for the speech evidence, to derive posterior odds which can then be combined with other evidence.

In this chapter, we will demonstrate the ideas behind the LR with some real speech data. This will also help to make some of the current limitations on the LR approach clearer. We will start with a very simple example with categorical phonological values to remind the reader of the general approach, and then tackle some more complicated examples with continuous variables.

Illegal high vowel devoicing in Japanese

The first example concerns what has been termed *illegal vowel devoicing* in Standard Japanese (Kinoshita 2001). In Standard Japanese, it is well known that the front and back high vowels /i/ and /ɯ/ can regularly disappear under certain phonologically well-defined circumstances, for example when they are in syllables that do not carry a pitch accent, and are surrounded by voiceless consonants (Vance 1987: 48–55). So, for example, the Japanese word for *person* is /hitoˀ/. It has a high front vowel /i/ in the first syllable surrounded by voiceless consonants h and t, and, since the second syllable carries the pitch accent (marked by the ˀ), the syllable in which the /i/ appears is unaccented (Japanese pitch accent was explained in Chapter 6). If pronounced carefully, *person* is [çito], with an [i], but in normal speech, the [i] disappears: [çto]. This is known as vowel devoicing. Now, since this is a normal part of Standard Japanese phonology, everybody can be expected to do it and so there is no potential for speaker identification.

However, some speakers devoice high vowels in the 'wrong' environments. For example, the proper name *Tokushima* should be said [tokɯˀçima] with a [ɯ], but some speakers say [tokçima] without a [ɯ]. The /ɯ/ should not be devoiced because, although it occurs between two voiceless consonants, [k] and [ç], it also occurs in a syllable that carries a pitch accent. Another example is the word for *zoo*, /doobɯtsɯˀen/, normally said [doːbɯtsɯɛn] with a second [ɯ]. Some speakers illegally devoice the second /ɯ/ and say [doːbɯtsɛn]. This should not happen because the second /ɯ/ does not occur between two voiceless consonants. Since some speakers do this and some do not (i.e. there is between-speaker variation), it has potential as a forensic-phonetic parameter, depending of course on how consistent individual speakers are in their illegal devoicing behaviour.

Suppose that two Japanese speech samples under forensic comparison agreed in the amount of illegal devoicing in words like *Tokushima* and *doobutuen*. Perhaps all words in both samples showed illegal devoicing, or only 50%, or none. How could that similarity be evaluated?

Kinoshita (2001: 133–43) noted and calculated the incidence of illegal devoicing in the two words *Tokushima* and *doobutuen* for 11 Japanese males across two recording sessions. This is shown in Table 11.1. The two recording sessions were separated by about a fortnight, and in each recording session the words were elicited in natural speech twice. Thus, for example, 0% means that no tokens within a session had devoiced /ɯ/; 25% means that two out of the four tokens were illegally devoiced; 100% means that all four tokens in the sample had devoiced /ɯ/.

Table 11.1 Frequency of illegal devoicing (%) for 11 Japanese males across two recording sessions (after Kinoshita (2001: 135))

Speaker	Recording session 1	Recording session 2
AA	25	25
HA	100	100
JN	100	75
KA	100	75
KF	50	75
KH	0	0
KO	50	50
MN	0	0
TN	75	75
TS	100	75
TY	75	75

It can be seen from Table 11.1 that there are between-speaker differences in devoicing behaviour, and that there is also a certain amount of consistency within a speaker: 7 out of the 11 speakers show the same rate across both recordings.

These figures, together with the formula for the likelihood ratio, can be now used to give an indication of the answer to the forensic-phonetic question posed above: how to evaluate a case where both samples agree in the amount of illegal devoicing. Using the formula for the likelihood ratio, we have: LR = (probability of observing the same amount of illegal devoicing in both samples assuming they are from the same speaker) / (probability of observing agreement in illegal devoicing assuming that samples come from different speakers).

Table 11.1 shows that 7 out of the 11 speakers were consistent in the amount of illegal devoicing, which gives a probability for the numerator of 64%. With 11 speakers and two sessions there are 220 different-speaker pairings; 34 of these, or about 15%, agreed in amount of illegal devoicing (Kinoshita 2001: 141, 142). Therefore the probability of observing agreement in illegal devoicing assuming different speakers is about 15%. The LR for the data is then 64%/15% = 4.3. One would be, on average, 4.3 times more likely to observe agreement in illegal devoicing if both samples came from the same speaker than if they came from different speakers.

The LR for observing a difference in amount of devoicing between samples can be estimated in the same way. The probability of observing disagreement in illegal devoicing assuming same speakers is 4/11 = 36%; the probability of observing disagreement assuming different speakers is 186/220 = 85%. The LR for two samples showing differing degrees of illegal devoicing is then 36%/85% = 0.42. One would be, on average, 1/0.42 = about 2.4 times more likely to observe disagreement in illegal devoicing if both samples came from different speakers than if they came from the same speaker.

According to these figures, then, observing agreement in illegal devoicing between two samples would offer limited support for the prosecution hypothesis (a LR of 4.3),

Forensic Speaker Identification

whereas observing disagreement in illegal devoicing would offer limited support for the defence (a LR of 0.4).

Calculation of likelihood ratio with continuous data

The first thing to understand in the calculation of a likelihood ratio is that it is necessary to take into account not just the size of the difference between the questioned and suspect samples – that is, how *similar* they are – but also how *typical* the questioned and suspect samples are of the reference population (the group of speakers to which they belong). The notions of similarity, typicality and reference population will be explained below, and then combined with the ideas of distributions and probability discussed in Chapter 8 to give an example of what this entails. The general idea will be outlined first, and then an example will be given using figures from actual data.

Three basic ideas

Assume that questioned and suspect samples have been compared with respect to several linguistic/phonetic and acoustic features, and the inevitable differences quantified. In order to evaluate these differences in terms of a likelihood ratio, it is necessary to know two things. One is a measure of *similarity*: how big are the differences between the two samples? The other is a measure of *typicality*: how typical of the group of speakers as a whole is the combined sample from questioned and suspect? The necessity for both types of information, and not just the magnitude of the difference between the samples, can be seen from the following scenario.

Assume that the questioned and suspect voice samples have been measured for an acoustic parameter, for example long-term mean F0 (LTF0), and the difference between the LTF0 of the two samples found to be 5 Hz. Assume further that we know that the average LTF0 for Broad Australian is 120 Hz, and that LTF0 values are normally distributed, with a standard deviation of 20 Hz. These figures would have been obtained by measuring the mean LTF0 of each of a very large sample of Broad male speakers, and then calculating the mean and standard deviation of those means. The mean LTF0 of each speaker would also have had to be based on speech made at several different times in order to correctly reflect within-speaker variation.

The large set of LTF0 measurements from the many Broad speakers we will call the *reference sample*, because it is with reference to them that the differences between the questioned and suspect samples have to be evaluated. Provided the reference *sample* is large enough, statistical theory tells us that it will provide good estimates of the important statistical parameters – mean, standard deviation, etc. – of the reference *population* of all mean LTF0 measurements of all Broad Australian male speakers. This is why, given the fact that it is usually unrealistic to imagine measuring a whole population (statistical or actual), a large sample is taken from the population. Although *sample* and *population* strictly speaking mean different things, to avoid unnecessary confusion I have not distinguished between *reference sample* and *reference population* below and have simply used the first term.

It will be recalled from Chapter 8 that one standard deviation above and below the mean of a normally distributed population can be expected to contain about 68% of the observations, so about 68% of speakers in the reference sample would have LTF0 values around the mean between 100 Hz and 140 Hz. In other words, if we were to pick an individual speaker at random from this sample, we would expect that 68 times out of one hundred he would have a LTF0 between 100 Hz and 140 Hz. Another 13% of speakers would have values below the mean between 100 Hz and 80 Hz, and another 13% would have values above the mean between 140 Hz and 160 Hz. Thirteen times out of a hundred we would expect to pick a speaker at random with a LTF0 between 100 Hz and 80 Hz and thirteen times out of a hundred we would expect a randomly chosen speaker between 140 Hz and 160 Hz. A few individual speakers in the reference sample would be found with LTF0 measurements lower than 80 Hz, and some with measurements higher than 160 Hz: we would expect them to be picked very infrequently at random.

Now, in terms of the range of values of the Broad-speaking reference sample, the observed difference of 5 Hz between questioned and suspect samples appears very small. Taking two standard deviations above and below the mean as a suitable indicator of range, the 5 Hz difference is 5/80 = ca. 6% of the sample range. Questioned and suspect samples can therefore be said to be very similar in terms of their LTF0. However, if the questioned sample was 120 Hz and the suspect 125 Hz, their average value would be ca. 123 Hz, which is, assuming an average for the sample of 120 Hz, by definition very *typical* for the reference sample. It is a difference that you would expect to occur very very often if samples were taken at random. Clearly, even though the questioned and suspect samples are exceedingly similar, their very typicality makes the strength of the evidence on its own against the suspect weaker.

Now assume that the values for the questioned and suspect samples are 80 Hz and 85 Hz. Given our reference sample with an average of 120 Hz and a standard deviation of 20 Hz, these two values of 80 Hz and 85 Hz are clearly at one extreme of the distribution. In probability terms, it would now be highly unlikely for two values of this size to be drawn at random from the population. Now the evidence against the suspect is much stronger. Clearly, then, *the magnitude of the difference between questioned and suspect samples is not enough: the typicality of the two samples measured against a reference sample needs also to be taken into account.* These ideas of similarity and typicality can now be explained quantitatively, and in greater detail, and the notion of reference sample also requires some additional discussion.

Assume that the voice of a male offender with a Broad Australian accent has been recorded and measured with respect to a particular acoustic feature. These measurements will have a mean and a standard deviation. Assume too that a suspect voice, likewise Broad and male, is also measured for these features, yielding a suspect mean and standard deviation. Finally, assume that the mean and standard deviation is known for a representative reference sample of male speakers of Broad Australian English. As mentioned above, these would have been calculated from many individual speakers' mean values. The questioned and suspect samples are then assessed using the two criteria of similarity and typicality. These will be discussed in turn.

Similarity

Firstly, and intuitively, the size of the difference between the questioned and suspect samples will be important – the larger the difference between the mean values of the

Forensic Speaker Identification

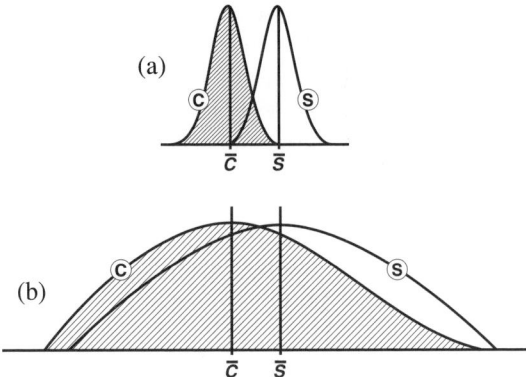

Figure 11.1 Two pairs of samples (a and b) whose means ($\bar{c}$ and $\bar{s}$) are separated by the same amount, but which differ in the spread of their distribution (standard deviation)

two samples, *ceteris paribus*, the more likely it is that two different speakers are involved. Basic statistical theory tells us, however, that *the assessment of similarity also involves taking the samples' distribution into account, since two samples that are separated by a given amount and have largely overlapping distributions are more similar than two samples that are separated by the same amount and have non-overlapping distributions.*

This situation is shown in Figures 11.1a and b. Figure 11.1 shows the distribution of two pairs of questioned and suspect samples: one questioned–suspect pair on the top (a) and one questioned–suspect pair on the bottom (b). The outline of the distribution of each sample is indicated by a curved line marked c (for questioned sample) and s (for suspect sample), and the samples' mean values are shown by $\bar{c}$ (for mean of questioned sample) and $\bar{s}$ (for mean of suspect sample). The questioned and suspect means in the (a) pair have been drawn to be separated by the same amount as in the (b) pair, thus $(\bar{c} - \bar{s})_{\text{pair a}} = (\bar{c} - \bar{s})_{\text{pair b}}$, so from this point of view both pairs are equally similar. However, it is also clear that the (b) pair of samples has largely overlapping distributions, whereas the (a) pair does not.

One interpretation of the high degree of overlap in the (b) pair is that it is because both samples have been taken from the same underlying group, or population. For this reason the (b) pair is more similar than the (a) pair, even though they are separated by the same distance. The statistic that quantifies the spread of a distribution is its standard deviation (discussed in Chapter 8). In evaluating the degree of similarity between questioned and suspect samples, therefore, both the difference between the sample means, and the sample standard deviations have to be taken into account. The questioned and suspect samples in pair (b) in Figure 11.1 would be considered to be more similar than those in pair (a) for example.

The number of items in the respective samples is also important and enters into the estimation of the similarity between the samples. It was pointed out in Chapter 2 that this is because we can be more confident about what the true, i.e. population, mean and standard deviation are the more observations are available. It is doubly important given the phenomenon of within-speaker variation. Thus, the higher the number of items in the sample from which to calculate the sample's mean and standard

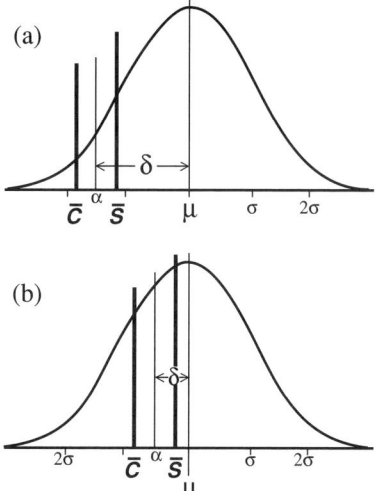

Figure 11.2 An illustration of the typicality criterion for the calculation of a likelihood ratio (a) = less typical, (b) = more typical

deviation the better: if we have, say, ten LTF0 values for the offender from different conversations, and ten from the suspect this is better than just one apiece.

Typicality

It is also necessary to know for the estimation of a LR how typical the questioned and suspect means are from the point of view of the reference sample of speakers. They are likely, of course, to have values that are near the mean value for the population, since that is where the majority of the population lies. However, it might be the case that they are less typical, and lie farther from the population mean.

This idea of typicality is illustrated in Figure 11.2a and b. The top (a) and bottom (b) parts of Figure 11.2 show two normal curves which both represent the same distribution of a particular acoustic parameter, say LTF0, in the reference sample. The mean value for the reference sample is marked by μ, and there are ticks above and below this mean at one standard deviation intervals (points one standard deviation away from the mean are shown where possible by σ; points two standard deviations from the mean are marked with 2σ).

The position of the questioned and suspect means with respect to the distribution of the reference sample is indicated by thick vertical lines. As in the previous example, the mean of the questioned sample is marked by $\bar{c}$ and the mean of the suspect sample by $\bar{s}$. In both (a) and (b) curves, the difference between the questioned and suspect means (the thick vertical lines) can be seen to be the same. However, in the top curve both questioned and suspect means lie nearer the lower tail of the distribution of the reference sample. In the (a) curve the questioned and suspect means lie between two standard deviations and one standard deviation below the mean value of the reference sample, whereas in the (b) curve, the questioned and suspect means lie within one standard deviation below the reference sample mean. A value α is shown intermediate between the questioned and suspect means, and the distance δ between α and the reference sample mean μ is a measure of the typicality of the questioned and suspect means: a small value for δ means more typical, a large value means less typical. In (a) the value for δ is greater than for (b), so the combined questioned and suspect samples are less typical of the reference sample.

Reference sample

In the preceding section was introduced the idea of a reference sample, against whose distributional statistics the typicality and similarity of the questioned and suspect speech samples are to be evaluated. It is important to realise that what constitutes the reference sample is not invariant, and will depend on the case. In particular, it will depend on the nature of the defence, or alternative, hypothesis. For example, a plausible narrowing in the defence account of the voice evidence will often be from 'It was the voice of (an Australian male) other than the accused' to 'It was the voice of some other Australian male who *sounds like* the accused'. Under the first hypothesis, the reference sample will be all other male Australian speakers; under the second, it will be all male Australian speakers who sound like the accused.

Such a change in the defence hypothesis can have a considerable effect on the LR. For example, the probability of observing a high degree of similarity between two voice samples is going to be very high under the prosecution hypothesis that they come from the same speaker. However, the probability of observing a high degree of similarity is also likely to be high under the defence hypothesis that the samples come from two speakers that sound similar – certainly higher than if the defence were simply that it was just someone else speaking. Thus the value for the LR $[p(E \mid H_p)/p(E \mid H_d)]$ will tend towards unity under the similar-sounding speaker defence, and will consequently contribute less support for the assertion that the same speaker was involved.

This does not mean that the similar-sounding speaker hypothesis will automatically work in favour of the defence, however, for two reasons. Firstly, changing the hypothesis changes the prior odds. If the (Australian male) questioned voice belonged to someone else in Canberra, say, the prior odds would be ca. 144 000 to 1 against it being the accused (assuming 360 000 Canberrans, two-fifths of whom are adult males). These odds would shorten considerably if they reflected only the number of adult males in Canberra who actually sound like the accused. The second reason – a rather important one which was explained in the previous chapter asking 'What is a voice?' – is that *just because two voices sound similar to a lay ear does not necessarily mean that they are similar in all their phonetic characteristics that a trained ear will register*. Thus *it is not necessarily the case that the LR will be considerably higher for sound-alikes than for non-sound-alikes*.

A likelihood ratio formula

A formula for the calculation of the likelihood ratio from continuous data of the type often illustrated in this book (for example F0 or formant frequencies) is given at (11.1). This is the formula used in Aitken's 1995 book *Statistics and the Evaluation of*

Evidence for Forensic Scientists to demonstrate the calculation of a LR to compare refractive indices of glass fragments from a window broken by an offender with fragments found on the suspect. The formula appears daunting, and the reader is not expected to understand it in its entirety. However, it has been included so that the reader can see how it is constituted, and in particular which parts contribute to the notions of similarity and typicality just discussed.

In this formula, the strength of the evidence is reflected in how much the LR is greater or lesser than unity. As already explained in Chapter 4, a LR value of 20, say, means that one would be 20 times more likely to observe the difference between questioned and suspect samples if they came from the same speaker than from different speakers. A LR value less than 1 would indicate that one is more likely to observe the difference if different speakers were involved. A value of unity for the LR means one would be just as likely to observe the difference between the samples if the same speaker were involved as if different speakers were involved, and the evidence is valueless.

Formula for the Likelihood Ratio with continuous data (after Aitken 1995: 180).

$$\text{LR} \approx \frac{\tau}{a\sigma} \times \underbrace{e^{\left[-\frac{(\bar{x}-\bar{y})^2}{2a^2\sigma^2}\right]}}_{\text{similarity term}} \times \underbrace{e^{\left[-\frac{(w-\mu)^2}{2\tau^2} + \frac{(z-\mu)^2}{\tau^2}\right]}}_{\text{typicality term}} \quad (11.1)$$

$\bar{x}$ = mean of questioned sample; $\bar{y}$ = mean of suspect sample
μ = mean of reference sample
σ = standard deviation of questioned and suspect samples
τ = standard deviation of reference sample
$z = (\bar{x} + \bar{y})/2$
$w = (m\bar{x} + n\bar{y})/(m + n)$
m = number in questioned sample
n = number in suspect sample
$a = \sqrt{1/m + 1/n}$

It can be seen that the formula to the right of the approximate equality sign ($\approx$) is made up of three parts multiplied together. Thus the value for the LR is the product of three quantities. The last two quantities, which each consist of e raised to a fraction, represent the terms for similarity and typicality, and are so marked. Parts of the formula can be easily related to the explanations of similarity and typicality just discussed. Thus the ($\bar{x} - \bar{y}$) part of the similarity term represents the difference between the questioned and the suspect means, and the parts of the typicality term that relate to the distance of both questioned and sample means from the overall mean are ($w - \mu$) and ($z - \mu$) (z is the average of questioned and suspect means, and μ is the mean of the reference sample.)

The first item after the approximate equality sign shows how much bigger the standard deviation of the reference sample (τ) is than the standard deviation of the questioned and suspect samples (σ). The numerator τ reflects the amount of variation present in the reference sample, and can be visualised in Figure 11.2, where the distance between μ and σ is one standard deviation. Larger values for τ indicate a wider spread and greater variation within the reference sample for this parameter. The variation in the questioned and suspect samples denoted by the denominator σ can be visualised in Figure 11.1a and b. In Figure 11.1a, the variation around the mean is very small for both questioned and suspect samples compared to the variation in Figure 11.1b. The a in the denominator $a\sigma$ is a term to adjust for the number of items in the questioned and suspect samples.

The calculation of the $\tau/a\sigma$ term can be illustrated with actual values for the refractive indices of glass. If the standard deviation of the reference sample τ was 0.004, and the standard deviation of the questioned and suspect samples σ was 0.00004, assuming a value for a of 0.5477, the value for the first term $\tau/a\sigma$ is $0.004/(0.00004 \times 0.5477) = 182.6$. This means the variation in the reference sample is about 180 times greater than in the questioned and suspect samples, and the LR for evidence with this type of glass would include a multiplication of the product of the similarity and typicality terms by a (large) factor of ca. 180.

This ratio of variation of the reference sample to that of the questioned and suspect samples is an important term, because it controls to a large extent the range of possible LRs, that is, how powerful the LR can get for a particular parameter. It can be appreciated that if the spread of the distribution in the reference sample is large relative to the spread of the individual questioned and suspect samples under comparison, this will result in the possibility of a large value for the first term $\tau/a\sigma$ and a large LR.

An important point arises from consideration of the first term in the LR equation. It is this. The nature of acoustic speech parameters like formant frequencies and F0 is such that the spread of the reference sample is usually not very much greater than that of individual samples. For example, the population standard deviation for a male F2 value in /ʉː/ may be ca. 150 Hz, compared to ca. 100 Hz for an individual male. Thus the contribution to the magnitude of the LR from the first term in the LR formula is not likely to be very big, owing to the nature of speech acoustics. Disregarding the adjustment a for the number of items in the samples, the contribution of this term for the F2 example would be $150/100 = 1.5$, which is smaller by an order of magnitude than the glass fragment example. It is to be expected therefore that the range of possible LRs from speech acoustics will thus be consequently constrained, and the maximum strength of evidence from one parameter limited. It must not be forgotten, however, that one of the advantages of the LR is its combinability, and, given the inherent multidimensionality of voices, even if the LRs from several individual parameters are low, their *combined* LR might not be. That this is so will be demonstrated in the following section.

Applications

Two examples will now be given of the calculation of an overall LR with actual data. The first, using the word *hello*, is an extremely restricted example involving few speakers, few comparisons and few parameters. It will help the reader to follow the second, more extensive, example from Japanese. This uses more parameters, with twice the number of speakers and hence very many more comparisons. The idea in both these examples is to simulate a situation where a suspect sample is being compared acoustically, using the LR formula discussed above, to a questioned sample.

A demonstration of the method

It is of interest to see two things from these examples. Firstly, the demonstrations will show how the LR is actually calculated, and what its magnitude might be if the figures were *from particular cases*. This is because the LR is the figure that would allow us to quantify the strength of the evidence *in a particular case*. Secondly, and probably of more immediate interest for the legal profession, it is also intended to see to what extent the LR actually reflects reality. *If the LR approach works, likelihood ratios greater than unity should occur for the same-speaker pairs, and LRs of less than unity should occur for the different-speaker pairs*. We shall see that this is to a large extent true of the data, and that the LR can thus be used as a measure to discriminate between same-speaker and different-speaker pairs.

For ease of presentation, the results of the calculations will be given for both *hello* and Japanese examples first. Then the limitations and shortcomings of the demonstration, and current LR approaches in general, will be discussed.

Calculating likelihood ratios with hello

This example uses two acoustic parameters from the word *hello* where the author's research on similar-sounding speakers (Rose 1999a,b) has suggested that the ratio of between-speaker to within-speaker variance is relatively high, and where the highest realistic discrimination rate occurs. *Hello*s from two same-speaker pairs and from two different-speaker pairs were compared. The same-speaker *hello*s were taken from recording sessions separated by at least one year. Thus we are truly dealing with non-contemporaneous within-speaker variation, although it is perhaps unrealistic to have such a long hiatus separating questioned and suspect samples.

The constitution of the corpus in terms of speaker, recording session and number of *hello*s is given in Table 11.2. Table 11.2 shows, for example, that one of the same-speaker pairs uses speaker RS, who has 7 *hello*s from his recording session 1 tested against 9 *hello*s from his recording session 2.1. (RS's recording session 2.1 was carried out one year after his session 1.) One of the different-speaker pairs consists of 7 *hello*s from speaker EM's recording session 2.1 compared against 9 *hello*s from speaker PS's recording session 2.1.

As mentioned, the *hello*s were compared with respect to LRs derived from two acoustic parameters. Both of these involve the second formant centre frequency. One is the F2 at the second target in the /aʉ/ diphthong ('F2 /aʉ/'), and the other is the F2 in the /l/ ('F2 /l/'). The values for these two parameters in a single token of *hello* can be seen in the spectrogram of a *hello* in panel (c) of Figure 8.25, where the centre frequency of F2 in /l/ is ca. 900 Hz and the centre frequency of F2 at the second diphthongal target looks to be about 1650 Hz (panel (d) shows that when estimated digitally it is 1701 Hz).

Table 11.2 Speakers, recording sessions and number of *hello* tokens used to demonstrate calculation of LR

Same-speaker pairs	Different-speaker pairs
RS 1 (7) vs. RS 2.1 (9)	EM 2.1 (7) vs. PS 2.1 (9)
MD 1 (6) vs. MD 2.2 (9)	JM 1 (5) vs. DM 2.2 (9)

The pairs of speakers, and the particular recording sessions, were deliberately chosen, on the basis of similarity and difference in their mean F2 /aʉ/ values, to yield probabilities that are likely to both agree and be in conflict with the known reality. It is often a good idea to bias an experiment against the hoped-for outcome, thus making possible an *a fortiori* argument. Here I have biased the experiment both for and against. Thus one same-speaker pair (speaker RS) was chosen because RS has very similar mean values for the F2/aʉ/ parameter across recording sessions (1293 Hz and 1230 Hz), and one same-speaker pair (speaker MD) was chosen because MD has very different values (1694 Hz and 1553 Hz). This is because MD typically has rather large within-speaker variation.

The same-speaker comparison with RS is therefore biased to show a LR greater than unity in accordance with the facts; the same-speaker comparison with MD is biased against the LR showing the true relationship. Likewise, one different-speaker pair (EM, PS) has fairly different mean values for F2 /aʉ/, and one different-speaker pair (JM, DM) has exceedingly similar values (1494 Hz and 1490 Hz). The different-speaker comparison with EM and PS is therefore biased to show a LR less than unity in accordance with the facts; the different-speaker comparison with JM and DM is biased against the LR showing the true relationship.

To summarise so far, we are going to see a demonstration of how the LR is applied in four separate comparisons. Two of the comparisons involve non-contemporaneous samples from the same speaker, and two involve samples from different speakers. The samples themselves all consist of several tokens of the word *hello*, and the LR will be estimated using two acoustic parameters from these *hello*s.

Results of likelihood ratio comparison on hello

Table 11.3 shows the results of the four paired comparisons based on LRs. In the column headed *condition*, the type of comparison – whether same-speaker or different-speaker pair – and the names of the samples are given. The next column shows the parameter: F2 /aʉ/ (the frequency of the second formant at the second diphthongal target) or F2 in /l/ (the frequency of the second formant in /l/). The next three columns, headed $\bar{x}$, SD and n, show the values for the designated questioned sample: mean, standard deviation and number of *hello* tokens, and the next three show the corresponding values for the designated suspect sample. Thus it can be seen that RS's first sample (RS 1) contained seven *hello*s, and that they had a mean of 1293 Hz for F2 /aʉ/, compared to the mean value of 1230 Hz for his second set of nine *hello*s one year later (RS 2.1). His values for F2 /l/ are in the row immediately below this. The mean and standard deviation values for the reference sample (μ, σ in the LR formula at (11.1)) are given in the caption at the top. (The reference sample consisted of data from the six speakers being compared. This is of course both unrealistic and entirely inappropriate, since test and reference data should be from different speakers.)

Same-speaker pair RS/RS

The likelihood ratios for the two parameters are in the rightmost-but-one column. Thus in the first same-speaker comparison (RS 1 vs. RS 2.1) the LR for F2 /aʉ/ was 3.32. This means that, using the parameters from the reference sample, one would be

A demonstration of the method

Table 11.3 Evaluation of likelihood ratios for same-speaker and different-speaker comparisons using two acoustic parameters in *hello*. Reference sample mean and standard deviation = 1464 Hz and 142 Hz (F2 /aʉ/); 966 Hz and 95 Hz (F2 /l/)

Condition	Parameter	'Questioned sample'			'Suspect sample'			LR	Combined LR
		$\bar{x}$	SD	n	$\bar{x}$	SD	n		
1 Offender, suspect Same speaker (*RS 1 vs. RS 2.1*)	F2 /aʉ/	1293	104	7	1230	75	9	3.32	5.34
	F2 /l/	1011	77	7	1056	85	9	1.61	
2 Offender, suspect Different speaker (*EM 2.1 vs. PS 2.1*)	F2 /aʉ/	1533	94	7	1415	69	9	0.07	0.06
	F2 /l/	1040	40	6	1090	57	10	0.86	
3 Offender, suspect Same speaker (*MD 1 vs. MD 2.2*)	F2 /aʉ/	1694	87	6	1553	116	9	0.30	0.06
	F2 /l/	851	46	6	917	44	9	0.20	
4 Offender, suspect Different speaker (*JM 1 vs. DM 2.2*)	F2 /aʉ/	1494	41	5	1490	59	9	4.5	2.5×10^{-6}
	F2 /l/	776	25	5	967	67	9	5.7×10^{-7}	

just over three times more likely to observe this difference in F2 /aʉ/ if the values came from the same speaker than if they came from different speakers. The LR for F2 /l/ in this comparison was 1.61. The combined LR has been calculated as the product of the two LRs on the assumption that they are independent (more on this below). Thus assuming effective independence of the parameters, their combined LR will be 3.32 × 1.61 = 5.34, and this is given in the rightmost column.

Taking both parameters into account, then, one would be just over five times more likely to observe the differences if this was a same-speaker rather than a different-speaker pair. If these were data from a real case, and ignoring the prior odds, this LR of 5.3 would constitute weak support for the prosecution. It can also be noted that, as this is indeed a same-speaker pair and the LR is bigger than 1, the LR agrees with the reality.

Different-speaker pair EM/PS

The second comparison, with the different-speaker pair of EM and PS, has a combined LR of considerably less than unity, at 0.06. This means that in a real case the observed differences would be (1/0.06 =) 16.7 times more likely assuming different speakers: weak support for the defence. Again, note the LR is less than 1 and in accord with the actual event (*hello*s from different speakers).

Forensic Speaker Identification

Same-speaker pair MD/MD

As intended, the case with MD is not so good: his same-speaker comparison has the same LR as the EM/PS comparison: nearly 17 times more likely if the speakers were different! This is clearly because MD has a wide range of values for these parameters, especially F2 /au̶/. This is a nice example of the dramatic effect of between-speaker differences in variation mentioned in Chapter 2. Since the LR is less than 1, it is not in accord with reality (*hello*s both from MD).

Different-speaker pair JM/DM

The final comparison, with different speakers JM and DM, was also intended to yield a high LR because of the almost identical value for F2 /au̶/ of the two speakers (1494 Hz vs. 1490 Hz). Indeed, it can be seen that the LR for this parameter is 4.5, showing that one would be between four and five times more likely to observe this difference *were the samples from the same speaker*! (The relatively low LR value, given the exiguous difference of 4 Hz between the two speakers' F2 /au̶/ values, is because the means are typical of the reference population.)

However, JM has a uvularised lateral, which is reflected in a very low F2. He also tends to show relatively little within-speaker variation across recordings. Together, these factors mean that the LR for the F2 in /l/ for the DM/JM comparison is astronomically low: 5.7×10^{-7}, and dramatically reverses the 4.5 LR from the F2 /au̶/ parameter. One is now $1/(2.5 \times 10^{-6})$, or about 400 000 times more likely to observe this difference if the samples were from different speakers. Ignoring prior odds, this would of course constitute extremely strong evidence for the defence. This is a good example of the necessity of combining LRs from different parameters in forensic phonetics: just considering the LR based on the F2 in /au̶/ would have been extremely misleading.

These *hello* examples have shown how LRs can be calculated from some actual pairs of samples to yield LRs of very different magnitudes that could then be interpreted as giving varying degrees of support for the defence or prosecution hypothesis. In three out of the four cases, the LR also corresponded to the reality.

Calculating likelihood ratios with Japanese vowel formant data

The Japanese example is like the *hello* example just discussed but it is more complicated, involving more speakers and more parameters. It is taken from an actual controlled experiment in Kinoshita (2001) designed to investigate how well same-speaker and different-speaker pairs could be discriminated using formants. In this experiment, care was taken to simulate typical forensic conditions, with natural speech recorded at different times and including some comparable vowels in different words, and one example of the same word.

Ten male Japanese speakers were recorded twice, the recording sessions being separated by about two weeks. In both recording sessions, the speakers interacted with the investigator by carrying out verbal tasks like answering where a bus stopped on a particular map of a town they were given. In this way, several tokens of many words containing selected vowels could be obtained from natural speech. The

A demonstration of the method

recordings were then analysed acoustically, many parameters were measured and analysed statistically, and six acoustic parameters were selected from these on the basis of their high F-ratios (high ratio of between-speaker to within-speaker variance) and lack of correlation.

Three of these parameters were taken from the single Japanese word *moshimoshi*. *Moshimoshi* is used to answer the phone in Japanese, much like *hello*. The moshimoshi parameters were all third formant frequencies: in the first o, the second m, and the second sh (thus: m**o**shi**m**o**shi**). The other three parameters selected were F2 in /i/, and F2 and F3 in /e/. Unlike the *moshimoshi* parameters, however, these parameters were simply measured from vowels in different words, like n**e**moto *name*, t**e**rebi *television*, sush**i**ya *sushi shop*, j**i**nja *shrine*. As can be seen, these vowels all occurred in different phonological environments. Thus the *e* in the first word occurred between *n* and *m*, but in the second word it occurred between *t* and *r*. All the vowels did share the characteristic of occurring in the same prosodic environment, however, in that they all occurred on pitch-accented syllables. The overall situation was then similar to that normally found in the acoustic comparison of forensic samples, where vowels from the same word can be compared and also vowels from different words, but in the same prosodic environment can be compared.

All same-speaker and different-speaker pairs in the data were examined, and overall LRs were calculated. This was done in the same way as with the two same-speaker and two different-speaker pairs in the *hello* example above, except that now the overall LR was calculated as the product of the individual LRs for not just two, but each of the six parameters. A further, and very important, difference from the *hello* example was that, in the Japanese case, the reference sample was always independent of the test sample. That is, every same-speaker and every different-speaker pair of Japanese samples was evaluated against a reference sample that did not contain data from the speakers being compared.

A very few results – for two same-speaker pairs and two different-speaker pairs, as for the *hello*s – are shown in Table 11.4. In Table 11.4, the speaker pairs are shown

Table 11.4 Examples of individual and overall likelihood ratios for same and different pairs of Japanese speakers. Pairs with the highest and lowest correct overall LR are shown (after Kinoshita 2001)

Pair	Other words			*Moshimoshi*			Overall LR
	i F2	e F2	e F3	o F3	m F3	sh F3	
				Individual LRs			
Same-speaker							
HA/HA	33.911	9.102	31.342	3.571	0.739	8.188	**209 034**
TS/TS	2.667	4.042	0.965	7.396	0.015	1.236	**1.4**
Different-speaker							
TN/KF	1.982	4.299	0.995	0.497	0.144	0.324	**0.2**
JN/HA	1.4^{-20}	5.6^{-15}	1.0^{-107}	0.53	8.3^{-16}	9.79^{-5}	**3.38^{-161}**

down the left, divided into same-speaker and different-speaker. Along the top are shown the six acoustic parameters, divided into the three from the single word *moshimoshi* and the three from the vowels of different words. Under each parameter is its corresponding LR, and at the extreme right is the overall LR, which is the product of the six individual LRs. Thus it can be seen that when HA's first sample was compared with his second sample two weeks later (HA/HA), the overall LR was ca. 209 000. If this were a real case, it would constitute strong supporting evidence for the prosecution (209 000 times more likely to observe the difference between the samples if they were from the same speaker than if they were from different speakers.) When one of TN's samples was compared with one of KF's samples, however, the overall LR was 0.2: a LR of (1/0.2 =) 5 times more likely if the samples were from different speakers.

Several things are very clear in Table 11.4. For a start, *it is clear that the acoustic parameters of speech are capable of yielding some extremely large overall LR values.* For example, the overall LR for the JN/HA comparison, 3.38^{-161}, is an infinitesimally small figure. Because the pairs with the highest and lowest overall LR in the experiment were included in Table 11.4, an idea of the maximum range can be obtained.

Secondly, it is also clear that overall LRs from speech can vary considerably. For example, in same-speaker comparisons, speaker HA had the greatest overall LR of ca. 209 000, and speaker TS had a very low overall LR of 1.4. Likewise, *mutatis mutandis*, for the different-speaker pairs: 0.2 for the comparison between TN and KF is a very different figure from 3.38^{-161} for JN and HA!

Thirdly, it can be seen that, just as with the *hello* examples, the contribution of the individual parameters to the overall LR can be quite different. For HA's same-speaker comparison, for example, the F2 in /i/ and F3 in /e/ contributed the most to his overall LR, whereas one parameter, F3 in /m/, shows a LR of *less* than 1 – that is, typical of a different-speaker pair. For TS, F3 in /e/ is associated with a LR lower than 1. This all points to the well-known fact, already mentioned in Chapter 2, that not all parameters behave homogeneously in a given speaker, or across speakers. One parameter in a speaker might have a very typical value for the population, and not be very similar across repeats, yielding a relatively low LR. Another might show a greater similarity and be less typical, yielding a relatively high LR. Again it is clear that many parameters more than just one need to be considered.

The likelihood ratio as a discriminant distance

What will probably be of greatest interest to the reader is the extent to which the LRs correctly discriminated same-speaker and different-speaker pairs. It did this rather well, in fact. In this experiment with Japanese, 90 same-speaker comparisons were tested and 180 different-speaker comparisons. Out of the 90 same-speaker comparisons, 83 had LRs greater than 1, which means that 92% were correctly discriminated. And 174 of the 180 different-speaker comparisons had LRs smaller than 1, which is 97% correct. These are both quite good rates. It is interesting to note that it seems to be slightly more difficult to tell when you have two samples from the same speaker than when you have two samples from different speakers. If an equal error rate estimate is needed for the test, a slight adjustment to the threshold of 1 will yield 6% (that is a rate of 94% correct).

A demonstration of the method

That the use of the LR as a distance measure works rather well on forensically realistic speech material to discriminate same-speaker from different-speaker pairs is a very important result. This is because up to now, there has been an enormous amount of theoretical discussion on how a forensic-phonetic analysis *could* work, but very little demonstration that it actually *can*. There is to be sure often indirect evidence of the validity of the approach when the prosecution's forensic-phonetic identification goes unchallenged, or the prosecution is dropped on the basis of the defence's report. However, as was made clear in the US Supreme Court decision in *Daubert v. Merrell Dow*, it should be criterial for the admissibility of scientific evidence to know to what extent the method can be, and has been, tested.

The results of this experiment do give some confidence in the method, especially since it will be remembered that the results were obtained with only six acoustic-linguistic parameters. As explained in Chapter 3 on different types of parameters, this actually constitutes only perhaps a quarter of the evidence that can be expected to inform a forensic-phonetic judgement. There will be auditory-linguistic and non-linguistic parameters as well as non-linguistic acoustic parameters to consider.

A speechalyser?

In their book *Interpreting Evidence*, Robertson and Vignaux introduce the concept of the LR by describing a hypothetical breathalyser that is calibrated to be correct 95% of the time when the driver is actually over the limit (pp. 13,14). They then derive a LR for a positive breathalyser sample. A positive sample is 95%/5% = 19 times more likely if the driver is over the limit.

Why not a 'speechalyser' then? If we were able to calibrate tests like the Japanese example above, it would be possible to quote a LR for the test in the same way as the hypothetical breathalyser. It is possible to imagine a test for two speech samples where, just like the Japanese example, they are compared by calculating their LR and seeing if it is greater than or less than 1. If this test is calibrated to an error rate of, say, 8% for same-speaker pairs, then assuming that the samples come from the same speaker, the probability of observing a LR greater than 1 is 92%, and assuming that the samples come from different speakers the probability of a LR greater than 1 is 8%. This gives a LR *for the test* of 92/8 = 11.5. One would then be between 11 and 12 times more likely to observe the result with this test if the samples were from the same speaker than if they were from different speakers.

This approach seems promising, and it is in fact often used when background statistics are not available to calculate a proper LR for the data (which will often be the case in forensic-phonetic comparison). However, it has its own problems, and it is important to be aware of them in case this approach is used. These are discussed below.

Problems and limitations

Once again, a reality check is in order. Both the *hello* and the Japanese demonstrations of the calculation and use of the LR contain problems and shortcomings that must be pointed out. To do this, it is important to keep separate the two uses to which the

Forensic Speaker Identification

LR was put: its intended, *bona fide* use as a means to quantify the strength of evidence, and its use as a distance measure to discriminate same-speaker and different-speaker speech samples. We examine the limitations of the *bona fide* use first.

Limitations of bona fide *likelihood ratio example*

There is clearly a big difference for the court between the overall LRs of 209 000, as for example with HA's same-speaker comparison in Table 11.4 above, and the 1.4 that was TS's same-speaker LR. The former LR gives considerable support to the prosecution hypothesis; the latter, as near unity as makes no difference, is effectively worthless. The magnitude of the LR matters, and therefore a good estimate is vital.

The LR approach instantiates a theorem. If the formula is correct and the values that are plugged into it are correct then the output, i.e. the LR, will also be correct. The first point to be made, then, is that the estimate of the LR is only as good as the method used to derive it. Since there were some deficiencies in the methods used in the examples above, the actual LR values will not be correct, although given the results of the discrimination test, they are unlikely to be very far out.

The deficiencies in the method involve three things: the reference sample, the formula used, and the combination of individual LRs. The shortcomings associated with each of these will be discussed in turn.

Reference sample

The reference sample is the sample against which the similarity and typicality of the questioned and suspect samples is assessed. It is also called the background sample, or background data. The main problem with the reference sample is its representativeness. For the LR estimate to be good, the reference sample needs of course to reflect the relevant population. What the relevant population is, and therefore its size, depends on the defence hypothesis, but it is likely to be big. As explained in Chapter 2 in the section on means and small samples, if you want to obtain a reliable estimate of a parameter in the male population of broad Australian speakers you need to sample many speakers, not just 6 as in the *hello* example, or even 10 or 11 as in the Japanese examples.

Formula

An important problem is the estimation of the variance in the reference sample (symbolised as τ in the formula). If this is inaccurate, the magnitude of the LR will be incorrect. The reality of the situation is that forensic speech samples are compared from different occasions – if they were not it would be known who the offender was. As should be perfectly clear by now, there will be differences between the voice of the same speaker on different occasions. Therefore, the estimate of the variance for the reference sample must include a component that accounts for the within-speaker variation that arises from the same speaker speaking on different occasions.

As already pointed out, the LR estimates quoted above are obtained from a formula designed for the refractive indices of glass fragments, but a pane of glass is not like a speaker. Whereas it is not clear what kind of variation you get for the mean refractive index of a pane of glass from one situation to the next, there can be considerable within-speaker variation for a particular phonetic parameter. The variance within speakers needs to be incorporated, therefore. However, pooling all tokens for the relevant population of speakers to get an estimate of the typicality would improperly combine single-sample with long-term variation. The typicality term in the LR formula therefore needs to be modified so as to take into account such significant long-term within-speaker variation.

This is not totally straightforward. One complication is that the amount of variance can differ considerably for different occasions. For many parameters, the amount of variation in a sample of speech on a single occasion, say a short telephone call, is less than the amount of variation present in two samples of speech separated by a short period (e.g. two consecutive telephone calls), and this in turn is less than the variation in two samples of speech separated by a longer period, say two telephone calls separated by a month. Another complication is that there is between-speaker variation in the variation between samples on different occasions, so that there is between-speaker variation in the within-speaker variation.

The primary significance of this complex variation structure is that it would require a considerable amount of sophisticated statistical research, presumably using estimates of variance components (e.g. Box *et al.* 1978: 571–83) to determine the best way to model and incorporate such variance into a LR formula for speech. This has not yet been properly attempted.

A second important point is that the formula assumes that the parameters are normally distributed. This is not necessarily correct for speech. In the Japanese comparison, for example, F2 and F3 in /e/ were normally distributed (as were in fact most of the formants); however, F3 in /m/ was heavily negatively skewed and F3 in /o/ was clearly bimodal (Kinoshita 2001: 294, 295). This means that the LRs for F3 in /m/ and /o/ will not be correct, and also that more complicated statistical modelling, for example mixed-Gaussian, is required for the distributions. Bimodal distribution has also been shown to characterise the long-term F0 of some Japanese females (Yamazawa and Hollien 1992).

Combination of parameters

It was pointed out in Chapter 4 when introducing the likelihood ratio that the combination of LRs, and therefore the overall LR, depends on the extent to which the items of evidence on which they are based can be considered mutually independent. An example of two items of dependent evidence might be that both questioned and suspect are of aboriginal appearance (first item of evidence), and there is a match in suspect's and offender's ABO blood group (second item). The probability of a match in blood group depends to a certain extent on race – among native Australians the percentage of A group individuals is about 37%, compared for example to 25% for caucasians – therefore the two pieces of evidence are not independent.

The parameters involved in the Japanese and *hello* demonstrations are not of the same nature as in the above example: they are not categorical, as blood group, but continuous. In order to investigate dependence with respect to such continuous parameters, it needs to be determined to what extent they correlate. If one were to examine the degree to which F2 in /l/ and F2 in /aʉ/ are correlated in *hello*, one would find that it is not possible to demonstrate a significant correlation within a speaker, possibly because the ranges of the parameters are so small. Across speakers, however, there is a very slight negative correlation between the two parameters: speakers with a lower frequency for F2 in /l/ tend to have a slightly higher frequency for F2 in the second diphthongal target of /aʉ/, and vice versa. This in itself is strange, since one might expect that the correlation would be positive, reflecting overall vocal tract length. One possible interpretation, suggested by F. Nolan, is that the negative correlation reflects broadness of accent. Broader speakers might have a more strongly uvularised or velarised lateral, with its expected lower F2, and more fronted offglide to the diphthong, with its higher F2.

In the Japanese example, the levels of correlation between the six different parameters were also very low indeed – that is one reason why these six parameters were chosen in the first place – and they were therefore ignored (Kinoshita 2001: 269–77). Nevertheless, an accurate estimate of the LR in a particular case will need to take the actual degree of correlation into account, whether significant or not. This will almost certainly have an affect on the magnitude of the LR.

We now turn to the limitations in the example of the use of the LR as a discriminant distance.

Limitations with the likelihood ratio as a discriminant distance

The Japanese comparison above showed that, despite its theoretical limitations, the LR based on the glass refractive index formula could function well as a distance measure in a test to discriminate same-speaker and different-speaker pairs in forensically realistic speech. What are the limitations here?

Paucity of data

In the foregoing sections it was pointed out that the method suffered from inadequacies in reference sample. The main limitations in using the LR as a discriminant distance, too, have to do with lack of data. 'More data' can be understood in two senses. Firstly, although successful, the Japanese example is still just one test. Many more similar tests are needed to obtain an idea of the actual performance of the method and its associated confidence limits. This would then enable us to say, for example, that we would be 95% confident that the actual performance of the test lies between an error rate of 10% and 4%.

It is necessary to see how the test performs on more data, but not just in the sense of its replicability. More data also means more subjects. More subjects are needed because, although as pointed out in Chapter 5 verification performance approaches a constant with additional speakers, it is highly unlikely that that point has been reached with only 10! To the extent that these 10 speakers distribute in their parameters like the population, then the results may be comparable, and the performance figures of 92% and 97% for same-speaker and different-speaker pairs realistic. But we cannot know until we know how the population distributes from a large sample. It may be the case that, if 50 speakers had to be discriminated, the performance with six parameters would drop: perhaps only 60% of all pairs could be correctly discriminated.

A demonstration of the method

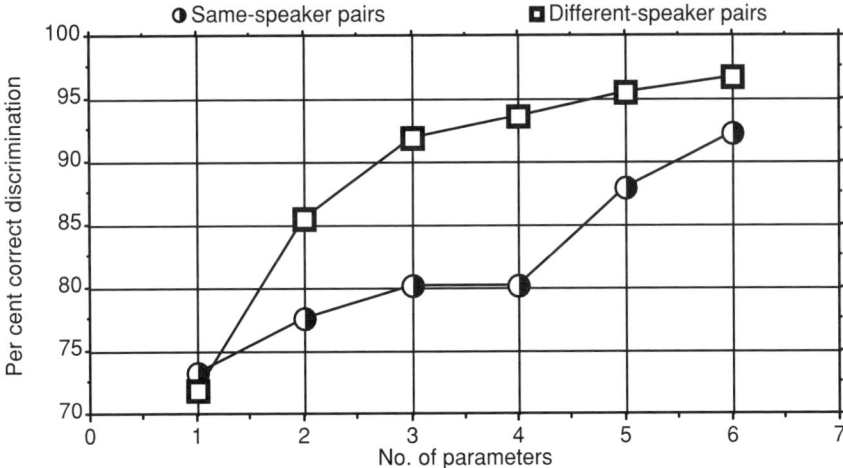

Figure 11.3 Mean per cent correct discrimination of same- and different-speaker pairs as a function of number of parameters used (adapted from Kinoshita 2001: 322)

One of the beauties of the LR approach, however, and one that is due to the multiparametric nature of the human voice, is that there is no theoretical upper limit on the number of parameters that can be combined. It can be shown with the Japanese example that increasing the number of parameters from one to six results in a gradual improvement in discriminant performance. That is, the more parameters that are considered, the more the LR is forced lower than a value of unity for different speaker pairs, and forced higher than a value of unity for same-speaker pairs.

This is shown in Figure 11.3, where the percentage of correct discriminations for both same-speaker and different-speaker pairs in the Japanese example is plotted against the number of parameters used. This plot was obtained by calculating average values for all possible discrimination performances with a given number of parameters. Thus the figures of 72% or 73% for the correct performance of a single parameter were obtained from the average performance for all six individual parameters (the average of p(arameter)1, p2, p3, ..., p6), and the discrimination performance figures for two parameters were obtained from the average of all possible permutations of two parameters (average of p1 and p2, p1 and p3, p1 and p4, p2 and p3, ..., etc).

It can be seen in Figure 11.3 that with only one parameter only about 72% or 73% of all pairs can be correctly discriminated, but that, with the exception of the hiccough between three and four parameters in the same-speaker pairs, the overall performance increases to above 90% for six parameters. (The slightly lower performance for same-speaker data already mentioned is also clear.)

Not only do the graphs increase, neither do they appear to have flattened out, so perhaps *seven* parameters on average might have discriminated the speakers even better than six. The message here, then, is that if more speakers need to be discriminated, it may be simply that more parameters are needed to achieve the same rates as with the six shown here. This is for future research to determine.

The shape of the graphs in Figure 11.3 also of course suggests another question. Is it possible that there is actually an optimum number of parameters, such that looking

323

at more than that number will not change the result? If this were so, forensic comparison might not involve the combination of so very many parameters after all, and not be so time-consuming as was the lament in Chapter 2.

Now, if the same-speaker curve is extrapolated further to the right using any of the normal techniques it will soon have intercepted the 100% line. This suggests that it might not be necessary to look at more than, say, ten parameters to be confident that a LR of more than one would not be reversed. On the other hand, no straightforward extrapolation of the different-speaker curve will intercept the 100% line: perhaps this indicates that on average one cannot achieve 100% correct discrimination rate, no matter how many LRs are combined. Whether there is in fact a lily-gilding effect with the number of LRs to be combined is yet another question for future research.

As always, caution is needed in the interpretation of these data. It must not be forgotten that the graphs in Figure 11.3 represent averages, so they do not show the worst possible rate that could be obtained. If instead of the average one took the *worst single* parameter, then the *worst two* parameters then the *worst three*, and so on, then the performance would not be so good. For example with same-speaker comparisons, the correct discrimination rate remains below 60% for four parameters, and only becomes useful (ca. 80%) after five have been combined. By the same token, however, neither do the average parameters represent the *best possible* performance. An 80% correct discrimination of same-speaker pairs has already been achieved with the two best parameters, for example, and only one best parameter (F3 in /s/) is needed to successfully discriminate *all* 180 different-speaker pairs.

Average probablities

In view of the lack of data on background statistics generally available in voice comparisons, a hypothetical 'speechalyser' was envisaged above that would enable us to derive a LR for the actual test that could then be used to quantify the strength of the evidence in a particular case (the LR for the test was 11.5.) What are the limitations involved here?

The speechalyser approach involves what is known as an *average probability* (Aitken 1991: 70–7) It lets us know how the test would evaluate pairs of speakers *on the average*. Averages characterise a set of figures by quoting a typical one – a figure in some sense in the middle, which can stand for all of the figures. The test therefore tells us that 'if you apply this test a lot of the results will cluster around an LR of 11.5'.

Averages can be very useful, but they do not give any idea of the range of figures they represent, or the figure *in a particular case*. This can be easily seen with the Japanese data. Table 11.4 showed the actual LR for HA's same-speaker comparison was about 209 000. For TS, it was just over 1. In HA's case, there is very strong support for the prosecution; in TS's the evidence is equivocal. But both these cases would have to be evaluated with the LR for the test (calculated above) of 11.5. In both cases, it can be seen that a considerable distortion can be involved (although admittedly this is somewhat exaggerated because HA and TS are speakers with extreme overall within-speaker LRs.) In HA's case the strength of the evidence is emasculated from 209 000 to 11.5; TS's case is more serious: support for the prosecution now appears where there really isn't any.

A demonstration of the method

Probably the greatest caveat concerning this type of approach is that it is only as good as the material used to calibrate the test. It was mentioned above that average probabilities are usually used when the background statistical data to compute a proper LR are lacking. In such cases, tests are conducted and discriminant performances are determined with whatever data are to hand (hair samples from your friends, for example, or Japanese vowel data to evaluate differences in Australian *hello*).

Such a cavalier *faut de mieux* attitude to the choice of background data will not do where speech is concerned. It would almost certainly lead to highly misleading and unreliable results. It is known from research comparing the distributional properties of acoustic parameters that these can differ considerably between languages and dialects. For example, there are different amounts of variance in the formant frequencies for the Japanese mid vowel /e/ and for the phonetically corresponding vowel in English (Kinoshita 2001: 155–7.). The same will almost certainly apply even for different styles of the same language, for example formal vs. informal, or spoken vs. read-out material.

Given what is known about the differences in variability of acoustic parameters in different kinds of speech data, then, the forensic phonetician has to be very careful in selecting data from which to derive average probabilities. Only material that closely matches the questioned and suspect samples in as many linguistic features as possible can be countenanced, so that the calibration material bears close resemblance to the data being tested. Given these conditions, it might be just as sensible to go ahead and use the data to compute an approximate *bona fide* LR in the first case.

Chapter summary

Three examples, one using *hello* and two using Japanese data, have been given of how a LR is calculated from real speech data. The last example, with Japanese vowel formants, demonstrated that *the approach clearly works, and its theoretical predictions are born out*. Same-subject data are evaluated with LRs larger than 1, different-subject data are evaluated with LRs less than 1. It can be noted that implementation of this approach seems to require phonological knowledge: to be able to identify a word's pitch-accent carrying syllable, for example, in order to be able to measure its vowel formants. A further important point demonstrated is that, although the magnitude of individual LRs from acoustic speech parameters is likely to be small, *the magnitude of overall LRs from speech can in fact be very large if several, ideally independent, parameters are combined*. It was also shown how the overall LR can function as a distance measure to successfully discriminate same-speaker and different-speaker pairs. For the latter use, it is then possible to quote a LR for the test.

The shortcomings of both the *bona fide* approach and the discriminant approach were discussed. These have to do with paucity of data, not yet fully adequate statistical models, and the recourse to average probabilities. These limitations should of course always be made clear to the court. The paucity of data is partly because the use of LRs, as part of the Bayesian approach, is still new in forensic phonetics, although it is becoming standard procedure in some other areas of forensic science. We need to find out more about reference samples of many important parameters. More tests remain to be done, especially with many speakers, to enable precise calibration of the tests. Improvements in the statistical modelling of parameters are needed. As knowledge in these areas increases, so will the accuracy of LR estimation improve. This is guaranteed, given the status of the LR as a theorem.

In the absence of fully adequate background data, much smaller reference samples have to be used, as in the *hello* and Japanese examples above, and it is conceivable that the investigator will either have such samples at their disposal, or be able to collect them at reasonably short notice. Therefore *the approximate nature of the LR quoted should always be made clear, and an attempt should be made to estimate how approximate the estimate is.*

As constantly emphasised above, the LR is the proper way to present scientific forensic evidence, and there is no reason why forensic phonetics should be exempt. Indeed, *it is always sensible to inquire what the LR is for the evidence in a speaker identification case.*

12

Summary and envoi

This chapter brings together the main points made in preceding chapters. It does this by first summarising them, and then considering what the requirements are for a successful forensic speaker identification. Finally it looks to the future.

Probably the most important point that the reader can draw from this book is that the forensic comparison of voice samples is complicated but perfectly possible. Its complexity is one of the reasons why it is difficult. I hope too that the book has shown that forensic speaker identification is not just a question of going and measuring some speech acoustics with a computer. It should be clear from the many examples given of the ways speakers can differ that reference needs to be made basically to all levels of linguistic structure and how they function. Not only that, but knowledge of linguistic typology is also required, for what could be an important parameter in one language might not be in another. So of course you have to quantify the acoustics, but this is done in the context of the linguistic structure of the samples, and this structure in turn is interpreted in the context of the use to which it is being put in the real world.

Forensic speaker identification is about trying to discriminate between speech samples from the same speaker and speech samples from different speakers. In order to do this, use is made of the fact that, as explained in the 'Basic ideas' part of the book, a voice is a multidimensional object that can be quantified along many different dimensions, or parameters. The parameters in which speaker identity is encoded are of many different types. They can be discrete or continuous; auditory or acoustic; linguistic or non-linguistic; and they are describable and quantifiable by experts within the various disciplines that contribute to speech-science, especially linguistics. This book has given examples of some of the many different types of parameters that can be used, from aspects of phonemic structure through formant frequencies of a particular sound, to long-term fundamental frequency and spectrum.

Speech samples must be compared with respect to as many as possible of the strongest parameters that are available in the data. A strong parameter is a parameter that is known to have high between-speaker variation and low within-speaker variation. The comparison is carried out for each parameter by first quantifying the difference between the two samples in that parameter. The task confronting the expert is then to say how likely they would be to observe that difference if the two samples were from the same speaker, and if the two samples were from different speakers. In order to calculate these two probabilities, it is not enough to quantify how *similar* the two samples are. It must also be determined how *typical* both of them are of the general population.

The ratio of these two probabilities is called the Likelihood Ratio, and quantifies the strength of the evidence for that parameter. A LR much greater than 1 means that the probability of observing the difference assuming the speaker is the same is much bigger than the probability assuming different speakers, and thus gives support for the prosecution hypothesis. Likewise, a LR much smaller than 1 gives support for the defence. The LRs for the individual parameters are then combined to give an overall LR for all the speech evidence, and this is the information that the expert should aim to provide the court with.

If LRs from many independent parameters are combined, an overall LR that will be big enough to have probative value will usually be obtained. This overall LR must then be evaluated in the light of the prior odds – the odds for or against the guilt of the accused before the voice evidence is adduced. Although the term *identification* is often used, this is misleading, since the aim is not to identify a particular speaker or speakers. Trying to give an estimate of how likely it is that the same speaker is involved is the job of the court, and outside the competence of the forensic speaker identification expert.

It is probably true, albeit inductively so, as with fingerprints, that we do all have different voices, or idiolects. A voice is a very complicated object, and can be best envisaged semiotically, that is, in terms of the different types of information it signals. The number of parameters necessary to uniquely identify a voice is probably very large. On its own, this should not constitute in principle any more of an impediment to speaker verification than occurs with fingerprints, which also need to be characterisable in many dimensions, albeit very many less than is assumed necessary for a voice. As pointed out in Chapter 2, however, the first thing that compromises speaker verification is the reduction in dimensionality that occurs as a result of practical limitations in real-word forensic comparison.

To be sure, a speaker's acoustic output is uniquely determined by their vocal tract, and extraction and quantification of both acoustic and auditory parameters is relatively easy even for medium quality recordings. This makes characterisation of a speaker as they are speaking on a particular occasion a relatively easy, if somewhat laborious task. However, it is the existence of within-speaker variation arising from the interaction of communicative intent and individual vocal tract shapes that creates the greatest problems. This gives rise to a situation where there will always be differences between voice samples, and where it is necessary to evaluate them as either within-speaker differences or between-speaker differences.

The consequences of this are that with a typical case involving an effectively open class verification, probabalistic conclusions are necessary:

> ... it has to be stated unambiguously that phonetic science cannot at the present time offer a universally applicable theory or automatic procedure which will provide a reliable identification of any speaker under each and every circumstance ...
>
> Braun and Künzel (1998: 16)

Summary and envoi

As pointed out above, these conclusions are properly expressed in terms of an approach that estimates with what probability the inevitable differences would be observable if they came from the same voice, and with what probability they would be observable if they came from different voices. The information necessary for an *accurate* calculation of these two probabilities, and hence their LR, will not usually be available owing to a lack both of adequate background statistical information and of adequately refined statistical models. This constitutes another limitation, and means that the estimate of the LR cannot normally pretend to perfect accuracy.

Nevertheless, forensic phoneticians will normally still be capable of providing an overall LR estimate from various combinations of LRs based on relatively 'hard' and 'soft' probabilities. Provided the limitations are understood, and made explicit, it is clear that forensic phoneticians are able to supply the court with highly useful evidence based as it is on their extensive linguistic, phonetic, and phonemic knowledge, conceptualised within a LR approach.

Requirements for successful forensic speaker identification

Desiderata for a successful forensic speaker identification can be considered from four aspects: data, method, investigator and language.

Data

In order to carry out forensic speaker identification effectively, several criteria must be met. Although all cases will differ in details, three criteria relating to data must always be fulfilled: quantity, quality and comparability. The speech samples must be copious and of adequate quality. A large amount of data is required in order to be able to make a good estimate of two of the things necessary for computing the LR: the average values of the samples, and the magnitude of variation in the samples. The quality of the data also determines the accuracy of the measurements and descriptions.

In addition to these two desiderata, *the samples to be compared must be capable of being compared, and comparability is essential*. Speech samples can be comparable in two different senses, called *situational comparability* and *linguistic comparability*. The values of forensic-phonetic parameters – their means and variances – vary with different non-linguistic circumstances, and samples taken under differing circumstances are then situationally non-comparable. For example, it would normally not be possible to compare a sample of shouted speech recorded on a video surveillance system during a hold-up with a sample of subdued speech from a suspect badly recorded during a police interview, even if there were a lot of data for analysis.

Linguistic comparability has to do with the amount of data that is comparable from a linguistic point of view. Samples containing the same words or phrases in the same prosodic context, for example *hello* at the beginning of a telephone conversation, are highly comparable. If the same words are not present, recourse has to be had to comparable segments in different words in the same prosodic context, for example /aː/ in *car* or *Parkes* (a place name). But in this case, careful control has to be exercised, since the differences in adjacent segments can increase the amount of variance to be expected. At the very least, the prosodic context has to be preserved. Thus one could not compare /aː/ in an unstressed syllable with /aː/ in a stressed syllable, as for example in the words *Parkes* and *car* in the two sentences *He's gone to PARKES*, and *It was HIM in the car, not ME*.

In order to accurately evaluate the inevitably occurring differences between the samples in terms of a LR, adequate background data from a suitable reference population must be available. Whereas it is quite possible that the samples will be of sufficient quantity and quality, and be comparable, it is highly unlikely that truly adequate background data will be available to furnish an accurate LR. Under these circumstances, the expert must use what is available, and try to estimate how this compromises the accuracy of the LR.

Because it will usually not be clear to the interested parties whether the data meet the three criteria of quantity, quality and comparability, it is not possible to know the potential of forensic-phonetic samples until an expert has listened to them, and has made a quick appraisal of their acoustics. It is usually sensible, therefore, to ask an expert to undertake a preliminary assessment of the data in order to evaluate whether or not it can be used for the intended purpose. If the data are adequate, then by definition there will be lots of it, and processing it will take time. The expert can try to estimate how long it will take. Since each case is different, the data may also contain other relevant information that is not immediately obvious to the untrained ear that will help the expert advise the parties. For example, both samples may have indistinguishable voice quality but clearly differ in the phonetic quality of the realisation of a particular allophone. Also, of course, the expert will be able to assess the degree to which the samples contain material that is linguistically comparable.

Method

Although each case is different, there will usually be some common points in the methodology. Both auditory and acoustic comparisons are necessary, and complementary. An auditory analysis is necessary to decide what is comparable, and to assess auditory linguistic differences. An acoustic analysis will usually supply the bulk of the quantifiable material, and will also act as a check against problems arising from possible perceptual integration. Steps should be taken to minimise expectation effects. Both these topics were discussed in Chapter 9 on 'Speech perception'.

Investigator

Of course, a successful outcome will not be guaranteed by the adequacy of the data alone – the ability of the practitioner will have something to do with it also. A brief look at the requirements for practitioners is therefore in order.

An expert is someone who knows a lot about their area of expertise. More importantly, perhaps, an expert also knows what they don't know, and is able to appreciate the ways in which the analytical methods they use are limited by the nature of the data.

Qualifications

The *International Association for Forensic Phonetics* stipulates in its code of practice that practitioners must have *an academic training and qualifications in phonetics/speech science*. The level of qualification is not specified, although in the IAFP's draft prerequisites for accreditation it is set at master's degree (or equivalent) or above. Since some practitioners will have acquired at least some of their practical expertise as an undergraduate, the training can be the result of either undergraduate or postgraduate study. The (now defunct) Forensic Speaker Identification Standards Committee of the *Australian Speech Science and Technology Association*, on the other hand, was being deliberately more restrictive than the IAFP by stating in its code of practice that at least a PhD is required of practitioners. The committee assumed that the complexity in the ways in which identity is encoded in speech was such that a PhD was required. Nowadays, as universities compete in a venal marketplace, a master's degree can imply very different lengths of study and research, and there can also be variation in the amount of time universities require for a PhD. Therefore it is perhaps worth inquiring into the nature of the qualification of the practitioner in each case.

Area of expertise

In order to discuss this point, it is necessary to clarify some distinctions, notably between *phonetics*, *speech science*, and *speech technology*. As characterised in the introduction, phonetics is a subject that investigates speech from many angles, asking especially: how people speak, how the speech is transmitted acoustically, and how it is perceived. This is capable of a broad construal to include all aspects of human vocalisations, and in this sense it is used synonymously with speech science. However, phonetics can also be understood in a much narrower sense to refer to speech as realisation of linguistic structure. This would exclude the examination of the differences between speakers in the production of the same sound, for example, or aspects of vocalisations that do not realise speech sounds, for example how someone sounds when they are angry, how they laugh, clear their throat, or hiccough. When considered in this narrow way, reference is often made to linguistic phonetics. Some researchers, e.g. Ladefoged (1997: 589–90) appear to have an even narrower concept of linguistic phonetics that is restricted to aspects of sounds that signal differences between words in a given language, but not, say, to aspects of sounds that signal the difference between two dialects, or sociolects.

A further distinction is usually drawn between *speech science* and *speech technology*. The latter has to do with technological aspects of speech such as noise reduction, or speech encoding, and focuses on the implementation of scientific ideas rather than issues relating to their nature, like their coherence, falsifiability or justification. The difference between speech science and speech technology can be illustrated with the minimum of parody by the concept of the formant, discussed in detail in the acoustics chapter. Despite their unquestionable theoretical status as vocal tract resonances, formants, especially the higher formants, are notoriously difficult to track automatically. They are therefore not of any central interest in speech technology, and are generally not used either for commercial speaker recognition or speech recognition (although research is of course still being done with them; see e.g. Hansen *et al.* 2001). In speech science and phonetics, however, which are not wedded to automaticity, formants remain important constructs that, as explained in Chapter 8 on 'Speech acoustics', and demonstrated in Chapter 11, relate a speaker's acoustic output to their linguistic vowel system, and also their identity.

As far as the general area of qualification for forensic phonetics is concerned, the IAFP, as just mentioned, specifies *phonetics or speech science*. From its accreditation requirements, however, which insist on a comprehensive knowledge of forensic phonetics, it is clear that the emphasis is on narrow phonetic and linguistic expertise. Thus the IAFP requires for accreditation: proficiency in auditory, articulatory and acoustic phonetics; knowledge of social and regional variation in the main languages in which the candidate is to work; proficiency in use of up-to-date analysis equipment; and understanding of the law as it relates to expert evidence. The Australian Forensic Speaker Identification Standards Committee also stipulates phonetics or speech science, but would allow also speech technology, as areas of qualification.

Now, it should be clear from all of the previous discussion in this section that the evaluation of between-sample differences in forensic speaker identification demands a combination of specialist skills on the part of the ideal practitioner. This will be someone who, firstly, can demonstrate understanding of the complexity of the relationship between the output of the vocal mechanism and the various types of information signalled in it. This information contains both linguistic and non-linguistic parameters with individual identifying potential. This in turn means that the expert must be proficient in the description and transcription of both voice quality and segmental and suprasegmental aspects of phonetic quality; they must be able to recognise the potential comparative parameters, both acoustic and auditory, and know how best to extract and quantify them. Most importantly, they must know what parts of samples are linguistically comparable. This seems to require expertise in phonetics broadly construed, i.e. speech science, and not just linguistic phonetics: 'First of all, the expert really has to be an expert in speech science/phonetics' (Braun and Künzel 1998: 17). Hollien (1990: 335) has more to add: 'The relevant and necessary background . . . includes a rigorous education in experimental phonetics, psychoacoustics, engineering, computer science, linguistics, statistics, physiology, and the like'.

It is hard to avoid the centrality of linguistic knowledge in all this, however – the knowledge of how speech relates to language. Because linguistics forms such a central part of the methodology, expertise in speech technology on its own is clearly not enough.

A final point concerning qualifications for forensic phoneticians relates to the forensic rather than the phonetic expertise needed. It has been pointed out that forensic science is an autonomous discipline concerned with evaluating evidence. Therefore, the fact that someone can demonstrate expertise in their own field, say phonetics, does not mean that they have the requisite forensic expertise (Robertson and Vignaux 1995: 199–201). One of the aims of this book has been to show what is necessary, in addition to phonetic expertise, in order to assess and present the value of forensic-phonetic evidence.

Language

A constraint on both data and practitioner is the language of the samples, and how it relates to the investigator's native language. IAFP enjoins in its Code of Practice that *members should approach with particular caution forensic work on speech samples in languages of which they are not native speakers.* This is weak, since it omits to specify what *particular caution* entails.

Being a native speaker of the language of the forensic samples confers enormous advantages on the investigator. Native speaker–listeners are able, as part of their natural linguistic competence, to correctly interpret the enormously complex variation in the sounds they hear in speech. This is important, given the linguistic fact that the same sound can realise a linguistic unit in one language, and signal individual differences in another. Thus a native listener knows automatically which (parts of which) sounds realise which phonemes and which (parts of which) sounds signal individual information. They also know what constitutes typical speech behaviour in their speech community, and what idiosyncratic. Thus they can easily recognise idiosyncratic articulations. Given an adequate description of the language, and very few languages are in fact described well enough for this, non-native-speaking phoneticians can do all this too, but certainly nowhere near as easily.

Because of this, in comparison of languages that are not the investigator's native language, it is clear that investigators must have comprehensive knowledge of or access to the phonological (phonetic and phonemic) linguistic and sociolinguistic structure of the variety in question. They must also work in conjunction with a native speaker whom they can interrogate and whose comments the investigator can interpret linguistically. If the native speaker is themself a linguist, so much the better. It is the author's opinion, not necessarily shared by others, that one should only undertake forensic speaker identification in a language other than one's own if one already speaks the language well, and then only together with a native speaker/linguist.

In this conjunction it can be noted that, although professional linguistic phoneticians are often polyglots, and will thus have been exposed to a wider variety of speech sounds as they actually occur in different languages, the ability to speak a lot of languages is not, as is quite often supposed, on its own an automatic qualification for a forensic phonetician.

The language of the samples also plays a role in that the set of parameters that will be of use in forensic-phonetic comparison will vary depending on the language. Sometimes, a forensic comparison is requested between samples in different languages. It is known that languages can differ in potentially important forensic parameters. For example, native Mandarin speakers employ a wider pitch range than native speakers of American English (Chen 1974). A little research has been done into the speech of bilinguals, and bidialectals, which shows that they tend to preserve these linguistic differences. Unfortunately, not enough is known yet about bilingual speakers to say whether any voice quality remains the same across two samples of the same speaker speaking in two different languages or dialects. Most likely it will depend on how good a command the speaker has of both varieties. Until we have a much better knowledge of this area, cross-linguistic forensic comparison is clearly counter-indicated.

In the future?

Advances in computer technology and signal processing have brought about mind-boggling improvements both in the speed and feasibility of acoustic speech analysis and statistical processing. These are particularly tangible for a researcher like the author who has experienced the leap from analogue to digital investigation. For example, a long-term F0 distribution that used to take at least a day or two to compile can now be computed almost instantaneously, although not necessarily as accurately (Rose 1991: 237). Some analyses, for example the computation of a long-term average spectrum, are now commonplace that were simply not feasible before. As mentioned above, it is now relatively easy to quantify acoustically how a speaker is speaking on a particular occasion even from moderate-quality speech. One of the obvious spin-offs of this is the ability to do automatic speaker verification and/or recognition. Whether we like it or not, we are obviously in for more technological advances. It is often asked whether this will result in improved forensic speaker verification.

The answer to this question depends on what you understand by *improvement*. Technology will enable us to do it quicker – of that there is no doubt. For some, improvement means full automation. As was made clear in Chapter 5, very few are optimistic that forensic speaker identification will eventually emulate its presentation in the movies and American TV crime shows and become fully automatic. The consensus is that, because of the nature of human voices and the sound systems of language, and the lack of control over forensic speech samples, there will never be a fully automated system for forensic speaker identification, no matter what the technological advances. The very fact that two identical twins can have very similar acoustics but differ in the implementation of a single segment of their linguistic system (Nolan and Oh 1996) means that it is likely that there will always be a need for careful auditory analysis before instrumental analysis, which will also usually be selective, and not automatic.

But the real meaning of improvement in the context of forensic speaker identification can only be in terms of the evaluation constraints discussed. Namely, real improvement can only make sense in terms of the *accuracy and magnitude of the LR estimate*. As made clear above, the accuracy of the LR will improve with better knowledge of background data statistics. For example, what is the between-speaker and within-speaker variation in the acoustics of forensically common words like *hello*, *O.K. fuck(in')*, *yeah*? This is clearly not primarily a technological issue, although technology will certainly help us to get and process forensically realistic data to determine the answer. It is much more a question of shoe leather and *Sitzfleisch*. In addition to this, the accuracy of the LR is also a function of the appropriate statistical modelling of the data, and this is still to be addressed. Once again, the kind of advances in computer technology that have revolutionised statistical analyses will speed up the process, but will not supplant analytical thinking.

As far as the magnitude of the LR is concerned, perhaps this is where the clearest contribution from technology can come from. As mentioned above, speech technology works with very different parameters from phonetics and speech science. We know that the automatic parameters of speech technology like the *cepstrum* or *delta cepstrum* perform exceptionally well in automatic speaker recognition and verification. Very little research has been done on how well these automatic parameters perform on forensically realistic data. The fact that they are not easily relatable to phonetic quality is not an excuse to ignore them, and their contribution needs to be assessed (Rose and Clermont 2001: 34–5). That is, it needs to be seen whether automatic parameters can deliver LRs of greater magnitude than those currently used. This is still some way off, however, since the background data for the distributions of the automatic parameters also need to be determined first, itself a daunting proposition given the mathematical complexity of the task.

Summary and envoi

Finally it is worth pointing out that technological improvement does not always represent the *summum bonum*. The introduction of digital telephones, for example, means that we now have to know to what extent the original signal is distorted by digital encoding, and what remains comparable.

One should not invest too much hope in advances from technological progress therefore. Rather, improvement in forensic speaker identification will come from better knowledge: better knowledge of background data and how to process it adequately; better knowledge on the part of the legal professions and law enforcement agencies of what the forensic-phonetic expert is trying to do. But above all better knowledge and understanding of how individuality is encoded in speech: knowledge of what the best parameters are, where to look for them, and under what real-world circumstances they are comparable.

Glossary

Glossaries differ in their exhaustiveness. For example, a glossary may simply be a more or less exhaustive list of technical terms with simple definitions. Or it may be used to explain just those technical terms that the writer assumes some readers might not be familiar with. The glossary below is intended as a kind of a quick-reference list for important technical terms that may occur in forensic-phonetic reports.

(Arithmetical) mean The proper statistical term for *average*.
(Vowel) backness An important descriptive parameter for vowels. Refers to how far forward or back the body of the tongue is.
(Vowel) height An important descriptive parameter for vowels. Refers to how high or low the body of the tongue is.
Accent The pronunciation used by a speaker (as opposed to other things like choice of words or syntax) that is characteristic of a particular area, or social group.
Acoustic forensic analysis The expert use of acoustic, as opposed to auditory, information to compare forensic speech samples.
Acoustic phonetics (or **speech acoustics**) That part of phonetics that deals with the acoustic properties of speech sounds, and how they are transmitted between speaker and hearer.
Allomorph The realisation of a morpheme. Two allomorphs of the plural morpheme in English for example are *s*, as in *gnats* and *es* as in *horses*.
Allophone A speech sound functioning as the realisation of a phoneme.
Articulation rate A measure of how fast someone speaks, usually quantified in terms of syllables per second, exclusive of pauses.
Articulatory phonetics The study of how speech sounds are made by the speaker.
Auditory forensic analysis (or **technical speaker recognition by listening**) The expert use of auditory, as opposed to acoustic, information to compare forensic speech samples.
Aural-spectrographic identification Highly controversial method of speaker identification using both visual examination of spectrograms and listening.
Between-speaker variation The fact that different speakers of the same language differ in some aspects of their speech. One of the conditions that makes forensic speaker identification possible.
Centisecond (or **csec** or **cs**) Unit for quantifying duration in acoustic phonetics: one-hundredth of a second.
Cepstrum A very common parameter used in automatic speaker recognition, one effect of which is to smooth the spectrum.

Glossary

Closed set comparison An unusual situation in forensic speaker identification where it is known that the offender is present among the suspects.
Continuity A technical term for how fluently someone speaks.
Convergence The tendency for two participants in a conversation to become more similar in their speech behaviour to signal in-group membership. Speakers can also diverge from one another.
Conversation analysis The study of how conversation is structured and regulated.
Decibel (or **dB**) Unit for quantifying amplitude in acoustic phonetics.
Defence fallacy An error in logical reasoning that assumes (1) the probability of the evidence given the hypothesis of innocence, $p(E \mid H_i)$ is the same as the probability of the hypothesis of innocence given the evidence, $p(H_i \mid E)$; and (2) ignores how probable the evidence is under assumption of guilt, $p(E \mid H_g)$.
Dialectology The study of how language varies with geographical location.
Digitising The process of converting an analogue speech signal, e.g. from a cassette recorder, into a digital form that can be used by a computer for speech analysis.
Diphthong A vowel in a single syllable that involves a change in quality from one target to another, as in *how*, or *high*.
Expectation effect A well-known perceptual phenomenon whereby one hears what, or whom, one expects to hear.
F- (or **Formant**) **pattern** The ensemble of formant frequencies in a given sound or word.
False negative In speaker recognition, deciding that two speech samples have come from different speakers when in fact they are from the same speaker.
False positive In speaker recognition, deciding that two speech samples have come from the same speaker when in fact they are from different speakers.
FFT (or **Fast Fourier Transform**) A common method of spectral analysis in acoustic phonetics.
Formant A very important acoustic parameter in forensic speaker identification. Formants reflect the size and shape of the speaker's vocal tract.
Formant bandwidth An acoustic parameter that reflects the degree to which acoustic energy is absorbed in the vocal tract during speech.
Fundamental frequency (or **F0**) A very important acoustic parameter in forensic speaker identification. F0 is the acoustic correlate of the rate of vibration of the vocal cords.
Hertz (or **Hz**) Unit for quantifying frequency: so many times per second. 100 Hz for example means one hundred times per second.
Incidential difference One of the ways in which speakers can differ in their phonemic structure.
Indexical information Information in speech that signals the speaker as belonging to a particular group, e.g. male, middle-class, with a cold, Vietnamese immigrant, and so on.
Intonation The use of pitch to signal things like questions or statements, or the emotional attitude of the speaker.
Kilohertz (or **kHz**) Unit for quantifying frequency: so many thousand times per second. 1 kHz for example means one thousand times per second.
Likelihood ratio (**LR**) A number that quantifies the strength of the forensic evidence, and that is thus an absolutely crucial concept in forensic identification. The strength of the evidence is reflected in the magnitude of the LR. LR values greater than 1 give

support to the prosecution hypothesis that a single speaker is involved; LRs less than 1 give support to the defence hypothesis that different speakers are involved.

The LR is the ratio of two probabilities. In forensic speaker identification these are: the probability of observing the differences between the offender and suspect speech samples assuming they have come from the same speaker; and the probability of observing the differences between the suspect and offender speech samples assuming they have come from different speakers.

Linear prediction A commonly used method of digital speech analysis.

Long-term A common type of quantification in forensic speaker identification whereby a parameter, usually fundamental frequency, is measured over a long stretch of speech rather than a single speech sound or word.

Manner (of articulation) The type of obstruction in the vocal tract used in making a consonant, e.g. fricative, or stop.

Millisecond (or **msec** or **ms**) Common unit for quantifying duration in acoustic phonetics: one-thousandth of a second.

Monophthong A vowel, in a single syllable, that does not change in quality.

Morpheme A unit of linguistic analysis used in describing the structure of words: the smallest meaningful unit in a language. For example, the word *dogs* consists of two morphemes: {dog} and {plural}.

Naive speaker recognition When an untrained listener attempts to recognise a speaker, as in voice line-ups, etc.

Open set comparison The usual situation in forensic speaker identification where it is not known whether the offender is present among the suspects.

Parameter (or **dimension**, or **feature**) A generic term for anything used to compare forensic speech samples, e.g. mean fundamental frequency, articulation rate, phonation type.

Phonation type The way the vocal cords vibrate, giving rise to auditorily different qualities, e.g. creaky voice, or breathy voice.

Phone A technical name for speech sound.

Phoneme A unit of linguistic analysis: the name for a contrastive sound in a language. For example, *bat* and *pat* begin with two different phonemes.

Phonemics The study of how speech sounds function contrastively, to distinguish words in a given language. Phonemics is an important conceptual framework for the comparison of forensic speech samples.

Phonetic quality One of two very important descriptive components of a voice, the other being voice quality. Describes those aspects of a voice that have to do with the realisation of speech sounds.

Phonetics The study of all aspects of speech, but especially how speech sounds are made, their acoustic properties, and how the acoustic properties of speech sounds are perceived as speech by listeners. Phonetic expertise is an important prerequisite for forensic phonetics.

Phonology One of the main sub-areas in linguistics. Phonology studies the function and organisation of speech sounds, both within a particular language, and in languages in general.

Phonotactic difference One of the ways in which speakers can differ in their phonemic structure.

Pitch (1) An important auditory property of speech. Pitch functions primarily to signal linguistic categories of intonation tone and stress, but overall pitch and pitch range can be used to characterise an individual's voice. (2) Another term for fundamental frequency.

Pitch accent The use of pitch to signal differences between words that is partly like tone and partly like stress. Japanese is a pitch-accent language.

Place (of articulation) Where in the vocal tract a consonantal sound is made.

Population A statistical term referring to the totality of something. For example, the population of male speakers of General American under 25 years of age means all male speakers of General American under 25 years of age.

Posterior odds In forensic speaker identification, the odds in favour of the hypothesis of common origin for two or more speech samples after the forensic-phonetic evidence, in the form of the likelihood ratio, is taken into account. The posterior odds are the product of prior odds and LR.

Prior odds In forensic speaker identification, the odds in favour of the hypothesis of common origin for two or more speech samples before the forensic-phonetic evidence is taken into account.

Probability A number between 0 and 1 (or 0% and 100%) quantifying one of two things: (1) The degree of belief in a particular hypothesis, such as *these two speech samples have come from the same speaker*. (2) The frequency of occurrence of an event, for example the number of times two samples from the same speaker have the same quality for a particular vowel. In forensic speaker identification, type (2) probabilities should be used to assess the probability of the evidence under competing prosecution and defence hypotheses. This then facilitates the evaluation, by the court, of the type (1) probability (i.e. the probability of the hypothesis that the speech samples come from the same speaker).

It is logically and legally incorrect for the forensic phonetician to try to assess the type (1) probability of the hypothesis that two or more speech samples come from the same/different speakers.

Prosecution fallacy An error in logical reasoning that assumes that (1) the probability of the evidence given the hypothesis of guilt, $p(E \mid H_g)$ is the same as the probability of the hypothesis of guilt given the evidence, $p(H_g \mid E)$; and (2) ignores how probable the evidence is under assumption of innocence, $p(E \mid H_i)$.

Realisational difference One of the ways in which speakers can differ in their phonemic structure.

Sample A statistical term referring to a sub-part of a population. A speech sample is thus a part of the speech of a particular person. Large samples give a good idea of the statistical properties of the population they have been taken from.

Segmentals A generic term for vowels and consonants.

Sociolect A way of talking that is typical of a particular social group.

Sociolinguistics The study of how language varies with sociological variables like age, sex, income, education, etc.

Spectral slope An acoustic parameter that relates to the way the vocal cords vibrate.

Spectrogram A picture of the distribution of acoustic energy in speech. It normally shows how frequency varies with time. Spectrograms are often used to illustrate an acoustic feature or features of importance. To be distinguished from *spectrograph*, which is the name of the analogue instrument on which spectrograms used to be made. Nowadays they are made by computer.

Spectrum The result of an acoustic analysis showing how much energy is present at what frequencies in a given amount of speech.
Speech perception That part of phonetics that studies how the acoustic properties of speech sounds are perceived by the listener.
Standard deviation A statistical measure quantifying the spread of a variable around a mean value.
Stress Prominence of one syllable in a word used to signal linguistic information, like the difference between *implant* (noun) and *implant* (verb) in English.
Subglottal resonance A frequency in speech attributable to structures below the vocal cords, e.g. the trachea.
Suprasegmentals A generic term for tone, stress and intonation.
Syllable (or **speaking**) **rate** A measure of how fast someone speaks, usually quantified in terms of syllables per second, inclusive of pauses.
Systemic difference One of the ways in which speakers can differ in their phonemic structure.
Technical speaker recognition Attempts to recognise a speaker informed by theory, as in forensic speaker identification, automatic speaker recognition, etc.
Tone The use of pitch to signal different words, as in tone languages like Chinese.
Transposing the conditional A common error in reasoning that involves assuming that the probability of the evidence given the hypothesis is the same as the probability of the hypothesis given the evidence.
Variance A statistical measure quantifying the variability of a variable; the square of the standard deviation.
Voice quality One of two very important descriptive components of a voice, the other being phonetic quality. Describes those long-term or short-term aspects of a voice that do not have to do with the realisation of speech sounds.
Voiceprint Another name for spectrogram. Usually avoided because of its association with voiceprint identification.
Voiceprint identification Highly controversial method of speaker identification exclusively using visual examination of spectrograms.
Voicing/phonation Refers to activity of the vocal cords.
Within-speaker variation The fact that the same speaker can differ in some aspects of their speech on different occasions, or under different conditions. One of the conditions that makes forensic speaker identification difficult.

References[1]

Abe, I. (1972) 'Intonation', in Bolinger, D. (ed.) *Intonation*, Harmondsworth: Penguin Books.
AFTI (n.d.) *Voice Print Identification*, Applied Forensic Technologies International, Inc., HTTP://www.aftiinc.co./voice.htm (accessed 27/08/2001).
Aitken, C. G. G. (1991) 'Populations and Samples', in Aitken and Stoney (eds) (1991): 51–82.
Aitken, C. G. G. (1995) *Statistics and the Evaluation of Evidence for Forensic Scientists*, Chichester: Wiley.
Aitken, C. G. G. and Stoney, D. A. (eds) (1991) *The Use of Statistics in Forensic Science*, Chichester: Ellis Horwood.
Anderson, S. R. (1985) *Phonology in the Twentieth Century*, Chicago: University of Chicago Press.
Atal, B. S. (1974) 'Effectiveness of linear predication characteristics of the speech wave for automatic speaker identification and verification', *JASA* 55: 1304–12.
Atal, B. S. (1976) 'Automatic recognition of speakers from their voices', *Proc. IEEE* 64/4: 460–75.
Baldwin, J. (1979) 'Phonetics and speaker identification', *Medicine, Science and the Law* 19: 231–2.
Baldwin, J. and French, P. (1990) *Forensic Phonetics*, London: Pinter.
Barlow, M. (ed.) (2000) *Proc. 8th Australian International Conference on Speech Science and Technology*, Canberra: ASSTA.
Barlow, M., Clermont, F. and Mokhtari, P. (2001) 'A methodology for modelling and interactively visualising the human vocal tract in 3D space', *Acoustics Australia* 29/1: 5–8.
Barry, W. J., Hoequist, C. E. and Nolan, F. J. (1989) 'An approach to the problem of regional accent in automatic speech recognition', *Computer Speech and Language* 3: 355–66.
Beck, J. M. (1997) 'Organic variation of the vocal apparatus', in Hardcastle and Laver (eds) 256–97.
Bernard, J. R. L. (1967) *Some measurements of some sounds of Australian English*, unpublished PhD thesis, Sydney University.
Blakemore, C. (1977) *Mechanics of the Mind*, Cambridge: Cambridge University Press.
Bolt, R. H., Cooper, F. S., Green, D. M., Hamlet, S. L., Hogan, D. L., McKnight, J. G., Pickett, J. M., Tosi, O. and Underwood, B. D. (1979) *On the Theory and Practice of Voice Identification*, Washington DC: National Academy of Sciences.
Bower, B. (2001) 'Faces of perception', *Science News* 160/1: 10–12.

---

[1] *Abbreviations*: AJL, *Australian Journal of Linguistics*; ASSTA, *Australian Speech Science and Technology Association*; FL, *Forensic Linguistics*; JASA, *Journal of the Acoustical Society of America*; JIPA, *Journal of the International Phonetics Association*; Proc. IEEE, *Proceedings of the Institute of Electrical and Electronics Engineers*; SC, *Speech Communication*; UCLA WPP, *Working papers in phonetics, University of California at Los Angeles*.

References

Box, G. E. P., Hunter, W. G. and Hunter, J. S. (1978) *Statistics for Experimenters*, New York: Wiley.
Braun, A. (1995) 'Fundamental frequency – how speaker-specific is it?', in Braun and Köster (eds) (1995): 9–23.
Braun, A. and Köster, J.-P. (eds) (1995) *Studies in Forensic Phonetics*, Beiträge zur Phonetik und Linguistik 64, Trier: Wissenschaftlicher Verlag.
Braun, A. and Künzel, H. J. (1998) 'Is forensic speaker identification unethical? – or can it be unethical not to do it?', *FL* 5/1: 10–21.
Bricker, P. D. and Pruzansky, S. (1976) 'Speaker recognition', in N. J. Lass (ed.) (1976): 295–326.
Broeders, A. P. A. (1995) 'The role of automatic speaker recognition techniques in forensic investigations', in *Proc. Intl. Congress Phonetic Sciences* 3: 154–61.
Broeders, A. P. A. (1999) 'Some observations on the use of probability scales in forensic identification', *FL* 6/2: 228–41.
Broeders, A. P. A. (2001) 'Forensic speech and audio analysis forensic linguistics 1998 to 2001: a review', Paper at the 13th INTERPOL Forensic Science Symposium.
Broeders, A. P. A. and Rietveld, A. C. M. (1995) 'Speaker identification by earwitnesses', in Braun and Köster (eds) (1995): 24–40.
Brown, K. (1996) *Evidential value of elemental analysis of glass fragments*, unpublished First Class Honours Thesis, University of Edinburgh.
Bull, R. and Clifford, B. (1999) *New Law Journal* Expert Witness Supplement (Feb.): 216–20.
Butcher, A. (1981) *Aspects of the speech pause: phonetic correlates and communicative functions*, PhD Thesis, University of Kiel (published in *Arbeitsberichte* 15, Institut für Phonetik der Universität Kiel).
Butcher, A. (1996) 'Getting the voice line-up right: analysis of a multiple auditory confrontation', in McCormak and Russell (eds) (1996): 97–102.
Calvert, D. R. (1986) *Descriptive Phonetics*, 2nd ed, New York: Thieme.
Chambers, J. K. and Trudgill, P. (1980) *Dialectology*, Cambridge: Cambridge University Press.
Champod, C. and Evett, I. (2000) Commentary on Broeders (1999), *FL* 7/2: 238–43.
Champod, C. and Meuwly, D. (2000) 'The inference of identity in forensic speaker recognition', *SC* 31: 193–203.
Chen, G. T. (1974) 'The pitch range of English and Chinese speakers', *Journal of Chinese Linguistics* 2/2: 159–71.
Clark, G. M. (1998) 'Cochlear implants in the second and third millennia', in Mannell and Robert-Ribes (eds) (1998), vol. 2: 1–6.
Clark, J. and Yallop, C. (1990) *An Introduction to Phonetics and Phonology*, Oxford: Blackwell.
Clermont, F. (1996) 'Multi-speaker formant data on the Australian English vowels: a tribute to J. R. L. Bernard's (1967) pioneering research', in McCormak and Russell (eds) (1996): 145–50.
Clermont, F. and Itahashi, S. (1999) 'Monophthongal and diphthongal evidence of isomorphism between formant and cepstral spaces', *Proc. Spring Meeting of the Acoustical Society of Japan*, Meiji University Press: 205–6.
Clermont, F. and Itahashi, S. (2000) 'Static and dynamic vowels in a "cepstro-phonetic" subspace', acoustical letter, *Journal of the Acoustical Society of Japan* 21/4: 221–3.
Clifford, B. R., Rathborn, H. and Bull, R. (1981) 'The effects of delay on voice recognition accuracy', *Law and Human Behaviour* 5: 201–8.
Cole, S. A. (2001) *Suspect Identities: A History of Fingerprinting and Criminal Identification*, Cambridge, MA: Harvard University Press.
Collins, B. (1998) 'Convergence of fundamental frequency in conversation. If it happens, does it matter?', in Manell and Robert-Ribes (eds) (1998), vol. 3: 579–82.
Cox, F. (1998) 'The Bernard data revisited', *AJL* 18: 29–55.

Crowley, T., Lynch, J., Siegel, J. and Piau, J. (1995) *The Design of Language*, Auckland: Longman.
Cruttendon, A. (1986) *Intonation*, Cambridge: Cambridge University Press.
Crystal, D. (1987) *The Cambridge Encyclopedia of Language*, Cambridge: Cambridge University Press.
Daniloff, R., Schuckers, G. and Feth, L. (1980) *The Physiology of Speech and Hearing: An Introduction*, Englewood Cliffs: Prentice-Hall.
Daubert (1993) *Daubert v. Merrell Dow Pharmaceuticals, Inc.* (1993) 113 S Ct 2786.
Davies, S. B. and Mermelstein, P. (1980) 'Comparison of parametric representations for monosyllabic word recognition in continuously spoken sentences', *IEEE Transactions on Acoustics, Speech, and Signal Processing* 28: 357–66.
Deffenbacher, K. A. *et al.* (1989) 'Relevance of voice identification research to criteria for evaluating reliability of an identification', *Journal of Psychology* 123: 109–19.
Delgutte, B. (1997) 'Auditory neural processing of speech', in Hardcastle and Laver (eds) (1997): 507–38.
Diller, A. V. N. (1987) 'Reflections on Tai diglossic mixing', *Orbis* 32/1,2: 147–66.
Dixon, R. M. W. (1983) *Searching for Aboriginal Languages: Memoirs of a Field-Worker*, St Lucia: University of Queensland Press.
Doddington, G. R. (1985) 'Speaker recognition – identifying people by their voices', *Proc. IEEE* 73/11: 1651–64.
Durie, M. and Hajek, J. (1994) 'A revised standard phonemic Orthography for Australian English vowels', *AJL* 14: 93–107.
Elliott, J. R. (2001) 'Auditory and F-pattern variation in Australian *Okay*: a forensic investigation', *Acoustics Australia* 29/1: 37–41.
Elzey, F. (1987) *Introductory Statistics: A Microcomputer Approach*, Monterey: Brooks/Cole Publishing.
Emmorey, K., Van Lancker, D. and Kreiman, J. (1984) 'Recognition of famous voices given vowels, words, and two-second texts', *UCLA WPP*: 120–4.
Evett, I. W. (1991) 'Interpretation: A personal odyssey', in Aitken and Stoney (eds) (1991): 9–22.
Evett, I. W., Scrange, J. and Pinchin, R. (1993) 'An illustration of the advantages of efficient statistical methods for RFLP analysis in forensic science', *American Journal of Human Genetics* 52: 498–505.
Fant, G. (1960) *Acoustic Theory of Speech Production*, The Hague: Mouton.
Fant, G. (1973) *Speech Sounds and Features*, Cambridge, MA: MIT Press.
Finegan, E., Besnier, N., Blair, D. and Collins, P. (1992) *Language: Its Structure and Use*, Marrickville: Harcourt Brace Jovanovich.
Foster, K.R., Bernstein, D.E. and Huber, P.W. (1993) 'Science and the toxic tort', *Science* 261: 1509–614.
Foulkes, P. and Barron, A. (2000) 'Telephone speaker recognition amongst members of a close social network', *FL* 7/2: 181–98.
France, D. J., Shiavi, R. G., Silverman, S., Silverman, M. and Wilkes, D. M. (2000) 'Acoustical properties of speech as indicators of depression and suicidal risk', *IEEE Transactions on Biomedical Engineering* 4: 829–37.
French, J. P. (1990a) 'Acoustic phonetics', in Baldwin and French (eds) (1990): 42–64.
French, J. P. (1990b) 'Analytic procedures for the determination of disputed utterances', in Kniffka (ed.) (1990): 201–13.
French. J. P. (1994) 'An overview of forensic phonetics with particular reference to speaker identification', *FL* 1/2: 169–84.
Fromkin, V. A. (ed.) (1978) *Tone: A Linguistic Survey*, London: Academic Press.
Fromkin, V., Rodman, R., Collins, P. and Blair, D. (1996) *An Introduction to Language*, 3rd edn, Marrickville: Harcourt Brace and Company.

References

Fromkin, V., Blair, D. and Collins, P. (1999) *An Introduction to Language*, 4th edn, Marrickville: Harcourt.

Frye (1923) Frye vs. United States (1923), 293 Federal Reports (1st series) 1013, 1014 (CA).

Furui, S. (1981) 'Cepstral analysis technique for automatic speaker verification', *IEEE Transactions on Acoustics, Speech and Signal Processing* ASSP 29/2: 254–72.

Furui, S. (1994) 'An overview of speaker recognition technology', *Proc. ESCA Workshop on Automatic Speaker Recognition Identification Verification*: 1–8.

Furui, S. (2000) 'Steps towards flexible speech recognition: recent progress at Tokyo Institute of Technology', in Barlow (ed.) (2000): 19–29.

Furui, S. and Matsui, T. (1994) 'Phoneme-level voice individuality used in speaker recognition', *Proc. 3rd International Conference on Spoken Language Processing*: 1463–6.

Furui, S., Itakura, F. and Saito, S. (1972) 'Talker recognition by longtime averaged speech spectrum', *Electronics and Communications in Japan*, 55A/10: 54–61.

Gfroerer, S. and Wagner, I. (1995) 'Fundamental frequency in forensic speech samples', in Braun and Köster (eds) (1995): 41–8.

Gibbons, J. (ed.) (1994) *Language and the Law*, London: Longman.

Gigerenzer, G., Swijtink, Z., Porter, T., Daston, L., Beatty, J. and Krüger, L. (1989) *The Empire of Chance*, Cambridge: Cambridge University Press.

Gish, H. and Schmidt, M. (1994) 'Text-independent speaker identification', *IEEE Signal Processing Magazine* 11/4: 18–32.

Goddard, C. (1996) 'Can linguists help judges know what they mean? Linguistic semantics in the court-room', *FL* 3/2: 250–72.

Goddard, C. (2001) 'Thinking across languages and cultures: six dimensions of variation,' paper given at the Linguistic Institute, University of California at Santa Barbara.

Goggin, J. P., Thompson, C. P., Strube, G. and Simental, L. R. (1991) 'The role of language familiarity in voice identification', *Memory and Cognition*, 19: 448–58.

Goldstein, A. G., Knight, P., Bailis, K. and Connover, J. (1981) 'Recognition memory for accented and unaccented voices', *Bulletin of the Psychonomic Society* 17: 217–20.

Gonzalez-Rodriguez, J., Ortega-Garcia, J. and Lucena-Molina, J. J. (2001) 'On the application of the Bayesian framework to real forensic conditions with GMM-based systems', *Proc. 2001 Speaker Odyssey Speaker Recognition Workshop*: 000–00.

Good, I. J. (1991) 'Weight of evidence and the Bayesian likelihood ratio', in Aitken and Stoney (eds) (1991): 85–106.

Gould, S. J. (1996) *Life's Grandeur*, London: Vintage.

Gould, S. J. (2000) 'A tale of two work sites', in S. J. Gould *The Lying Stones of Marrakech: Penultimate Reflections in Natural History*, London: Jonathan Cape: 251–67.

Greisbach, R. (1999) 'Estimation of speaker height from formant frequencies', *FL* 6/2: 265–77.

Greisbach, R., Esser, O. and Weinstock, C. (1995) 'Speaker identification by formant contours', in Braun and Köster (eds) (1995): 49–55.

Gruber, J. S. and Poza, F. T. (1995) Voicegram identification evidence, *American Jurisprudence Trials* 54, Lawyers Cooperative Publishing.

Gussenhoven, C. and Jacobs, H. (1998) *Understanding Phonology*, London: Arnold.

Hansen, E. G., Slyh, R. E. and Anderson, T. R. (2001) 'Formant and F0 features for speaker verification', *Proc. 2001 Speaker Odyssey Speaker Recognition Workshop*: 25–9.

Hardcastle, W. J. and Laver, J. (eds) (1997) *The Handbook of Phonetic Sciences*, Oxford: Blackwell.

Harrington, J., Cox, F. and Evans, Z. (1997) 'An acoustic phonetic study of Broad, General, and Cultivated Australian vowels', *AJL* 17: 155–84.

Hayne, J. A. and Crockett, A. J. A. (1995) Supreme Court of Victoria Court of Appeal Report 325/94 (R v. Bell).

Hirson, A. (1995) 'Human laughter – a forensic phonetic perspective', in Braun and Köster (eds) (1995): 77–86.
Hirson, A., French, P. and Howard, D. (1995) 'Speech fundamental frequency over the telephone and face-to-face: some implications for forensic phonetics', in Lewis (ed.) (1995): 230–40.
Hirst, D. and Di Cristo, A. (eds) (1998) *Intonation Systems: A Survey of Twenty Languages*, Cambridge: Cambridge University Press.
Hollien, H. (1990) *The Acoustics of Crime*, New York: Plenum.
Hollien, H. (1995) 'Consideration of guidelines for earwitness lineups', *FL* 3: 14–23.
Hollien, H., Huntley, R., Künzel, H. and Hollien, P. A. (1995) 'Criteria for earwitness lineups', *FL* 2: 143–53.
Hombert, J.-M. (1977) 'A model of tone systems', *UCLA WPP* 36: 20–32.
Hombert, J.-M. (1978) 'Consonant types, vowel quality, and tone', in Fromkin (ed.) (1978): 77–111.
Horvarth, B. (1985) *The Sociolects of Sydney*, Cambridge: Cambridge University Press.
Hyman, L. (1977) 'On the nature of linguistic stress', in L. Hyman (ed.) (1977) *Studies in Stress and Accent*, Southern California Occasional Papers in Linguistics No. 4: 37–82, Los Angeles: University of Southern California.
Ingram, J., Prandolini, R. and Ong, S. (1996) 'Formant trajectories as indices of phonetic variation for speaker identification', *FL* 3: 129–45.
Inoue, Y. (1979) 'Japanese: A story of language and people', in T. Shopen (ed.) (1979) *Languages and Their Status*: 241–300, Cambridge, MA: Winthrop.
IPA (1999) *Handbook of the International Phonetic Association*, Cambridge: Cambridge University Press.
Jakobson, R., Fant, G. and Halle, M. (1952) *Preliminaries to Speech Analysis* (tenth reprint 1972), Cambridge, MA: MIT Press.
Jassem, W. (1971) 'Pitch and compass of the speaking voice', *JIPA* 1/2: 59–68.
Jassem, W. and Nolan, F. (1984) 'Speech sounds and languages', in G. Bristow (ed.) (1984) *Speech Synthesis: Techniques, Technology and Applications*: 19–47, London: Granada.
Jassem, W., Steffen-Batog, S. and Czajka, M. (1973) 'Statistical characteristics of short-term average F0 distributions as personal voice features', in W. Jassem (ed.) (1973) *Speech Analysis and Synthesis* vol. 3: 209–25, Warsaw: Polish Academy of Science.
Jefferson, G. (1989) 'Preliminary notes on a possible metric which provides for a "standard maximum" of silence of approximately one second in conversation', in D. Roger and P. Bull (eds) *Conversation*, Philadelphia: Multilingual Matters.
Johnson, K. (1997) *Auditory and Acoustic Phonetics*, Oxford: Blackwell.
Jones, Alex (1994) 'The limitations of voice identification', in Gibbons (ed.) (1994): 346–61.
Kersta, L. G. (n.d.) 'Voiceprint identification', manual, Somerville, NJ: Voiceprint Laboratories Corporation.
Kersta, L. G. (1962) 'Voiceprint identification', *Nature* 196: 1253–7.
Kieser, J. A., Buckingham, D. M. and Firth, N. A (2001) 'A Bayesian approach to bitemark analysis', paper at the *7th Indo-Pacific Congress on Legal Medicine and Forensic Sciences, Congress Program*: 217.
Kinoshita, Y. (2001) *Testing realistic forensic speaker identification in Japanese: a likelihood ratio based approach using formants*, unpublished PhD Thesis, the Australian National University.
Kniffka, H. (ed.) (1990) *Texte zur Theorie and Praxis forensischer Linguistik*, Tübingen: Max Niemayer Verlag.
Koenig, B. J. (1986) 'Spectrographic voice identification: a forensic survey', letter to the editor of *JASA* 79/6: 2088–90.
Kohler, K. J. (2000) 'The future of phonetics', *JIPA* 30/1,2: 1–24.
Koolwaaij, J. and Boves, L. (1999) 'On decision making in forensic casework', *FL* 6/2: 242–64.

References

Kotz, S., Johnson, N. L. and Read, C. B. (eds) (1982) *Encyclopedia of Statistical Sciences*, Wiley: New York.

Köster, O. and Schiller, N. (1997) 'Different influences of the native language of a listener on speaker recognition', *FL* 4/1: 18–27.

Köster, O., Schiller, N. and Künzel, H. J. (1995) 'The influence of native-language background on speaker recognition', in K. Elenius and P. Branderud (eds) *Proc. 13th Intl. Congress on Phonetic Sciences* 4: 306–9.

Kratochvil, P. (1968) *The Chinese Language Today*, London: Hutchinson.

Kratochvil, P. (1998) 'Intonation in Beijing Chinese', in Hirst and Di Cristo (eds) (1998): 417–31.

Kreiman, J. and Papçun, G. (1985) 'Voice discrimination by two listener populations', *UCLA WPP* 61: 45–50.

Kumar A. and Rose, P. (2000) 'Lexical evidence for early contact between Indonesian languages and Japanese', *Oceanic Linguistics* 39/2: 219–55.

Künzel, H. J. (1987) *Sprechererkennung: Grundzüge forensischer Sprachverarbeitung*, Heidelberg: Kriminalistik Verlag.

Künzel, H. J. (1994) 'Current approaches to forensic speaker recognition,' *Proc. ESCA Workshop on Automatic Speaker Recognition Identification Verification*: 135–41.

Künzel, H. J. (1995) 'Field procedures in forensic speaker recognition', in Lewis (ed.) (1995): 68–84.

Künzel, H. J. (1997) 'Some general phonetic and forensic aspects of speaking tempo', *FL* 4/1: 48–83.

Künzel, H. J. (2001) 'Beware of the telephone effect: the influence of transmission on the measurement of formant frequencies', *FL* 8/1: 80–99.

Labov, W. (1972) *Sociolinguistic Patterns*, Philadelphia: University of Pennsylvania Press.

Labov, W. (1986) 'Sources of inherent variation in the speech process,' in J. S. Perkell and D. H. Klatt (eds) *Invariance and Variability in Speech Processes*: 401–25, New Jersey: Lawrence Erlbaum Associates.

Labov, W. and Harris, W. A. (1994) 'Addressing social issues through linguistic evidence', in Gibbons (ed.) (1994): 287–302.

Ladd, D. R. (1996) *Intonational Phonology*, Cambridge: Cambridge University Press.

Ladefoged, P. (1962) *Elements of Acoustic Phonetics*, London: University of Chicago Press.

Ladefoged, P. (1971) *Preliminaries to Linguistic Phonetics*, Chicago: Chicago University Press.

Ladefoged, P. (1978) 'Expectation affects identification by listening', *Language and Speech* 21/4: 373–4.

Ladefoged, P. (1993) *A Course in Phonetics*, 3rd edn, Sydney: Harcourt Brace College Publishers.

Ladefoged, P. (1997) 'Linguistic phonetic descriptions', in Hardcastle and Laver (eds) (1997): 589–618.

Ladefoged, P. (2001) *Vowels and Consonants – An Introduction to the Sounds of Languages*, Oxford: Blackwell.

Ladefoged, J. and Ladefoged, P. (1980) 'The ability of listeners to identify voices', *UCLA WPP* 49: 43–51.

Ladefoged, P. and Maddieson, I. (1996) *Sounds of the World's Languages*, Oxford: Blackwell.

Langford, I. (2000) 'Forensic semantics: the meaning of *murder*, *manslaughter* and *homicide*', *FL* 7/1: 72–94.

LaRiviere, C. (1975) 'Contributions of fundamental frequency and formant frequencies to speaker identification', *Phonetica* 31: 185–97.

Lass, N. J. (ed.) (1976) *Contemporary Issues in Experimental Phonetics*, London: Academic Press.

Lass, R. (1984) *Phonology: An Introduction to Basic Concepts*, Cambridge: Cambridge University Press.

Laver, J. M. D. (1980) *The Phonetic Description of Voice Quality*, Cambridge: Cambridge University Press.
Laver, J. M. D. (1991a) *The Gift of Speech* Edinburgh: Edinburgh University Press.
Laver, J. M. D. (1991b) 'The description of voice quality in general phonetic theory', in Laver (1991a), Ch. 12: 184–208.
Laver, J. M. D. (1991c) 'The semiotic nature of phonetic data', in Laver (1991a), Ch. 10: 162–70.
Laver, J. M. D. (1991d) 'Voice quality and indexical information', in Laver (1991a), Ch. 9: 147–61.
Laver, J. M. D. (1991e) 'Describing the normal voice', in Laver (1991a), Ch. 13: 209–34.
Laver, J. M. D. (1991f) 'Labels for voices', in Laver (1991a), Ch. 11: 171–83.
Laver, J. M. D. (1994) *Principles of Phonetics*, Cambridge: Cambridge University Press.
Laver, J. M. D. (2000) 'The nature of phonetics', *JIPA* 30/1, 2: 31–6.
Laver, J. M. D. and Trudgill, P. (1991) 'Phonetic and linguistic markers in speech', in Laver (1991a), Ch. 14: 235–64.
Lehiste, I. (1970) *Suprasegmentals*, Cambridge, MA: MIT Press.
Lehman, W. P. (1975) *Language and Linguistics in the People's Republic of China*, Austin and London: University of Texas Press.
Lewis, J. W. (ed.) (1995) *Studies in General and English Phonetics – Essays in Honour of J.D. O'Connor*, London: Routledge.
Lieberman, P. (1992) 'Human speech and language', in S. Jones, R. Martin and D. Pilbeam (eds) *The Cambridge Encyclopedia of Human Evolution*: 134–8, Cambridge: Cambridge University Press.
Lieberman, P. and Blumstein, S. E. (1988) *Speech Physiology, Speech Perception and Acoustic Phonetics*, Cambridge: Cambrige University Press.
Lindley, D. V. (1982) 'Bayesian inference', in Kotz *et al.* (eds) (1982): 197–204.
Lindley, D. V. (1990) 'The present position in Bayesian statistics', *Statistical Science* 5/1: 44–65.
Lindsey, G. and Hirson, A. (1999) 'Variable robustness of non-standard /r/ in English: evidence from accent disguise', *FL* 6/2: 278–88.
Lisker, L. and Abramson, A. S. (1964) 'A cross-language study of voicing in initial stops: acoustical measurements', *Word* 20: 384–422.
Luksaneeyanawin, S. (1998) 'Intonation in Thai', in Hirst and Di Cristo (eds) (1998): 376–94.
Lyons, J. (1981) *Language and Linguistics: An Introduction*, Cambridge: Cambridge University Press.
Malakoff, D. (1999) 'Bayes offers a new way to make sense of numbers', *Science* 286: 1460–4.
Mannell, R. H. and Robert-Ribes, J. (eds) (1998) *Proc. 5th International Conference on Spoken Language Processing*, Canberra: ASSTA.
McCawley, J. D. (1978) 'What is a tone language?', in Fromkin (ed.) (1978): 113–31.
McClelland, E. (2000) 'Familial similarity in voices', paper presented at the BAAP Colloquium, Glasgow University.
McCormak, P. and Russell, A. (eds) (1996) *Proc. 6th Australian International Conference on Speech Science and Technology*, Canberra: ASSTA.
McDermott, M. C., Owen, T. and McDermott, F. M. (1996) *Voice Identification: The Aural Spectrographic Method*, HTTP://www.owlinvestigations.com.forensic_articles/aural_spectrographic/fulltext.html (accessed 27/08/2001).
McGehee, F. (1937) The reliability of the identification of the human voice', *Journal of General Psychology* 17: 249–71.
McQueen, J. M. and Cutler, A. (1997) 'Cognitive processes in speech perception', in Hardcastle and Laver (eds) (1997): 566–85.

References

Meuwly, D. and Drygajlo, A. (2001) 'Forensic speaker recognition based on a Bayesian framework and Gaussian mixture modelling (GMM)', *Proc. 2001 Speaker Odyssey Speaker Recognition Workshop*: 145–50.

Mitchell, A. G. and Delbridge, A. (1965) 'The speech of Australian adolescents', Sydney: Angus and Robertson.

Mokhtari, P. and Clermont, F. (1996) 'A methodology for investigating vowel–speaker interactions in the acoustic-phonetic domain', in McCormak and Russell (eds) (1996): 127–32.

Moore, B. C. J. (1997) 'Aspects of auditory processing related to speech perception', in Hardcastle and Laver (eds) (1997): 539–65.

Moroney, M. J. (1951) *Facts from Figures*, 2nd. edn, Harmondsworth: Pelican.

Morris, D. (1978) *Manwatching*, St. Albans: Triad/Panther.

Naik, J. (1994) 'Speaker verification over the telephone network: databases, algorithms and performance assessment', *Proc. ESCA Workshop on Automatic Speaker Recognition Identification Verification*: 31–8.

Nakasone, H. and Beck, S. D. (2001) 'Forensic automatic speaker recognition', *Proc. 2001 Speaker Odyssey Speaker Recognition Workshop*: 1–6.

Nolan, F. (1983) *The Phonetic Bases of Speaker Recognition*, Cambridge: Cambridge University Press.

Nolan, F. (1990) 'The limitations of auditory-phonetic speaker identification', in Kniffka (ed.) (1990): 457–79.

Nolan, F. (1991) 'Forensic phonetics', *Journal of Linguistics* 27: 483–93.

Nolan, F. (1994) 'Auditory and acoustic analysis in speaker recognition', in Gibbons (ed.) (1994): 326–45.

Nolan, F. (1996) 'Forensic Phonetics', notes distributed at the two-week course given at the 1996 *Australian Linguistics Institute*, Australian National University, Canberra.

Nolan, F. (1997) 'Speaker recognition and forensic phonetics', in Hardcastle and Laver (eds) (1997): 744–67.

Nolan, F. and Grabe, G. (1996) 'Preparing a voice lineup', *FL* 3: 74–94.

Nolan F. and Kenneally, C. (1996) 'A study of techniques for voice line-ups', unpublished report on IAFP research project.

Nolan, F. and Oh, T. (1996) 'Identical twins, different voices', *FL* 3: 39–49.

Noll, A. M. (1964) 'Short-time spectrum and *cepstrum* techniques for voiced pitch detection', *JASA* 36: 296–302.

Oasa, H. (1986) *A quantitative study of regional variations in Australian English*, unpublished MA Thesis, Australian National University.

Ohala, J. (2000) 'Phonetics in the free market of scientific ideas and results', *JIPA* 30/1,2: 25–9.

Osanai, T., Tanimoto, M., Kido, H. and Suzuki, T. (1995) 'Text-dependent speaker verification using isolated word utterances based on dynamic programming', *Reports of the National Research Institute of Police Science* 48: 15–19.

Papçun, G., Kreiman, J. and Davis, A (1989) 'Long-term memory for unfamiliar voices', *JASA* 85: 913–25.

Pruzansky, S. and Mathews, M. V. (1964) 'Talker-recognition procedure based on analysis of variance', *JASA* 36: 2041–7.

Q. v. Duncan Lam (1999) District Court of New South Wales 99-11-0711.

Rabiner, L. and Juang, B.-H. J. (1993) *Fundamentals of Speech Recognition*, Englewood Cliffs: Prentice-Hall.

Rietveld, A. C. M. and Broeders, A. P. A. (1991) 'Testing the fairness of voice identity parades: the similarity criterion', *Proc. XIII International Congress of Phonetic Sciences* 5: 46–9.

Robertson, B. and Vignaux, G. A. (1995) *Interpreting Evidence*, Chichester: Wiley.

Roca, I. and Johnson, W. (1999) *A Course in Phonology*, Oxford: Blackwell.

Rose, P. (1981) *An acoustically based phonetic description of the syllable in the Zhenhai dialect*, unpublished PhD Thesis, University of Cambridge.

Rose, P. (1987) 'Considerations in the normalisation of the fundamental frequency of linguistic tone', *SC* 6/4: 343–52.

Rose, P. (1989) 'On the non-equivalence of fundamental frequency and pitch in tonal description', in D. Bradley, E. Henderson and M. Mazaudon (eds) *Prosodic Analysis and Asian Linguistics: to honour R.K. Sprigg*: 55–82, Canberra: Pacific Linguistics.

Rose, P. (1990) 'Thai Phake tones: acoustic aerodynamic and perceptual data on a Tai dialect with contrastive creak', in R. Seidl (ed.) *Proc. 3rd Australian Intl. Conf. on Speech Science and Technology*: 394–9, Canberra: ASSTA.

Rose, P. (1991) 'How effective are long term mean and standard deviation as normalisation parameters for tonal fundamental frequency?', *SC* 10: 229–47.

Rose, P. (1994) 'Any advance on eleven? Linguistic tonetic contrasts in a bidialectal speaker', in Togneri (ed.) (1994): 132–7.

Rose, P. (1996a) 'Between- and within-speaker variation in the fundamental frequency of Cantonese citation tones', in P. J. Davis and N. Fletcher (eds) *Vocal Fold Physiology – Controlling Complexity and Chaos*: 307–24, San Diego: Singular Publishing Group.

Rose, P. (1996b) 'Speaker verification under realistic forensic conditions', in McCormak and Russell (eds) (1996): 109–14.

Rose, P. (1996c) 'Aerodynamic involvement in intrinsic F0 perturbations – evidence from Thai-Phake', in McCormak and Russell (eds) (1996): 593–8.

Rose, P. (1996d) 'Observations on forensic speaker recognition', invited paper at the 6th International Law Congress, Melbourne [distributed on diskette by Organising Committee, 6th ICLC].

Rose, P. (1997) 'A seven-tone dialect in Southern Thai with super-high: Pakphanang tonal acoustics and physiological inferences', in A. Abramson (ed.) *Southeast Asian Linguistic Studies in Honour of Vichin Panupong*: 191–208, Bangkok: Chulalongkorn University Press.

Rose, P. (1998a) 'A forensic phonetic investigation in non-contemporaneous variation in the F-pattern of similar-sounding speakers', in Mannell and Robert-Ribes (eds) (1998): 217–20.

Rose, P. (1998b) 'Tones of a tridialectal: tones of Standard Thai, Lao and Nyo – acoustic and perceptual data from a tridialectal speaker', in Mannell and Robert-Ribes (eds) (1998): 49–52.

Rose, P. (1999a) 'Differences and distinguishability in the acoustic characteristics of *Hello* in voices of similar-sounding speakers', *Australian Review of Applied Linguistics* 21/2: 1–42.

Rose, P. (1999b) 'Long- and short-term within-speaker differences in the formants of Australian *hello*', *JIPA* 29/1: 1–31.

Rose, P. (2000) 'Hong Kong Cantonese citation tone acoustics: a linguistic tonetic study', in Barlow (ed.) (2000): 198–203.

Rose P. and Clermont, F. (2001) 'A comparison of two acoustic methods for forensic discrimination', *Acoustics Australia* 29/1: 31–5.

Rose, P. and Duncan, S. (1995) 'Naive auditory identification and discrimination of similar voices by familiar listeners', *FL* 2/1: 1–17.

Rose, P. and Simmons, A. (1996) 'F-pattern variability in disguise and over the telephone – comparisons for forensic speaker identification', in McCormak and Russell (eds) (1996): 121–6.

Rycroft, D. (1963) 'Tone in Zulu nouns', *African Language Studies* 4: 43–68.

Schenker, H. (ed.) (1975) *Ludwig van Beethoven: Complete Piano Sonatas*, vol. 1, New York: Dover.

Schiller, N. O. and Köster, O. (1995) 'Comparison of four widely used F0-analysis systems in the forensic domain', in Braun and Köster (eds) (1995): 146–58.

Schiller, N. O., Köster, O. and Duckworth, M. (1997) 'The effect of removing linguistic information upon identifying speakers of a foreign language', *FL* 4/1: 1–17.

References

Shipp, T., Doherty, E. and Hollien, H. (1987) 'Some fundamental considerations regarding voice identification', letter to the editor of *JASA* 82: 687–8.

Simmons, A. (1997) *Differences and distinguishability: a Bayesian approach to forensic speaker identification*, unpublished First Class Honours Thesis, Australian National University.

Sjerps, M. and Biesheuvel, D. (1999) 'The interpretation of conventional and "Bayesian" verbal scales for expressing expert opinion: a small experiment among jurists', *FL* 6/2: 214–27.

Stevens, K. (1971) 'Sources of inter- and intra-speaker variability in the acoustic properties of speech sounds', *Proc. 7th Intl. Congress on Phonetic Sciences*, Montreal: 206–32.

Stevens, K. (2000) *Acoustic Phonetics*, Cambridge, MA: MIT Press.

Stoney, D. A. (1991) 'Transfer evidence', in Aitken and Stoney (eds) (1991): 107–38.

Studdert-Kennedy, M. (1974) 'The perception of speech', in T. A. Sebeok (ed.) *Current Trends in Linguistics* vol. 12: 2349–85, The Hague: Mouton.

Studdert-Kennedy, M. (1976) 'Speech perception', in Lass (ed.) (1976): 243–93.

Thompson, C. P. (1987) 'A language effect in voice identification', *Applied Cognitive Psychology* 1: 121–31.

Titze, I. R. (1994) *Principles of Voice Production*, Englewood Cliffs: Prentice Hall.

Togneri, R. (ed.) (1994) *Proc. Fifth Australian International Conference on Speech Science and Technology*, Canberra: ASSTA.

Tosi, O., Oyer, H. J., Lashbrook, W., Pedney, C., Nichol, J. and Nash, W. (1972) 'Experiment on voice identification', *JASA* 51: 2030–43.

Trudgill, P. (1978) *Sociolinguistics: An Introduction*, Harmondsworth: Penguin.

Trudgill, P. (1983) 'Acts of conflicting identity: the sociolinguistics of British pop-song pronunciation', in P. Trudgill (1983) *On Dialect: Social and Geographical Perspectives*: 141–160, Oxford: Blackwell.

Vance, T. J. (1987) *An Introduction to Japanese Phonology*, Albany: State University of New York Press.

van der Giet, G. (1987) 'Der Einsatz des Computers in der Sprechererkennung', in Künzel (1987): 121–32.

van Lancker, D. and Krelman, J. (1986) 'Unfamiliar voice discrimination and familiar voice recognition are independent and unordered abilities', *UCLA WPP* 63: 50–60.

VCS (1991) 'Voice comparison standards of the VIAAS of the IAI', *Journal of Forensic Identification* vol. 41.

VoiceID (2001) 'Technical Information about VoiceID', VoiceID Systems, Inc., HTTP://www.voicepass.com/techinfo.htm [accessed 27/08/2001].

Wakita, H. (1976) 'Instrumentation for the study of speech acoustics', in Lass (ed.) (1976): 3–40.

Warren, R. M. (1999) *Auditory Perception*, Cambridge: Cambridge University Press.

Wells, J. C. (1982) *Accents of English*, Cambridge: Cambridge University Press.

Wells, J. C. (1999) 'British English pronunciation preferences', *JIPA* 29/1: 33–50.

Wierzbicka, A. (1996) *Semantics – Primes and Universals*, Oxford: Oxford University Press.

Wolf, J. J. (1972) 'Efficient acoustic parameters for speaker recognition', *JASA* 51: 2044–56.

Xu Weiyuan (1989) *The sociolinguistic patterns of Pudonghua in Duhang*, unpublished Master's Thesis, the Australian National University.

Yamazawa, H. and Hollien, H. (1992) 'Speaking fundamental frequency patterns of Japanese women', *Phonetica* 49: 128–40.

Index

Note: page numbers in *italics* refer to figures and tables.

accents 144, 147, 337
   allophonic differences 189–90
   imitation 194
   regional 46
   social 46–8
acoustic analysis 35–6, 38–41
   see also speech acoustics
acoustic cues 41
acoustic energy 195–6
   acoustic theory of speech production 207
   fundamental frequency (F0) 199
   glottal volume-velocity wave 208, 209
   harmonics 206–7
   modification 269
   processing 269
      neuro-receptive stage 270
      organo-receptive stage 269
   source 207–9
acoustic features 26
   discrete parameters 51
   range of variation 301
acoustic forensic analysis *92*, *93*, 337
   speaker identification 330
   text dependency 94–5
acoustic output 221, 222
   frequency–amplitude distribution 222–3
   prediction 215
   vocal tract 38, 328
      characteristics 297
acoustic parameters 34–6, 38–41
   automatic 41, 42–3, 45
   long-term 45
   traditional 41–2
acoustic-phonetic form of language 287
acoustic phonetics 337
   quantification 200
   see also speech acoustics
acoustic resonators 133–4
acoustic signal amplification 269
acoustic theory of speech production 43, 195, 199, 207, 217, 218
   acoustic prediction of resonant frequencies 215
acoustic values 33
acoustic vowel plots 228–30, 283
   formant centre frequency 229
affective intent 292–3
   forensic significance 293

affricates 138, 139, *140*
   post-alveolar 142
airstream, pulmonic egressive 210
alcohol intoxication 299–300
allomorphic differences 193
allomorphs 337
   analysis of variation 193
   encoding meaning of plural morpheme 272
allophones 176–9, 337
   aspirated 180
   Australian accents 190–2
   Canadian English 179, 180
   diphthongal 231
   fronted velar 235
   phoneme realisation 176–7, 182, 185
      deviant 189
      forensic 188
      linguistic choice 188–9
      nasalised 185–6
   plosives 182
   target 188–9
   voiceless velar stops 178
allophonic differences 189–90
alternative hypothesis *see* defence hypothesis
alveolar place, consonantal manners 137–9, *140*
American English 20
   Black 187
   sociolinguistic variables 48
amplitude, speech perception 271–2
analysis of variance (ANOVA) 18
anticipatory assimilation 177–8
   in place of articulation 177
approximants 139
articulation
   anticipatory assimilation in place of 177
   manner 137, 153, 339
   place 137, 140–2, 153, 340
   rate 169, 173, 337
   secondary 143
   speed 167, 169
articulators
   active 140–1, *142*
   passive 140, 141–2
articulatory phonetics 5, 337
articulatory setting 279–80
assimilation, anticipatory 177–8

353

Index

auditory analysis 34–8
 forensic-phonetic investigation 275
 speaker identification 330
auditory forensic analysis *92*, 93, 337
 familiarisation 106–7
auditory parameters 34–8
auditory-phonetic analysis 36, 49
auditory-phonetic statements 36
auditory techniques, speech sample description/comparison 92–3
aural-spectographic identification 81, 107–8, *109*, 110–22, 337
 accuracy claims 122
 admissibility of evidence 120–1
 aural modality 114–15
 claims 117–20
 closed set identification 110–11
 combined aural-visual method 114–15
 criteria for decision 114
 definition 341
 duplication of questioned sample 112–13
 error rates 118, 119–20
 Frye test 121
 independence of modalities 114–15
 innocent suspect incrimination 112, 113
 instructions to suspect 112
 Koenig survey 119–20
 likelihood ratio 118–19
 linguistic controversy 111–20
 methodology 111–16
 open set identification 110–11
 parameters for comparison 113–14
 quantification lack 116
 sample comparison 113–14
 terminology 117
 theoretical basis lack 115–16, 122
 Tosi's study 118–19
 training requirements 116
 validation lack 117–18
 visual modality 114–15
Australian Aboriginal languages 294
Australian English 20, 150–1, *152*
 Broad accent 190–2, 306–10
 Cultivated 190–2
 differences in 190
 diphthongs 151, *152*
 ethnic accents 190
 General accent 190–2
 long vowels 151, *152*
 phonemes 183, *184*, 185
 speech rate 167, *168*
Australian Forensic Speaker Identification Standards Committee 331, 332

Bayes' theorem 66
 likelihood ratio 74
 mathematics 76
 practical applications 68
Bayesian inference 66–76
 arguments against 69, 73–6
 arguments for 69–73
 closed set comparison using fundamental frequency distributions 256–7
 complexity 76
 criticisms 73–6
 current acceptance 67–8

 empirical confirmation of theoretical predictions 69–70
 evidence combining 70
 explicitness 70–1
 individual case evaluation 73
 likelihood ratio 69–70
 speaker identification 89–90
 speaker verification 89–90
believe, meaning of 74–5
bell curve 195–268
between-speaker variation 5, 10–13, 19, 337
 allomorphic differences 193
 cognitive intent 291–2
 dimension power 18
 discrete parameters 51
 illegal vowel devoicing 305
 likelihood ratio 321
 phonemic structure 186–92
 sample number 30
bidialectals
 linguistic difference preservation 333
 within-speaker variation 187
bilabial place, consonantal manners 137, 139, *140*
bilingualism 103
 linguistic difference preservation 333
 within-speaker variation 187
Black American English 187
boundary tone 279
breathing pauses 171
British English, sociolinguistic variables 48

Canadian English, phonemes 179–80
Cantonese 38, 39
 pitch height 41–2
 speech rate 167, *168*
cepstral coefficients 42, 43, 262, *263*, 264
cepstrum 262–3, *264*, 265, 334, 337
 forensic significance 265
chi-square test 71
Chinese language
 Pudong dialect 47
 tone language 158–9
coarticulation 189
cognitive intent 291–2
communicative intent 284, *285*, 291–4
 affective intent 292–3
 cognitive intent 291–2
 regulatory intent 295–6
 self-presentational 296
 social intent 293–5
componentiality 125–73
computer technology 333–4
computerised analysis *92*, 93–6
confidence interval 27–8
consonantal sounds 136
consonants 5, 6
 apico-alveolar 141–2
 aspirated 163
 dorso-velar 142
 effects on vowels 235
 English 142–3
 length 150
 nasal 135
 phonemes 183, *184*, 185
 post-alveolar 142

Index

production 125–6, 136, 137–40
  symbols 37n
  tonal pitch 42
  voiced 163, 164
  voiceless unaspirated 163–4
conversation
  active participation 101
  analysis 295, 338
  interaction 295–6
  turn pauses 171
conversation analysis (CA) 269–75
COT–CAUGHT merger 44
creak 130, 131
creaky voice phonation type 243
crico-thyroid muscle 129

defence hypothesis 57, 58, 64–5, 310
  evidence probability 66
  fallacy 337
  ignorance 75–6
  likelihood ratio 64–5, 76
  probability 65, 75
dentition, effect on acoustic output 299
detection, missed 85
detection error trade-off (DET) 85–6, 87
devoicing, illegal in Japanese 304–6
  parameter number 323
dialectology 48, 293–4, 338
dialects 102, 103
  aural-spectographic identification 112–13
  phonemic systemic differences 187
  vowel quality 144
different-speaker pairs 18
dimensions
  discriminatory power 18
  mean values 25–6
  number 25, 28
  number of observations 28
  overlap 11–12
  power 17
  size 25, 26–8
    likelihood ratio 59
  speaker's mean position 26–7
diphthongs 149, 338
  English 151, *152*
  hello acoustic changes 241–2
  phonemic analysis 179, 180
  phonetic analysis 179, 180
discriminant analysis 17, 18
discrimination in forensic speaker identification 17–18
  threshold values 89
distance–pressure function 196
drug intoxication 299–300

ear tube 269
earwitnessing
  error proneness 100
  evidence 81, 97, 105–6
  voice distinctiveness for line-ups 101
  voice line-up 101, 106
elimination, false 85
emotions, linguistic signalling 292
energy
  mechanical 196
  see also acoustic energy

English
  allophones 179–80
  consonantal manners 137–9
  consonants 142–3
  diphthongs 151, *152*
  nasal sounds 134
  phonemes 183, *184*, 185
  pitch 156
  speech rate 167, *168*
  stress language 155
  vowels 150–1, *152*
    system 145, 147
epistemic verbs 74–5
equal error rate (EER) 13, 14, 87
etic-emic difference 175
Euclidean distance 15, 16
evidence
  aural-spectrographic identification 107–8
    admissibility 120–1
  Bayesian inference 66–76
  combining 70
  degree of similarity of samples 57–8
  earwitness 81, 97, 105–6
  independent for same speaker 30
  likelihood ratio 60–1, 63–4, 78, 110
  matching 72–3
  prior odds 63–4
  probability 14, 57–8, 70–1
    of hypothesis 56–7
  strength for speaker identification 88
  voice sample discrimination 13–14
expectation effect 273, 338
  forensic significance 273–4
  naive speaker recognition 104
expert witness, scientific validity determination 121

F-patterns *225*, 226–8, 338
  analysis of *hello* 239–43
  auditory and acoustic features 243–4
  higher frequency formants 237
  phonological environment of vowels 233–6
  software performance 266–7
F-ratio 18
  higher frequency formants 237, 238
  Japanese vowel formant data for likelihood ratio calculation 317
false alarms 85
false negative 9–31
false positive 9–31
familiar speaker identification 98–9
  error rate 99
familiarity, auditory forensic analysis 106–7
fast Fourier transform (FFT) 204, *205*, 338
FBI
  aural-spectrographic protocol 111, 113
  Koening survey 119
flaps 139, 140
fluency 170
forensic automatic speaker recognition (FASR)
  algorithm (FBI) 96
forensic data representativeness 22, *23*, 24–30
forensic-phonetic analysis 1–2
  identification outcome 5
  language of samples 333
  potential of samples 330

355

Index

forensic-phonetic investigation 55–79
  Bayesian inference 66–76
  defence/prosecution hypothesis 64–5
forensic-phonetic likelihood ratio 60
forensic-phonetic parameters 33–53
  continuous 50–1
  dependence evaluation 52
  discrete 50–1
  independence 53
  qualitative 34, 50
  quantitative 34, 50
  requirements 51–3
  types 34
forensic phonetics 2–3, 332
forensic science 332
forensic significance
  affective intent 293
  between-speaker variation 10
  cepstrum 265
  continuity 172–3
  expectation effect 273–4
  F-pattern variation 144
  formant frequencies 215
  fundamental frequency 246–8
  indexical factors, intrinsic 300–2
  linguistic analysis 48–9
  long-term spectrum 261–2
  nasal and nasalised sounds 135
  perceptual integration 275
  phonation type 131
  phonemic structure 185
  phonetic quality 281
  phonological environment 235–6
  pitch 131
  pitch accent 161–2
  rate of utterance 169
  social intent 294–5
  stress 156, 161
  suprasegmentals 161–2
  tone of voice 161, 282–3
  vocal tract length 215–16
  voice quality 281–2
  Vowel acoustics 237, 243–4
  Vowels 144
  within-speaker variation 10, 237
forensic speaker identification 2
  closed set/open set 110–11
  see also speaker identification
formality in language 294
formants 211–12, 217, 338
  acoustic vowel plot 229
  amplitude 211–12
  automatic tracking difficulties 331
  bandwidth 212, 338
  centre frequency 211, 213
    acoustic vowel plot 229
    determination 213–14
    software packages 267
    vowel differences 225
  frequencies 215–16, 219, 220, 231
    articulatory backness 233
    reference sample spread 311
    vocal tract influence 238
  higher frequency 237–8
  linear prediction analysis 224
  patterns 226

  peaks 217
  see also F-patterns
Fourier analysis 199–200, *201*, 202–3
  fast Fourier transform (FFT) 204, *205*, 338
frequency *see* fundamental frequency
frequentist approach 66–7
  hybrid 71
  statistical significance 72
  twin trace problem 73
fricatives 138, 139, *140*, 141
  alveolar 164
  English 143
  glottal 141
  post-alveolar 142
  voiced 164
  voiceless 164
    glottal 238, 239–40
frontal sinuses 133
  dimensions 298
Frye test 121
fundamental frequency (F0) 38–9, 40, 244–53, *254*, 255–62
  acoustic energy 199
  average 41
  closed set Bayesian comparison 256–7
  compass 251
  definition 338
  determinants 244–6
  distribution 253, 255–7
    long-term 248–53
    modelling 257–9
  forensic significance 246–8
  forensic speech sample comparison 162
  harmonics 205–6
  health state 253, *254*, 299
  in *hello* 246–8, 255
  kurtosis 252
  long-term 45
  long-term mean (LTF0) 306–7
  long-term spectrum 259–62
    forensic significance 261–2
  mean 44, 249–50, 259
  modal 252–3
  pitch 42, 161–2
    in *hello* 247
  probability 253, 255–7
    density 258–9
    statement 259
  range 41, 301
    individual speaker 251
  reference sample spread 311
  root-mean-square measure 251
  skew 251–2
  during speech 212, 220
  speech-specific speech perception 270–1
  speech waves 197–8
  standard deviation 250–1, 259
  vocal cord
    length 244–6
    mass 244–6
    vibration rate 209
  vowels 271

Gaussian mixture modelling 257
glottal fry 130

Index

glottal stops 141
glottal volume-velocity wave 208, *209*
　acoustic energy 208, 209
　energy content 209, *210*
　harmonics 208, 209
glottis 128, 141
　air flow 207, 208
　energy source 240
　spread segments 218
grammar 269–75

hard palate 132
harmonics 205–6
　acoustic energy of vowels 206–7
　amplitude–frequency plot 221
　glottal volume-velocity wave 208, 209
　spectra 216–17, 224–5, 239
　　smoothing 262
　during speech 212–13
health state
　fundamental frequency (F0) 253, *254*
　voice variation 20
hello acoustic changes 238–43
　creaky voice phonation type 243
　diphthong 241–2
　fundamental frequency (F0) 246–8
　intonation 242–3
　lateral 241
　pitch 242–3
　short low central vowel 240–1
　stress 242–3
　voiceless glottal fricative 238, 239–40
histogram 195–268
hypervolume 16

identification of speaker 9
illegal vowel devoicing 304–6
　forensic-phonetic parameter use 304
　parameter number 323
indexical factors, intrinsic 284, *285*, 296–302
　age effects on vocal tract 298–9
　extrinsic features 297
　forensic significance 300–2
　health 299–300
　indexical information 297
　intrinsic features 297–300
　limits to variation 300–1
　sex effects on vocal tract 298–9
information communication, constraints and choices 284
innocence presumption 74–5
inter-speaker variation *see* between-speaker variation
International Association for Forensic Phonetics (IAFP) 282
　accreditation 331, 332
　code of practice 330–1, 332–3
International Association for Identification (IAI), aural-spectrographic protocol 111, 113, 114, 122
International Phonetic Association (IPA) 36
interpretation principles 68
intonation 125, 153–4, 156–8, 338
　hello acoustic changes 242–3
　phonetic quality 279
　stress interaction 157–8

intoxication 299–300
Italian
　allophones 177–9
　phonemes 177–9
　velar stops 177, 179

Japanese
　illegal vowel devoicing 304–6
　vowel formant data for likelihood ratio calculation 316–18, 323
judiciary/jurors 68

kernel density function 195–268
Korean, non-pitch-accent language 161
kurtosis 252

labialised sounds 143
labio-dental approximant 188
labio-dental place 141
language 5, 7
　communicative intent 284, *285*, 291–4
　contrastive phonetic differences 182
　default values for parameters 301
　formality 294
　intrinsic indexical factors 284, *285*, 296–302
　listener's 104
　meaning 288
　morphology *286*, 287, 289
　naive speaker recognition 102–4, 105–6
　native 103, 104
　non-accentual differences 46
　non-pitch-accent 161
　phonation types 130–1
　phoneme composition 182–3
　phoneme number 176
　phonemic insight 179
　phonemic structure 181–2
　pitch 158
　pitch-accent 160
　range of sounds 41
　regional accents 46
　second 97, 103
　social accents 46–8
　speech samples 332–3
　speech sounds 175
　　association 33
　standard 102
　stress/stress-accent 155
　structure 286–90
　　phonemic level 179, 180
　symmetry 180
　target 103–4
　　competence 104
　see also tone languages
laryngeal activity, differential timing 166
laryngitis 299
　fundamental frequency 253, *254*, 299
larynx 127
larynx tube
　frequencies 238
　resonance frequency 221
lateral sounds *140*
leptokurtosis 252
lexical semantics 288, 290
lexicon 290, 294

357

Index

likelihood ratio 7, 57–64, 338
  accuracy 334
  acoustic parameters of speech 318
  applications 312–18
  approach to forensic phonetics 76, 77
  aural-spectographic identification 118–19
  automatic parameters 334
  Bayes' theorem 74
  Bayesian inference 69–70
  between-speaker variation 321
  calculation 303–4
    continuous data 306–10
    hello 313–16
  calibration of tests 319
  combination 60–4, 321–2, 323, 328
  continuous data
    calculation 306–10
    formula 311
  defence hypothesis 64–5, 76
  definition 338–9
  different-speaker comparison 313, 314, 315, 316
    Japanese vowel formants 317–18, 323–4
  dimension size 59
  discriminant distance 318–19
    lack of data 322–4
    limitations 322–5
  evidence 63–4, 110
  expectation effect 274
  familiar speaker recognition 99
  formula 310–12
    shortcomings 320–1
  hard 77, 78
  illegal vowel devoicing 305
  intuition-based 77, 78
  Japanese vowel formant data 316–18, 323–4
  limitations 319–25, 329
    average probabilities 324–5
  log scaling 62
  magnitude 334
  matching approach 72–3
  mean of reference sample 307
  mean of sample 309–10
  method 303–26
  naive speaker recognition 100–1, 105
  overall for forensic-phonetic evidence 60–1
  parameters
    combination 60–2, 321–2, 323, 328
    distribution 321
    optimum number 323–4
    strength 328
  recognition of familiar voices 100–1
  reference sample 306, 307, 310, 330
    mean 307
    shortcomings 320
  reflection of reality 313
  same-speaker comparison 313, 314–15, 316
    Japanese vowel formants 317–18, 323, 324
  sample
    mean 309–10
    size 308–9
  similarity measures 306, 307–9
  speaker identification/verification 90
  standard deviation 308, 311–12
    of reference sample 306–7, 311–12
  typicality measures 306, 309–10
  variability 318
  verbal scales 61–2
  within-speaker variation 321
linear discriminant analysis 14
linear prediction analysis 224, 338
linguistic analysis, forensic significance 48–9
linguistic comparability of speech samples 329–30
linguistic differences
  non-accentual 46
  signals 279
linguistic features, geographical/social background 293
linguistic mechanism 284, *285*, 286–90
linguistic message 45
  communicative intent 291
  pitch variation 19
  structure 46
linguistic parameters 34, 43–4
linguistic structure 286–90, 327
  tone of voice 290–1
linguistic typology 327
linguistic units 5
linguistics 3
  descriptive 37
  knowledge 332
lip-rounding 147, 149
  allophones 179
  anticipatory 189
lips
  activity 143
  position 147
  trills 140
listener ability
  accuracy 101
  linguistic competence 104
  variation 100–1
logarithms, likelihood ratio scaling 62

matching approaches 72–3
mean, arithmetical 25–6, 337
  degree of representativeness 27
  true 29–30
mechanical energy 196
modal voice 130, 131
monophthongs 149, 222, 338
morphemes 160, 289–90, 339
  allomorphic variation 193
  sound shape 192
  sounds 193
morphology of language *286*, *287*, *289*
morphophonemes 193
morphophonemics 192–3, 290
morphosyntax 295
mother-in-law language 294
multilingualism 103

naive speaker recognition 97–106, 339
  expectation effect 104
  familiar 98–9
  familiarity 97–100
  language 102–4, 105–6
    target 103–4
  likelihood ratio 100–1, 105
  listener ability 100–1
  non-linguistic properties of stimulus 102

Index

recording quality 102
time factors 102
unfamiliar 98, 99–100
voice variable distinctiveness 101
nasal cavity 132, 133–4
   dimensions 298
   resonant frequencies 238
nasal sinuses 133
   dimensions 298
nasal sounds 134–5, *140*
   forensic significance 135
   voiceless 134–5
nasal velopharyngeal setting 38
nasalisation 148, 153
   speech-specific perceptual compensation 271
nasalised sounds 134–5
   forensic significance 135
nasality, paralinguistic function 135
nasals 138
natural parameters *see* acoustic parameters, traditional
natural semantic metalanguage (NSM) 75
noise, extraneous 42
non-linguistic parameters 34, 43, 44
non-rhotic accents 47
non-rhoticity, variable 47
normal distribution 195–268
noun phrase 170

observations
   data *23*
   number 28
odds
   Bayesian inference 66
   posterior 63, 340
   see also prior odds
oral cavity 132–3
   proportions 298

paralinguistic information signalling 131
past performance 74
pauses 169
   filled 169, 170, 171–2, 173
   hesitation 170–1, 172
   juncture 170, 172
   percentage of overall speech 173
   respiration 171
   silent 169, 170
   turn 171
perceptual integration 274–5
   forensic significance 275
perceptual mechanisms 35
pharyngeal cavity 132–3
   proportions 298
phonation 341
   modal 130, 131
   see also voicing
phonation type 38, 130–1, 339
   creaky voice 130, 131, 243
   emotional signalling 292
   extralinguistic use 131
   forensic significance 131
   non-modal 131
phone 6, 175–94, 339
   distribution 180–1

phonemes 289, 339
   allophone distinction 176–9
   allophone number 178, 179–80
   alveolar nasal 177, 179
   Australian English 183, *184*, 185
   consonants 183, *184*, 185
   distribution 180–1
   English 183, *184*, 185
   realisation 176–7, 182, 185, 188–90
      nasalised 185–6
   reality 181–2
   systematic 193
   velar stop 235
   vowels 183, *184*, 185
phonemic analysis contrast 176
phonemic contrast 176
phonemic restoration 273
phonemic structure
   between-speaker differences 186–92
   forensic significance 185
   incidental differences 187–8
   phonotactic differences 187
   realisational differences 188–90
   systemic differences 186–7
   within-speaker differences 186–92
phonemics 5, 6, 175–94, 289, 339
   speech samples 185
phonetic-acoustic approach 35
phonetic features, automatic parameters 43
phonetic quality 278–9, 280, 339
   forensic significance 281
   voice quality difference 280
   see also speech sounds
phonetic transcription 36–7
phonetics 6, 48, 49, 287, 339
   articulatory 5, 337
   components *286*, 290
   forensic 2–3, 332
   speaker identification investigators 331
phonological environment 235–6
   speech perception 272
phonology 6, 48, *286*, 287, 289–90
   definition 339
phonotactics 182–3, 185
   differences 339
   structure/analysis 182–3
   within-speaker variation 187
pianists, cadenza speed 166–7
pitch 36, 130, 339–40
   average 10–11, 283
   contours 157
   contrastive 164–5
   convergence 19
   extralinguistic use 131
   forensic significance 131
   height 41–2
   hello acoustic changes 242–3
   intonational 154, 157–8
      linguistic differences 279
   language 158
   linguistic value 271
   overall 44
   phonetic 280–1
   production 38–9
   speech perception 271
   stress 37n, 155

359

Index

syntactic information signalling 156
tonal 42
tone languages 158, 159
   contrasts 165
tone of voice 36, 41–2, 290–1
transcription 38
   uses 156
variation 19–21
vocal cords 130
   size 283
vowels 220
   production 126–7
   see also fundamental frequency (F0)
pitch accent 153–4, 160–1, 340
   forensic significance 161–2
place–manner matrix 142–3
platykurtosis 252
plosives 137–8, 139, *140*, 141
   allophones 182
   English 143
population 22, 24, 340
post-alveolar approximant 188
pragmatic meaning 288
pressure fluctuations, time–pressure function/waves 196, *197*
presume, meaning of 74–5
Prinzivalli case 1, 44
   regional accents 46
prior odds 63–4, 340
   estimation 74
   incompatibility with presumption of innocence 75
   indeterminacy 73–4
   innocence presumption 74–5
   past performance 74
probability 14, 340
   average 324–5
   Bayes' theorem 66
   classical 66
   defence hypothesis 65, 75
   density 258–9
   estimates 56
   evidence 57–8, 70–1
      in support 56–7
   matching 72–3
   prior odds 63
   prosecution hypothesis 65
   statements of 78
   theory 68
pronunciation, within-speaker variation 47–8
prosecution hypothesis 57, 58, 310
   evidence probability 66
   fallacy 340
   prior odds 63
   probability 65
      conditions 71
Pudong Chinese dialect 47
Pythagoras' theorem 15–16

received pronunciation (RP) 190
receiver operating characteristic (ROC) 85
regional accents 46
regulatory intent 295–6
rejection, false 85, 86
resonant frequencies 211

rhotics 139, 140
   British pop music 187
rhythm, phonetic quality 279

same-speaker pairs 18
samples *see* speech samples
segmental sounds 125–73
self-presentational intent 296
semantic analysis 75
semantic structure 287–8
semantics *286*, 287–8
shunts 241
signal processing 333–4
similar-sounding speaker hypothesis 310
skewness 251–2
social accents 46–8
social intent 293–5
   forensic significance 294–5
social interaction, convergence/divergence 294–5
sociolects 144, 340
sociolinguistic variables 48, 51
   phonemic structure 188
   phonotactics 187
sociolinguistics 48, 340
soft palate 132, 133, 134, 135
   opening 238
sound spectrograph 228
   see also spectrograms
source–filter theory *see* acoustic theory of speech production
speaker(s)
   acoustic property estimation 89
   differing 3
   discrimination 5, 84–5
   height correlation with vowel formant frequencies 214
   multidimensional comparison 16, *17*
   paired 15–16
   profiling 2, 3
   recognition 40, 81–2
   similar-sounding 28–9
   space distribution 14
   true mean 29–30
   two-dimensional comparisons 15–16
   vocal tract characteristics 297–8
   see also between-speaker variation; naive speaker recognition; within-speaker variation
speaker identification 82, 83–4
   automatic systems 94, 95–6
   Bayesian approach 89–90
   categorical decisions 89–90
   closed sets 84
   comparison with forensic speaker identification 82–91
   control over samples 90–1
   data requirements 329–30
   decision types 85
   evidence strength 88
   false 85
   false acceptance 86, 87
   false rejection 85, 86, 87
   forensic-phonetic 96–7
   forensic speaker identification relationship 87–91
   improvement 334

investigators 330–2
    area of expertise 331–2
    language 332–3
    qualifications 330–1
likelihood ratio 90
method 330
open sets 84
parameter encoding 327
parameter performance 87
performance evaluation 85–7
reference samples 88
requirements 329–33
set membership 84
speaker pool size 87
strategies 85
threshold requirement 85–6, 90
voice distortion 87
speaker recognition 91, *92*
    acoustic parameters 95
    automatic 93, 94, 334
        acoustic parameters 244
        cepstrum 265
        forensic speaker identification 96
        performance of system 95–6
        text dependency 94–5
    cepstrum 334
    commercial 93
    naive 92
    technical 92, 341
        auditory forensic analysis 92–6
    text-dependency 95
    variation 100
    see also naive speaker recognition
speaker verification 82–4, 84–5, 87
    automatic 334
    Bayesian approach 89–90
    categorical decisions 89–90
    cepstrum 334
    comparison with forensic speaker identification 82–91
    control over samples 90–1
    detection error trade-off curves 86, 87
    dimensionality reduction 328
    forensic speaker identification relationship 87–91
    likelihood ratio 90
    reference samples 88
    speech properties of pool members 89
    system evaluation 87
    threshold requirement 85–6, 90
    voice distortion 87
speaking rate 169
speaking tempo 113
spectrograms 107, 220–5, 340
    formant structure *225*, 226–8
    inferences about speech production 224
    schwa 221–5
    software differences *266*
    speech sound rate 167, *168*
    visual comparison 108, *109*, 110, 113
    vowels 225–8
spectrum 341
    long-term 259–60
    LP-smoothed 221, 222, *223*, 224, 239
        cepstrum comparison 262–3
    short-term 260
    see also harmonics, spectra

speech
    acoustic energy 195–6
    articulatory mechanism 189
    chain 287
    changes with geographical location 293
    continuity 166, 169–73
        forensic significance 172–3
    defects 189
    deviant 189
    flow 170
    fluent 170, 171–2
    hesitant 171–2, 173
    non-linguistic temporal structure 173
    pattern recognition 42
    perception 148, 341
    range of variation of acoustic characteristics 301
    rate 166–7, *168*, 169
        paralinguistic 169
    recognition 40
    semantic structure 170
    spectrum 199
    structure 37
    syntactic structure 170
    units 40
    variation 19–21
    see also acoustic energy; acoustic theory of speech production; pauses
speech acoustics 3, 5, 6, 195, 209
    cepstrum 262–3, *264*, 265
        forensic significance 265
    changes 238
    fast Fourier transform (FFT) 204, *205*
    filter 209–11, 220
    formants 211–12, 213–14, 217
        frequencies 215–16, 219, 220
    frequency-domain representation 199
    likelihood ratio values 318
    perceptual decoding 148
    software performance 265–7
    source 207–9, 213, 220
        filter interactions 212–15
    source–filter theory 207
    spectral components 203
    spectral envelope *211*, 212, 217
    spectral representation 199–200, *201*, 202–7, 204
        harmonic spectrum 204–6
        sinusoidal components 205
        smoothed spectrum 206–7
    spectral slope 209
    spectrum 203–7
        harmonic 204–6
    speech waves 196–9
        pressure calculation *202*
    time-domain representation 199
    time-domain resynthesis 219
    variation 28–9
    vocal tract imprint 207
    see also acoustic theory of speech production; spectrograms
speech perception 6, 269–75
    amplitude 271–2
    bottom-up processing 272–3
    compensation for sexually different F0 271
    expectation effect 273–4
    nasalisation 271
    perceptual integration 274–5

361

Index

  phonological environment 272
  pitch 271
  plurals 272
  redundancies 273
  speech-specific 270–3
    forensic significance 273
  top-down processing 272–3
speech samples 5, 340
  allophonic evaluation 180
  between-sample comparability 185–6
  correct attribution 296
  discrimination between 327
  evaluation of forensic 3
  forensically important similarities/differences 77
  incriminating 88
  language 332–3
  linguistic comparability 329–30
  matching 72–3
  number 24, 28–30
  paired 58
  parameters for comparison 5, 327–8
  phonemics 185
  phonetic quality 92
  police interview 91
  pooling of unknown 30–1
  probability of evidence 57–9
  single suspect claim 88
  situational comparability 329
  voice quality 92
speech science, speaker identification investigators 331, 332
speech sounds 5, 195–6
  absence 44
  analysis 153
  components 152–3
  conditions of non-contrast for phonetically similar 181
  description 37, 153
  differences 40
  differential timing 162–4
  etic-emic difference 175
  languages 175
  phonemic level 175
  phonetic level 175
  presence 44
  production 125–6
  range 220
  rate 167
  segmental 126
  structure 152–3
  suprasegmentals 125, 153–62
    forensic significance 161–2
    typology 161
speech-specific perception 148
speech technology 334
  speaker identification investigators 331
speech waves 196–9
  amplitude 199
    maximum 202
    term 202–3
  complex 200, 202–3, 213, 219
  Fourier analysis 199–200, *201*, 202–3
  frequency 197–9
    term 202
  higher-frequency components 198–9
  phase term 204
  pressure calculation 200, 202–3
  radiated 220
  sinusoidal 202, 203
  time-pressure sine waves 200, *201*
  wave-form 197, 198, 199
speechalyser 319, 324
spread glottis segments 218
statistics 3, 7, 14
  analysis inferiority 71–2
  confidence interval 27–8
  frequentist approach 71–2
  see also mean, arithmetical
status relationships, encoding of perceived 295
stress 37n, 125, 153–6, 341
  forensic significance 156, 161
  hello acoustic changes 242–3
  intonation interaction 157–8
  linguistic difference signals 279
  phonetic quality 279
  pitch 155
  syllables 154–5
    length 155
stress-accent 155
  forensic significance 161
Student's *t*-test 71, 72
subglottal coupling 219
subglottal resonance/antiresonance 217–19, 220, 341
supraglottal resonance 240
supralaryngeal activity
  cepstrum 262
  timing 162–5
    differential 166
supralaryngeal articulations 127–9
supralaryngeal vocal tract (SLVT) 126, 131–5
  closure at one end 218
  energy modification 210
  filter function 209–11
  high-velocity air jets 207
  length 210, *211*
    forensic significance 215–16
  plasticity 216
  speech sound production 162–5
  squeezing 135–6, *137*
  transfer function 218
  vibration 210, 211
    filter function 212–15
    transfer function 212–15
  voice onset time (VOT) 163–4
suprasegmentals 125, 153–62, 341
  forensic significance 161–2
  phonetic quality 279
  typology 161
syllables
  intonational pitch 154, 157–8
  length 155
  prominence 155
  speaking rate 169, 341
  stress 154–5
syntactic units, pitch 156
syntax *286*, 287, 288–9
  constituents 289

taps 139, 140
target undershoot 233
technical speaker recognition by listening 93

Index

telephone voice 20, 21
  dimensionality reduction 21
text dependency 94–5
Thai language 158–9
theoretical predictions, empirical confirmation 69–70
tone languages 36, 38, 158–60, 161, 162
  forensic speech sample comparison 162
  pitch 158, 159
    contrasts 165
tone of voice 125, 153–4, 158–60, 282, 290–1
  acoustic values 39
  boundary 279
  comparisons 39
  definition 341
  forensic significance 161, 282–3
  linguistic mechanism 286
  normalisation of values 40–1
  phonetic quality 279
  pitch 36, 290–1
    height 41–2
  see also intonation
tongue
  body 144, 145
    height 147
    movement 233
    placement 147–8
    post-alveolar articulation 234
    vowel production 147–8, 149
  mechanical coupling with vocal cords 271
  position 144, 145
    control 147
  round trip 233
  trills 140
Tosi extrapolation 119
trachea, resonant frequency 218
transmission channel effect sensitivity 42
transposing the conditional 71, 341
trills 139, 140
triphthongs 149
twin trace problem 73

utterance-finality signalling 296
utterance rate 169
  forensic significance 169
uvula, trills 140

variance 341
variance, between-speaker:within-speaker 18
  see also F-ratio
velar, cardinal 235
velar stops 177
  Italian 177, 179
  voiceless *178*
velarised laterals 143
velum *see* soft palate
verb phrase 170
verbal scales for likelihood ratio 61–2
vocal cords 126, 127–9
  abduction 128
  activity 131
    timing 162–5
  adduction 128, 129
  air flow 240
  length 298
    fundamental frequency (F0) 244–6

mass 244–6
mechanical coupling with tongue 271
mucosal covering 298
oscillations 207
phonation type 130–1
pitch 130, 283
tensing 129, 246, 298
  pitch 130
vibration 33, 129–30, 242–3
  emotional signalling 292
  fundamental frequency determination 244–5
  fundamental frequency in *hello* 247
  pitch control 38–9, 130, 165
  pulmonic egressive airstream 210
  rate 165, 198, 209
  sex effects 298
  speech modulation 135
  as spring 245
  as string 245
voice onset time (VOT) 163–4
voicing 129–30
vocal fry 130
vocal mechanism 284, *285*, 286
vocal tract 125, 126, 144
  acoustic output 38, 297, 328
    prediction 215
  age effects 298–9
  air expulsion 210
  area function 215
  articulatory targets 238
  characteristics and acoustic output 297
  closure 241
  consonant production 137
  constriction 137
  dimensions 300
  formant frequencies 238
  genotype effects 300
  health effects 299–300
  imprint 207
  influences from movements of other structures 238
  length 215–19, 231, 298
    body size 298, 299
  radiation 213
  resonances 211
  resonant frequency prediction 215
  sex effects 298–9
  shape/size 41
  speaker characteristics 297–8
  tube constriction 144, 145
  wall compliancy 298
  see also supralaryngeal vocal tract (SLVT)
vocalic sounds 136
vocalis muscle 129
voice(s) 5, 7, 277–302
  choice 284, *285*
  cognitive processing 106
  comparisons 4, 7, 9
  complexity 4
  constraint 284, *285*
  dimensionality reduction 21–2
  discrimination 115–16
  disguise 97
  distinctiveness 97
  distortion 87
  entities 115, 116

363

Index

exposure 102
familiarisation 106–7
individual information 46
information content 46
inputs 284, *285*
linguistic content 46
meaning 277–8
mechanisms 284, *285*, 286
model 278, 283, 284, *285*
   communicative intent 284, *285*, 291–6
   intrinsic indexical factors 284, *285*, 296–302
   linguistic mechanism 284, *285*, 286–90
   tone of voice 284, *285*, 290–1
   vocal mechanism 284, *285*, 286
multidimensionality 14–16, *17*
non-linguistic aspects 37–8
quality 102, 278, 279–81, 341
   forensic significance 281–2
   organic component 279, 282
   phonetic quality difference 280, 281
   pitch 280–1
   setting component 279–80, 282
questioned 22, *23*, 24
similar sounding 35
speaking tempo 113
suspect 22, *23*, 24
tone 278
variable distinctiveness 101
see also between-speaker variation; within-speaker variation
voice line-up 101, 106
voice onset time (VOT) 163–4
voice samples
   differences 22
   discrimination 9, 10–13, 17–18
      average pitch 10–11
      paired observations 12–13, 15–16, 18
   forensic comparison 3
   multidimensional comparison 16, *17*
   number 24
   sources 22, *23*, 24
   two-dimensional comparisons 15–16
   see also dimensions
voice variation
   lack of control 19–21
   response to situations 20, 21
voiceprint identification *see* aural-spectographic identification
voicing 129–30, 142–3, 341
volume 16
vowel(s) 5, 6
   acoustic energy 206–7
   acoustic plots 228–30
   acoustic properties 220
   acoustic values 33
   articulation 280
   Australian 150–1, *152*
   backness 147–8, 153, 337
   central 235
   dynamicity 153
   effects on consonants 235
   F-pattern variation 235–6

forensic importance 144
formant frequencies 214
fundamental frequency 271
height 147–8, 153, 337
high front 235
length 149–50, 153
lip position 147
long 151, *152*
nasalisation 148
nasalised 135
occasion-to-occasion variation 233, 235–6
open 147
phonemes 183, *184*, 185
phonetic quality 198–9
phonological environment 233–5
   forensic significance 235–6
pitch 220
plots 235–6
primary correlates of quality 206
primary parameters of description 144–5
production 125–6, 126–7, 136
quality 43, 144, 220
rounding 147–8, 153
secondary parameters of description 148–50
short 150–1
sounds 136
spectrograms 225–8
steady-state 222
supralaryngeal vocal tract vibration 210–11
symbols 37n
tonal pitch 42
tongue position 145, *146*
   body placement 147–8
see also illegal vowel devoicing
vowel acoustics
   between-speaker variation 230–2
   within family 231–2
   forensic significance 243–4
   hello acoustic changes 238–43
   higher-frequency formants 237–8
   phonological environment 235–6
   speech perception 272
   within-speaker variation 232–5
   forensic significance 237

within-speaker variation 5, 10–13, 16, 19, 117, 341
   allomorphic differences 193
   cognitive intent 291–2
   dimension power 18
   discrete parameters 51
   occasion-to-occasion variation 235–6
   phonemic structure 186–92
   pronunciation 47–8
   sample number effects 29, 30
   sociologically defined 47
   voiceprints 117
   vowel acoustics forensic significance 237

z-score normalisation 40
zero 195–268
Zulu language 158–9